North Dakota's Geologic Legacy

North Dakota's Geologic Legacy

OUR LAND AND HOW IT FORMED

John P. Bluemle

NDSU NORTH DAKOTA STATE UNIVERSITY PRESS

Dept. 2360, P.O. Box 6050, Fargo, ND 58108-6050
www.ndsu.edu/ahss/ndirs

NDSU NORTH DAKOTA STATE UNIVERSITY PRESS

North Dakota's Geologic Legacy: Our Land and How It Formed
By John P. Bluemle, Ph.D.

Book design by Deb Tanner
Maps and illustrations by Up & Running Web Design, Bismarck, ND

International Standard Book Number (ISBN): 978-0-911042-82-5
Library of Congress Control Number: 2015953390
Printed in Canada

Bluemle, John P., author.
North Dakota's geologic legacy : our land and how it formed / John P. Bluemle.
pages cm
Includes bibliographical references and index.
LCCN 2015953390
ISBN 978-0-911042-82-5

1. Geology--North Dakota. I. Title.

QE149.B58 2016 557.84
QBI15-600240

For the people of North Dakota

Contents

Foreword vii

Preface xi

Introduction xvii

Chapter One: Geologic Origins of North Dakota's Landforms

- Geologic History 1
  - Deeply Buried Rocks 1
  - Geologic Time 2
  - Some Important Dates in Geologic Time 4
  - Rocks Found at the Surface 5
    - The Cretaceous Period 5
    - Geologic Formations Exposed at Surface in North Dakota 6
    - Tertiary Formations 12
  - Erosion of the Land 15
    - Amount of Erosion 15
    - North Dakota's Modern Landscape 19
  - Summary 21
- Geography and Climate 23
- Soils in North Dakota 28
- Landform Areas 31
  - Introduction 31
  - Physical Regions 32
    - The Great Plains 32
    - The Central Lowlands 36
    - The Red River Valley 39
  - Summary 39

Chapter Two: The Missouri Plateau

- Surface Materials 41
- Lignite and Petrified Wood 43
- Clinker 50
- Nodules and Concretions 57
- Badlands 63
  - Introduction 63
  - Age of the Badlands Materials 65
  - Why the Badlands Formed 68
  - Erosion in the Badlands 69
  - The Badlands Landscape 71
  - Summary 75

Buttes of Western North Dakota ... 76
Introduction ... 76
General Geology ... 76
Limestone and Sandstone Caprock ... 78
Other Sandstone-capped Buttes ... 88
Fox Hills Formation Sandstone Caprock ... 89
Hell Creek Formation Sandstone Caprock ... 89
Clinker Caprock ... 89
Silcrete Caprock ... 91
Sod-capped Buttes ... 93
Buttes Without Caprock ... 93
Buttes in Name Only ... 94
Glaciated Buttes ... 95
Summary ... 96

Chapter Three: The Missouri Valley

Geologic History of the Missouri River Valley ... 97
Other "Missouri" River Routes ... 106
Summary ... 107

Chapter Four: The Glaciated Plains and the Missouri Coteau

The Ice Age ... 109
Glaciers and Glacial Deposits ... 113
Landforms Shaped by Glaciers ... 125
Ground Moraine ... 125
Drumlin-like Landforms ... 129
End Moraine ... 132
Missouri Escarpment ... 132
Dead-ice Moraine ... 135
Ice-thrust Moraine ... 141
Landforms Shaped by Running Water ... 152
Eskers and Kames ... 152
Veblens ... 156
Tunnel Valleys ... 158
Outwash Plains ... 159
River Terraces ... 160
Abandoned River Valleys ... 160
Landforms Shaped by Glacial Lakes ... 163
Introduction ... 163
Glacial Lake Souris ... 164
Glacial Lake Dakota ... 166
Glacial Lakes Minnewaukan and Cando ... 167
Glacial Lake McKenzie ... 168

Chapter Five: Other Geologic Features

Wind-shaped Landforms 171
Sand Dunes 171
Loess 176
Wind-sculpted Features 178
Field Stones 180
Glacial Erratics 180
Siliceous Rocks 186
Silcrete 186
Knife River Flint 188
Chert 191
Stones Used in Construction 191
Coghlan Castle 194
Gold in North Dakota 195
Ransom County 196
Lake Souris–Denbigh Deposit 198
Bentley Lake Deposit 200
Hungry Gulch 201
Summary 202
Aquifers and Artesian Flows 203

Chapter Six: The Devils Lake Basin

Introduction 207
Late Preglacial History 209
Glacial History of the Devils Lake Basin 210
Summary of Circumstances Resulting in the Formation of the Devils Lake Basin 214
Fluctuations in the Level of Devils Lake 215
Puzzling Behavior of Devils Lake 218
Discussion 219
Summary 220

Chapter Seven: The Red River Valley

Glacial Lake Agassiz 223
History of Glacial Lake Agassiz 224
Lake Agassiz I 226
Lake Agassiz II 227
Surface Features 229
Grooves 229
Compaction Ridges 231
Spring Pits 234
Lake Shorelines 236
Scarps 236
Beaches 237
Bars 238
Deltas 238
Crustal Rebound 242
Pembina Escarpment 244

Chapter Eight: The Effect of Geology on the Human Environment
- Geologic Hazards ... 250
  - Introduction ... 250
  - Flooding ... 252
  - Landslides ... 253
  - Treppe ... 255
- Geology and Settlement Patterns ... 256
  - Summary ... 266

Chapter Nine: North Dakota's Energy Resources
  - Geologic Formations in the Williston Basin ... 268
- Lignite ... 269
  - Geology ... 269
  - Production History ... 270
  - Uranium Contained in Lignite ... 272
  - The Current Lignite Industry in North Dakota ... 274
  - Reclamation ... 275
  - The Future of Lignite ... 277
- Hydroelectric Power ... 278
- Geothermal Energy ... 279
- Wind Energy ... 281
- Solar Energy ... 284
- Oil and Gas ... 287
  - Early Oil and Gas Development in North Dakota ... 287
  - The Williston Basin ... 290
  - The Bakken Formation ... 291
    - Geologic Studies of the Bakken ... 291
    - Description of the Bakken ... 291
    - Formation of Bakken and Other Williston Basin Shales ... 293
    - Source Rocks ... 295
    - Reservoir Rocks ... 295
  - Considerations Relating to Oil and Gas Production in North Dakota ... 295
    - Regulation ... 295
    - Horizontal Drilling ... 296
    - Lineations ... 298
    - Hydraulic Fracturing ... 298
    - Pipelines ... 300
    - Natural Gas Description and Processing ... 300
  - Summary ... 302

Epilogue ... 305

Acknowledgments ... 307

Appendix A: Ice Age Chronology ... 311

Appendix B: Glossary ... 321

Notes 349
Sources and Recommended Reading 365
Compilation of Pertinent Resources 367
Index 377

Foreword

For me, one of the most fascinating aspects of natural North Dakota is the changes in the landscape that can be easily observed, even by someone without much in the way of geologic training, as one drives across our state. Consider, for example, a drive along Interstate 94 from Fargo to the Montana border. Fairly casual observations will reveal vast differences in the landscape as one drives through the very flat Red River Valley that is the home to Fargo, across the Prairie Pothole region which begins near Tower City and runs roughly to Bismarck, and through the flat-topped buttes of southwestern North Dakota.

Those who have been to north-central North Dakota may have had the opportunity to enjoy the Turtle Mountain region and the many recreational opportunities available; the curious may have also wondered why the landscape is elevated there relative to the surroundings. Most who are fairly familiar with our state have also likely taken note of the beautiful canyons and gullies of the badlands areas, which again are very different than other landscape regions in North Dakota. While limited in scope, as the home of Theodore Roosevelt National Park, the badlands are probably the best known landscape region in our state.

A resource that North Dakotans are very lucky to have is the North Dakota Geological Survey (NDGS). This group of scientists has produced a wealth of information about our state since the NDGS was created by the North Dakota Legislature in 1895. While they produce many reports, maps and other products that are highly useful for a professional geologist, the NDGS also engages in public outreach and has produced a number of products that are intended and suitable for the general public to help educate our state's citizens about North Dakota's geology. Their work has done much to enrich our state on many levels.

This book was born from two of the fundamental observations above: 1) the landscapes found in the different parts of North Dakota are highly varied and fascinating, and 2) the public outreach work of the NDGS. While this book is not a publication of the NDGS, Dr. John P. Bluemle enjoyed a 40-plus-year career with the NDGS that began in the early 1960s as an entry-level geologist and ran into the first decade of the 2000s; he concluded his career as the North Dakota state geologist.

Early in his career, John was assigned to study and map the surficial deposits of eastern North Dakota, which were left behind by vast glaciers that once covered the state. These sediments tell the story of days gone by when North Dakota was a very different place than today –colder even than the winters of the present, if one can imagine such conditions.

Over his career, John had the opportunity to work in every corner of the state, from the Minnesota border to Montana and from Canada to South Dakota. During these travels, he studied every major landscape system in the state. He also expanded his knowledge to the subsurface rocks and other geologic systems of the state that were not necessarily part of his original job duties. In addition to his work providing information utilized by professional geologists, John produced several NDGS publications intended to teach the general public about the state's geology. There probably isn't anyone alive today who knows more about North Dakota's geology or who has done more to reach out to the public to expand general knowledge of the state's geology.

In this book, John has summarized his vast knowledge of North Dakota's surficial geology. He takes the reader on a wonderful journey through landscapes dominated by deposits from glaciers, ancient lakes and running water, as well as landscapes dominated by erosion. The book is written at a level such that it is accessible to non-scientists, but the information will also be very useful to scientists who want to learn about the state's geology or expand their knowledge of it.

John is also an excellent and avid photographer, and the pages of this book are enriched with many of his photographs. These are supplemented with richly done color diagrams that help explain the many processes that shaped North Dakota's landforms. The writing is excellent, and clearly flows from topic to topic.

Interspersed with explanations of the geology readers will also find highly entertaining stories as John recounts some of the more humorous and/or unconventional experiences he had as he travelled the state working to understand its geologic story, as well as encounters he had with employees of various government agencies, contractors and even former North Dakota governors. As readers make their way through the pages, they will learn much about the science behind our state's geology, and they will also be exposed to social and political issues related to that geology.

While the focus of the book is primarily the landforms of the state, there is also an excellent discussion of the geologic energy resources that are present. This is fitting due to the importance of fossil fuels to North Dakota's modern economy and because of the potential for energy sources such as geothermal, wind and solar to be more fully developed in the future.

As a professor teaching geology at Dickinson State University, I have made extensive use of materials John has developed over his years of working in North Dakota to help my students better understand the state's geology. John has also been generous with his time and personal resources, sharing his photographs and on occasion spending time with my students as a guest speaker. In this book, he extends that generosity to any who desire to take advantage of it.

Since learning that John was working on a new book about North Dakota's geology, I have looked forward to the publication of this project with great pleasure and anticipation. I hope those who read it have as much fun delving through the pages of this book as I did reading the manuscript.

Eric C. Brevik
Belfield, N.D.

PREFACE

I arrived in Grand Forks in 1962, straight from Montana State College in Bozeman, to begin my job as a geologist with the North Dakota Geological Survey. One of my first assignments was to examine the fresh excavations for the Intercontinental Ballistic Missile sites then being installed in the eastern part of the state, a legacy of the Cold War. I traveled to all of the 150 sites, as they were excavated, and to another 15 launch-control site excavations, to examine the freshly exposed geology at each excavation.

The geologic insights I gained while visiting those 165 excavations were a remarkable introduction to the intricate glacial stratigraphy of eastern North Dakota. I saw buried layers of glacial lake sediment; beds of silt and sand, standing on end, contorted, twisted and left in the positions into which they had been shoved by glaciers 13,000 years earlier; exquisite boulder pavements embedded in ancient, buried soil zones; blocks of shale that had been pushed, broken, bent and mutilated by the glacial ice; and an array of other geologic phenomena, all freshly excavated and ready for my inspection. From my perspective as a geologist, the installation of the missile sites was North Dakota's most important role in the Cold War.

During my studies of the excavations, I experienced some of the absurdities of the Cold War mentality that pervaded our society in the 1960s. No one was authorized to tell me where the sites were. That information was classified, and I had no map showing their locations. So, every Monday morning, I'd stop at the Grand Forks Air Force Base to check in with the Corps of Engineers construction office. The Corps geologist would tell me the approximate locations of the sites being excavated that week or those that had been just recently excavated. He was not allowed to tell me the actual locations, so he'd say "one a few miles south of Langdon, one near Hoople," etc. Then I'd drive to the general area and, with a little hunting and some local inquiry, eventually arrive at the site, explain my purpose to whoever might be on the site (usually no one was there) and, depending upon the complexity of the geology, spend a few hours examining, studying and photographing the marvelous excavations: fresh, vertical walls, most of them 30 feet high in the actual missile excavations, but as high as 50 or 55 feet in the control sites. And because each excavation had four walls, I was able to see the geology, stunningly presented, in three dimensions. When I was finished with a site, I plotted its exact location on my map. So much for security regulations! I remember thinking that the Soviet Embassy probably had maps showing the locations of all the missile sites, but I was not bold enough, or quite foolish enough, to contact it and ask for one.

One day, while visiting with a construction superintendent, who was not a government employee, I mentioned my minor frustration at having to hunt for each site. He was amazed! He told me that maps showing the locations of all the sites were posted in every service station and every worker had a copy. The maps were posted by the construction company, not the Corps of Engineers. I had not noticed them. After all, the workers had to know where to report each day. The man I was visiting with could not give me one of the maps (against security regulations), but he did happen to lose one at my feet while we were talking.[1]

During my 42 years with the Geological Survey, I worked on every facet of North Dakota geology: the rocks that produce oil, gas, coal, gravel, ground water and all of our other mineral resources. My studies of the glacial sediments near Devils Lake helped me gain detailed insights into North Dakota's past climate changes. I was always most interested, though, in the origin of the hills and valleys I saw every day as I traveled around the state. I have spent a lifetime trying to understand how the land that is North Dakota came to be the way it is.

Through the years, I had the chance to work in every part of the state. A field season for me usually lasted from sometime in May, beginning when it was dry enough to get around, and ending in November, when the ground was frozen too hard to auger a hole.

My wife, Mary, and our three children, as they came along, lived with me in 25 or so North Dakota towns through the years. Our oldest son, Bill, was born in Park River while I was mapping Walsh County, and our daughter, Irene, arrived in Lisbon while I was mapping Ransom County. Paul, the youngest, was born in Grand Forks on the first day of January, when it was too cold to map anywhere.

Through the years, we lived, up to six months each, in places like Carrington, Cooperstown, Harvey, Hazen, Mayville, McClusky, Washburn, White Shield, Fort Totten, Fort Yates, Rock Lake and Turtle Lake. Our summer homes were in towns in 30 counties, and on four Indian reservations. We enjoyed every one of them.

Each place was special in some way. North Dakota people are open and friendly, and often interested in geology. When we arrived at a new place, we asked locally if anyone had an apartment for rent and, usually, someone did. In Harvey, we rented from the owner of the town bakery—a great choice! In Enderlin, our landlady's son, a hunter, kept us supplied with pheasants and geese that autumn. In another town, our landlady's son provided us with wild turkeys (some of them may have been poached—we didn't ask). In Fort Yates, the brand-new nursing home was not yet filled and still had room, so that became our home. Our little kids were a hit with the elderly residents.

Just before Thanksgiving, we would move back to our own home in Grand Forks. The day after we got home one year, I raked the yard and put up the storm windows. The next day, a blizzard blew, and the snow stayed until spring. Our neighbor, an elderly Norwegian man, commented in his wonderful accent, "that Bluemle, he always times things right." Well, I don't always time things right, but I was glad I had that year. For the 25 years that we spent our summers in the field, throughout North Dakota, we tended to visualize Grand Forks as a white and snowy winter wonderland because we were not around much to enjoy it in the summertime.

I claim some important knowledge and understanding of North Dakota, apart from the geology. While mapping, I noted stands of chokecherries, wild plums, buffalo berries and juneberries; that valuable information went on my field maps, right along with the geology, as did the locations of the best places to buy sausage and kuchen.

Most of the photos in this book illustrate landforms. They are intended to help readers appreciate and understand the geologic processes that shaped our modern landscape. I took most of them during the summers of 2009, 2010 and 2011 while traveling throughout the state, but all seasons are represented. Winter scenes can be interesting and I have a lot of them, but snow tends to obscure geologic features more often than it enhances them, so I did not use many of them in this book. Hiking to out-of-the-way locations, or perhaps climbing a butte to see the caprock, will reward you with spectacular views. I recommend it, but it is not really necessary to hike great distances to experience North Dakota's geology. I walked short distances to take some of the photos for this book, but many of the locations I visited could be viewed as well from a car window.

Writing this book, taking photos while traveling with Mary, reacquainting myself with places I had studied in the past, or discovering new places we had never been—these kinds of things made for enjoyable times. Sometimes we got a little more off the beaten path than I intended. One day, I drove along a road on the Missouri Coteau that became a trail, and eventually a narrow, mostly washed-out path with no place to turn around easily, so I kept going. Finally, we came to a barricade, so I stopped and walked around it to read the hand-printed sign: "Do Not Enter: Road Impassible." The seven miles I had just driven were impassible! I dug out a couple of the steel fence posts, drove to the good side of the barricade, and replaced the posts and sign. I think we may have gotten to some places and gotten some pictures of features that day that few readers of this book will ever see.

Another day, I walked over to a fence line, took a picture, headed back to the car, and promptly stepped in a badger hole and broke my foot. Still another time, we stopped to admire a herd of longhorn cattle, standing on the road, surrounding us. I thought one was particularly handsome, so I took his picture through the open window on the passenger side of our van. I felt something cold, turned around, and found myself nose to nose with a cow that had gotten her head in through the open window as far as her horns allowed. I presumed she probably wanted her picture taken, too, so I snapped her as well.

Some geologists move from country to country around the world;

we moved from county to county around the state of North Dakota. Unspoiled prairies and buttes, rivers and lakes, wildflowers and wild fruit are everywhere. Wildlife abounds. Gravel trails lead to broad horizons. Late-afternoon summer showers are followed by spectacular sunsets. The prairie landscapes are multidimensional; their breadths, elevations and depths reflect geologic events and processes I will explore in this book. And the geology is always there. Knowledge of the geology opens the door to another world just beneath the familiar scenes of our everyday lives. It takes us outdoors as we explore the intricacies of our Earth's history.

Mary and I have traveled beyond North Dakota—to most of the states and Canadian provinces and 20 other countries. Much of our travel has been to enjoy geology. We've seen wonderfully spectacular geology in places like Montana, Alberta and Sweden. Scotland is my favorite destination for geology and for the history of the science of geology. Our geology, here in North Dakota, may be more subtle than the places I just mentioned, but it is equally interesting.

My career has been satisfying, my work rewarding. Every day on the job was different for me. Whether it was the glorious summer days in the field, mapping geology, or wintry days I spent in my office, piecing together and trying to understand what I had mapped the previous summer, it was always fascinating.

North Dakota's Geologic Legacy is not meant to be a technical report. Nor is this book at all intended to be an exhaustive treatment of North Dakota's geology. Hundreds of technical reports have been published on North Dakota's geology. They deal with just about every aspect of the state's geology, and especially the formations that produce oil and gas and coal. I have written some of them, about the surface and ground water geology of North Dakota's counties, as well as the glacial landforms. This book, though, is devoted to landforms and how they originated.

INTRODUCTION

Some people visualize a mountain or an oil well, or perhaps a fossil dinosaur, when they think of geology. Or they may envision natural wonders, like the Grand Canyon or Yellowstone Park, or perhaps disasters such as volcanic eruptions and earthquakes. But what about less spectacular geologic features and events closer to home? Along with many of you, I have marveled at, and for 27 years, I lived on the remarkably flat plain of glacial Lake Agassiz. I have studied and enjoyed the rugged badlands along the Little Missouri River, climbed many of our buttes, and strolled through the dunes of the Sheyenne Grasslands and among the prairie potholes of the Missouri Coteau. All of these places owe their unique qualities to geology.

The North Dakota landscape can be divided into two main geologic parts: land that has been glaciated and land that has not. In western North Dakota, flat-lying beds of silt and clay cover the land. Petrified wood, clinker and concretions played vital roles in shaping the landscape which, for millions of years, has been carved by running water into its modern configuration.

Glacier ice shaped the eastern and northern parts of the state. When they came, the glaciers brought silt, clay, sand and boulders with them, and when they melted, the nutrients in these materials became the foundation for our rich prairie soils. The glaciated landscape is much younger, geologically, than is the land in the part of the state that was not glaciated. While they were here, the glaciers sculpted an intricate landscape, characterized by a variety of unique landforms. When the ice melted, water flowing from the melting ice carved valleys like

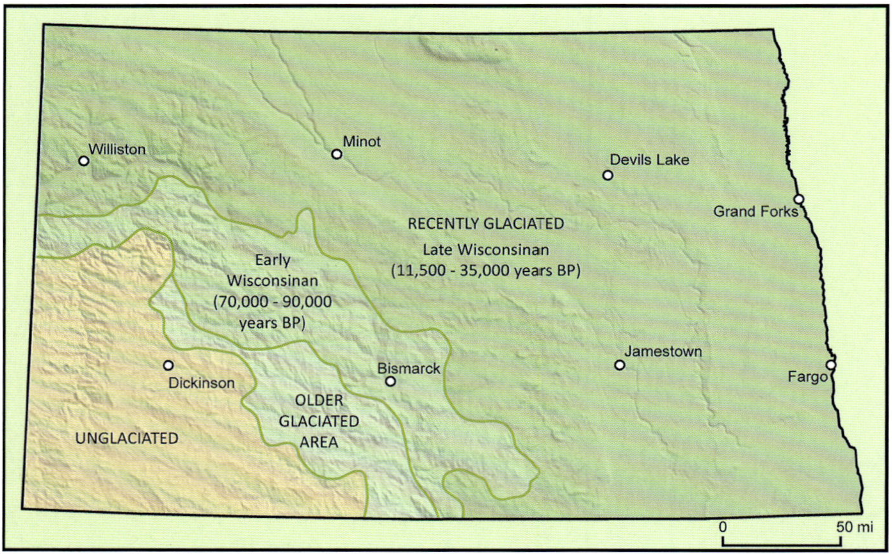

▲ This map of North Dakota shows the parts of the state in which the landscape is dominated by glacial landforms of various ages. Most of the area shown in green was glaciated twice during the past 90,000 years. The larger portion of the area ("Recently Glaciated"), shown in the darkest shade of green, was glaciated during Late Wisconsinan time, between 35,000 and 11,500 years ago (years BP = before present). The area shown in medium green ("Early Wisconsinan") was glaciated between 70,000 and 90,000 years BP. Late Wisconsinan glaciers did not advance over this area. Early Wisconsinan glacial deposits do, however, extend beneath the Late Wisconsinan deposits.

The distribution of known Early and Late Wisconsinan glacial deposits suggests that the Early Wisconsinan glaciation was more extensive in North Dakota than was the Late Wisconsinan glaciation.

The area shown in light green ("Older Glaciated Area") was glaciated, perhaps several times prior to Wisconsinan time (prior to 90,000 years ago). However, subsequent erosion has removed most of the evidence of glacial landforms except for scattered glacial erratics and some valleys that were probably carved by rivers of glacial meltwater. Glaciations occurring more than 90,000 years ago likely also covered some parts of North Dakota shown in orange. Deposits of such early glaciations likely extend north and eastward beneath the areas shown in darker green.

The area shown in orange ("Unglaciated") has generally been considered never to have been glaciated. I have, however, found some deposits that may be of glacial origin as far southwest as Rhame, in Bowman County, so it may be that early glaciers advanced over the entire state, perhaps several times.

those of the Sheyenne, James, Souris and Des Lacs rivers, and it deposited thick layers of gravel and sand, forming broad plains.

I wrote this book to describe and explain how geology shaped North Dakota's landscape. Why does the land look the way it does? What features identify each region of the state and set it apart from other places? Of what materials are the landforms made? How and when were they formed? I will try to answer these questions. I will explain how the geology had a role in determining where people settled, and why they chose to live where they did. And, I will discuss our abundant energy resources, a direct result of our geologic heritage.

The Foreword, Preface and Introduction of this book are followed by nine chapters. In chapter one, Geologic Origins of North Dakota's Landforms, I have emphasized the geologic events responsible for shaping our modern landscape. Chapters two, three and four are arranged by geographic regions. Chapter two, The Missouri Plateau, is a description of the geology west of the Missouri River. Following an explanation of the origin of the geologic materials responsible for our western landforms, I have detailed how those materials became the plains, badlands and buttes we see in western North Dakota today. Chapter three, The Missouri River Valley, is an accounting of how the river came to be, why it is where it is, why it separates such distinct western and eastern styles of geology. Chapter four, The Glaciated Plains and Missouri Coteau, describes the great array of landforms and materials that characterize the glaciated part of North Dakota.

Chapter five, Other Geologic Features, highlights a variety of features that otherwise do not seem to fit in neatly elsewhere: sand dunes, field stones, flint, aquifers—even gold occurrences in North Dakota. Chapters six, The Devils Lake Basin, and seven, The Red River Valley, deal specifically and in detail with two unique areas requiring special attention.

In chapter eight, The Effect of Geology on the Human Environment, I have dealt with some of the geologic hazards in North Dakota, how they relate to our lives, and how they helped influence the settling of the state. The final chapter, North Dakota's Energy Resources, provides some of the history of the geology and development of each of six energy resources.

The section titled Notes consists of explanatory information, as well as additional facts and anecdotes.

I have included a minimum of technical terms, but a few scientific terms were necessary to help explain certain geologic concepts. Most of the scientific terminology is briefly defined as I introduce it, but the glossary defines some relevant terms in more detail. Every term listed in the glossary is used in the text of the book and the definitions are specific to North Dakota geology.

Nearly all of the material in the book is based on my own studies and observations—knowledge I gained during the years I spent mapping the geology of North Dakota. Some facts, though, stem from other people's ideas and research and, where appropriate, I have noted sources on which I have relied. I have drawn upon, and cited, the work of other geologists, particularly when discussing ideas some of the earliest explorers and geologists had, as they first tried to explain the things they saw.

This book is not a technical or scientific paper and the referencing is not rigorous. I have kept references to a minimum. I have, however, included a listing of pertinent references, which includes some sources that I thought readers might find interesting—books or articles that provide additional information on a topic. In places where I have drawn on such materials, I have tried to identify the source. However, much of the knowledge that exists about North Dakota geology is difficult to pinpoint. Systematic citation of every source would have required extensive referencing that I considered inappropriate in this book.

Most of the photos are ground scenes that illustrate features readers can see while traveling around the state. I took most of them, but I have noted the sources of photos provided by other people. My choice of photos was based primarily on their geologic instructive value, not their aesthetic appeal. Even so, I hope the photos may help engender an appreciation of the beauty of North Dakota's landscape. I have included maps and diagrams to help explain the origin of certain landforms and to illustrate some geologic concepts.

CHAPTER ONE

GEOLOGIC ORIGINS OF NORTH DAKOTA'S LANDFORMS

Geologic History

Deeply Buried Rocks

During much of the past half-billion years, North Dakota was submerged beneath warm, shallow seas in which were deposited various kinds of sedimentary rocks. The weight and pressure of tremendous amounts of accumulating materials, as they built up through time, coupled with mechanical and chemical processes, transformed soft, loose sediments into hard rock. The great weight of the sediments caused them to sink to great depths, forming the Williston Basin.

The sedimentary rocks in the Williston Basin reach a thickness of more than three miles near Watford City. They constitute a vital part of our state's geology, as they contain great amounts of mineral resources. However, most of these rocks are deeply buried and played no direct role in the development of the modern landscape. For this reason, I will not discuss the deep sedimentary rocks or the early geologic history they record, except briefly in my discussion of North Dakota's energy resources.[2] Rather, I will deal with the materials and events that were directly involved in forming the landscape we see today.

Geologic Time

Geologic history records various geologic processes, such as (in order of occurrence): the accumulation of oil and gas in the Williston Basin, the rise and demise of the dinosaurs, the deposition of the lignite now being mined in the central part of the state, the erosion of the Little Missouri Badlands, the advancing and melting of the glaciers, and the Lake Agassiz flood. None of these past events can be directly observed, but we can assign relative ages to geologic layers associated with the events, and reliable radiometric methods make it possible to assign dates—actual ages in years before the present—to various geologic events.

I'll paraphrase an explanation of the great duration of geologic time provided to me several years ago by Dr. Donald Schwert, professor of Geology at North Dakota State University.

Most geologists accept that the Earth is about 4.6 billion years old. Let's travel the length of I-94, west to east, across North Dakota. Imagine that we start at the Montana line and call that 4.6 billion years ago. As we travel eastward, toward Minnesota, we pass upward through geologic time. Using the highway as a time-scale, the first life—primitive stuff: a few bacteria and algae—appeared on Earth at mile 84, as we pass the Abbey steeples at Richardton. There's nothing much more to note as we continue through Bismarck—just a few more amoeba-like organisms. The first land plants—conifers—appear as we enter the Red River Valley. Dinosaurs appear at Casselton, North Dakota's "hometown of governors." A little farther down the road—about 200 million years later—as we approach the Sheyenne River in West Fargo, they (dinosaurs, not governors) become extinct. At the edge of Lindenwood Park in Fargo (400 yards from the Red River) we meet the earliest hominids, our ancestors. Then, a few feet farther east, glaciers advance over North Dakota. They melted from the state less than four feet from the edge of the Red River (low-water stage). And when did North Dakota become a state? A half-inch before we reach the edge of the river.

The average person's frame of reference is restricted to the last quarter inch or so. This is why it is so hard to visualize geologic time.

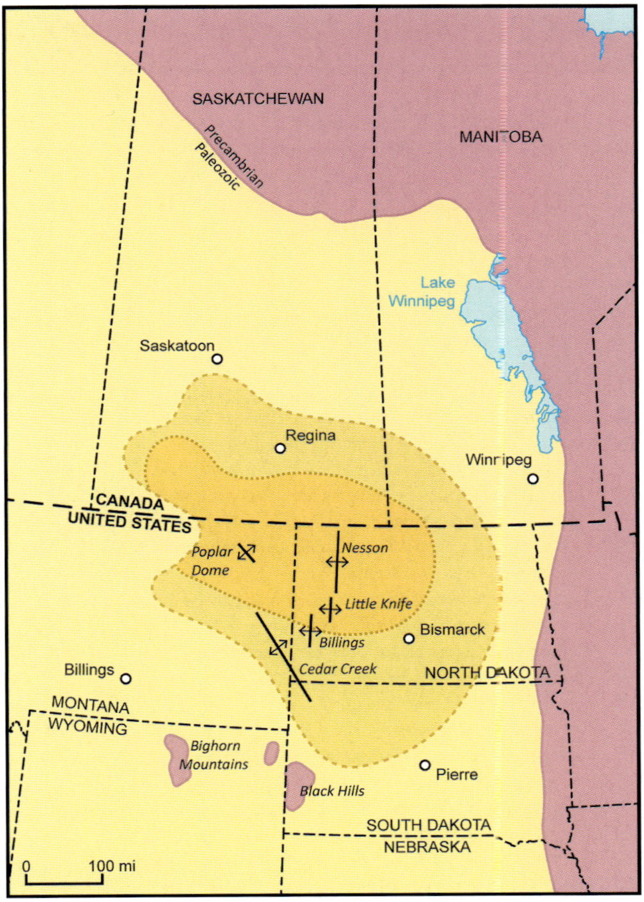

▲ This map shows the extent of the Williston Basin (area within dashed line). Several structural features are shown that occur within the Williston Basin. They include the Nesson Anticline, Little Knife Anticline, Billings Anticline and Cedar Creek Anticline in North Dakota, and the Poplar Dome in Montana. The darker-colored area within the Williston Basin is the approximate extent of the Bakken Formation.

The boundary separating Precambrian igneous and metamorphic rocks (the pink area to the north) from Paleozoic and younger sedimentary rocks (the yellow area to the south) extends through Saskatchewan, into Manitoba, and then south. Paleozoic and younger marine (materials deposited in seas: limestone, dolostone, shale, etc.), along with continental sediments (materials deposited on land: sandstone, etc.), are found south and west of the area covered by Precambrian rocks.

The small reddish areas in South Dakota and Wyoming are igneous intrusions (Black Hills, Big Horn Mountains and Devils Tower). Farther west, in Alberta, Montana and Wyoming, such intrusions are numerous.

Some Important Dates in Geologic Time

I have selected some representative events for this listing. The ages are approximate, and for some there is considerable disagreement. Don't take the listing too seriously! The dates and corresponding period names are taken mainly from the International Stratigraphic Chart, issued by the International Commission on Stratigraphy in 2009.

Years Ago	Event
13,810,000,001	End of last universe *(Big Crunch) (maybe!)*
13,810,000,000	Origin of current universe *(Big Bang)*
13,000,000,000	First stars formed
12,800,000,000	First galaxies formed
4,800,000,000	Solar system formed
4,600,000,000	Origin of the sun and Earth *(early Hadean time; no known rocks; Earth's crust still fluid)*
4,527,000,000	Earth collides with another planet *(moon forms from resulting debris)*
	Earliest single-celled life forms *(during Hadean time, 4.5 billion to 3.8 billion years ago)*
3,800,000,000	Oldest chemical evidence of single-celled life [found in Greenland] *(Eoarchean Era)*
3,480,000,000	First microbial communities [Australia] *(Paleoarchean Era)*
2,100,000,000	Earliest chemical evidence of multicellular life *(Rhyacian Period)*
2,000,000,000	Oxygen atmosphere/ozone layer forms; first photosynthesis *(Orosirian Period)*
1,000,000,000	First evidence for fossil metazoans (multicellular animals) *(Tonian Period)*
650,000,000	Snowball earth *(Cryogenian Period; worldwide Varangian Glaciation)*
600,000,000	First soft-bodied animals *(Ediacaran Period)*
570,000,000	First known arthropods (ancestors of insects, arachnids and crustaceans) *(Ediacaran Period)*
530,000,000	First fossil fish *(Cambrian Period)*
480,000,000	Mass extinction at beginning of Ordovician; glaciation; 60 percent of marine life destroyed
458,000,000	First land plants *(Ordovician Period)*
375,000,000	First amphibians *(Devonian Period)*
360,000,000	First conifer trees *(Devonian Period)*
359,000,000	Bakken Formation deposited *(Devonian/Mississippian Periods)*
340,000,000	First reptiles *(Mississippian Period)*
251,400,000	Mass extinction; 90 percent of all surface and marine species destroyed *(end of Permian Period)*
240,000,000	First dinosaurs *(Triassic Period)*; oldest known fossils found in Tanzania
195,000,000	First mammals *(Triassic Period)*; oldest known fossils found in China
160,000,000	First avian dinosaurs (birds) *(Jurassic Period)*
150,000,000	Archaeopteryx: pioneer bird *(Jurassic Period)*
145,000,000	Atlantic Ocean opens *(beginning of Cretaceous Period)*
130,000,000	Angiosperms (flowering plants) develop *(Cretaceous Period)*
65,500,000	Asteroid hits Mexico (Cretaceous Period ends); extinction of non-avian dinosaurs; 95 percent of all species destroyed
3,500,000	Ice Age begins *(late Pliocene)*
3,400,000	Australopithecus afarensis (Lucy) human ancestors in Ethiopia *(Pliocene Epoch)*
600,000	Homo erectus fossils from Ethiopia *(Pleistocene Epoch)*
200,000	Humans started looking like they do today
100,000	Homo sapiens (modern humans) appear in the fossil record
25,000	Neanderthal humans go extinct
13,800	Earliest flooding by Lake Agassiz; Lake Minnewaukan (Devils Lake) forms at 12,000
11,000	Last active glacial ice melts from North Dakota
9,000	Last stagnant glacial ice melts from North Dakota
8,300	Lake Agassiz drains from North Dakota for the last time
1,000	Icelander Leifur Ericksson lands in North America
208	Lewis and Clark spend the winter with the Mandan Indians at Fort Clark
?	You were born

Rocks Found at the Surface

Rocks and boulders scattered over the land surface are the oldest materials found in North Dakota. Most are glacial erratics (the word "erratic" is based on a Latin word for "wanderer"). The Ice Age glaciers carried erratics here from the places in Canada where they formed billions of years ago during a time when molten rock in the Earth's crust was still cooling and solidifying into igneous rock, such as granite. I will return to glacial erratics in chapter five, in my discussion of field stones.

THE CRETACEOUS PERIOD

Erratics aside, the oldest geologic materials at the surface in North Dakota, found in the same locations and positions in which they were originally deposited, are Cretaceous in age, ranging from 91 million to 65 million years old. During the Cretaceous Period, North Dakota was submerged beneath a shallow, subtropical sea. The 300- to 400-foot-deep sea was part of the Western Interior Seaway, a trough-like area that extended through North Dakota and connected the Arctic Ocean to the Gulf of Mexico. The fine-grained sediments that accumulated on the seaway floor included several shale formations: the Greenhorn, Carlile, Niobrara and Pierre, all of which can be seen in North Dakota today. Slightly coarser sediments—sandstone and siltstone—of the Fox Hills Formation were deposited near the shores of the sea, and in shallow estuaries and lagoons. All of these rock formations are marine, meaning they were deposited in seas.

By the time the Cretaceous Period came to a close, 65 million years ago, the Interior Seaway had become filled with sediment and a large delta had formed over western North Dakota. The sediments that make up this delta belong to the Hell Creek Formation, which consists of beds of sandstone, siltstone and mudstone; all materials deposited by running water, by wind, or in lakes or swamps.

The geologic history of the Cretaceous Period in North Dakota is known mainly through study of its fossils. The shale formations contain abundant, warm-water plant and animal fossils.[3] The plant and animal life, recorded by the fossils in the various Cretaceous formations, indicates a warm, tropical to semitropical climate. The youngest Cretaceous formation, the Hell Creek, contains small amounts of

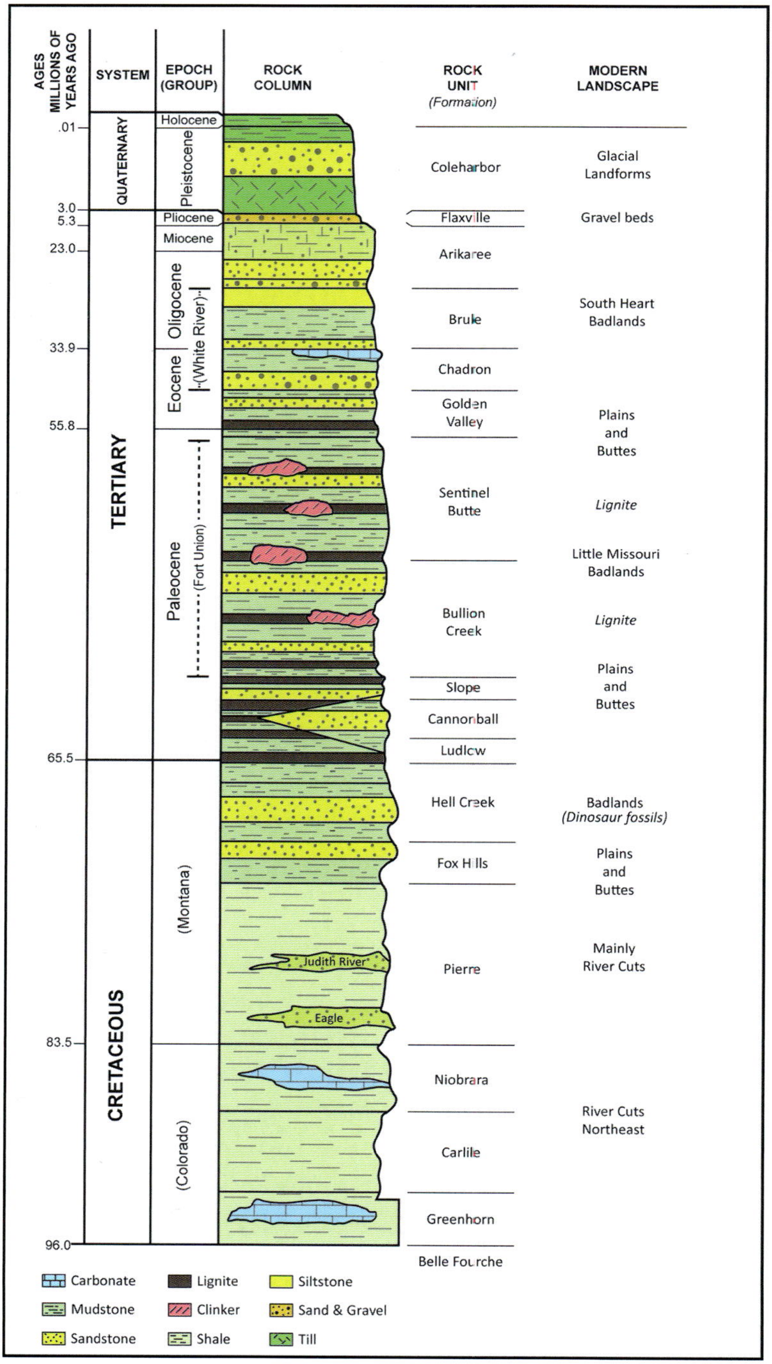

Geologic Formations Exposed at Surface in North Dakota

coal and petrified wood, and it is rich in dinosaur fossils. The Hell Creek dinosaur fauna was dominated by (or at least is best known for) large herbivores that thrived in woodlands, along with some carnivores. These North Dakota dinosaurs were among the last that lived on Earth.

The Greenhorn Formation

Cretaceous formations recognized in North Dakota include some small exposures of limey (calcareous) marine shale of the Cretaceous Greenhorn Formation. These patches of 91 million-year-old shale can be seen in only one small area in and along the streambed of the Pembina River in Pembina County.[4] Exposures of rocks belonging to Cretaceous formations younger than the Greenhorn are more widespread, and include, in order of decreasing age, the Carlile, Niobrara, Pierre, Fox Hills and Hell Creek formations. I have included photo examples of Carlile through Hell Creek formation exposures to show how they occur and how they differ in appearance from one another.

◀ This illustration shows the geologic materials that can be seen at the land surface in North Dakota. A stratigraphic column of all the rocks that occur in the state is shown on the inside back cover.

The **Rock Column** shows the lithology and texture of the various materials that occur (see legend of lithologies). Only some of the materials are likely to be found at any given location. The irregular line along the right side of the **Rock Column** indicates the relative resistance of materials to erosion—rock units extending farther to the right are more resistant; those that protrude less are more easily eroded.

In addition to the **Rock Column**, the names of recognized geologic formations are shown in the **Rock Unit** ***(Formation)*** column. Brief descriptions of the landscape developed on various rock units are shown in the **Modern Landscape** column.

To the left of the Rock Column, the ages, in millions of years, are shown for selected points of reference. The names of the major geologic systems (Cretaceous, Tertiary, Quaternary) are included, along with the epochs recognized by geologists. Also included are two traditional geologic groups—the Fort Union and White River.

This illustration is modified from part of the more-inclusive *North Dakota Stratigraphic Column,* published in 1986 by the North Dakota Geological Survey (NDGS Miscellaneous Series 66 by John P. Bluemle, Sidney B. Anderson, John A. Andrew, David W. Fischer and Julie A. LeFever).

Carlile and Niobrara Formations

The Carlile Formation (between 90 million and 87 million years old) and the Niobrara Formation (87 million to 84 million years old) are both marine shales, exposed only in valleys in the eastern part of the state. The Carlile is found only in the Pembina River Valley, and the Niobrara is found mainly in the Pembina and Sheyenne River valleys. The Carlile is a soft, dark gray calcareous shale, and the Niobrara is a lighter shade of gray due to its more limey content. The Niobrara tends to be yellowish in places where it is exposed and has been weathered.

▲ A road cut exposes the Cretaceous-age Niobrara Formation shale four miles north of Fort Ransom in Ransom County. The Niobrara Formation is limey (calcareous) shale that is exposed in several places along the Scenic Byway road following the Sheyenne River, as well as in river cuts along the Pembina River in northeastern North Dakota.

The golden color of the shale is the inspiration for the name of the Little Yellowstone Park, a few miles from this location. The shale weathers to the golden color in places where it is exposed at or near the ground surface. When it comes in contact with oxygen in the air, the limey shale weathers—oxidizes—to the golden color seen here. At depth, below the water table, the Niobrara shale is light gray in color. JPB Photo: 717/2009

▲ A river cut along the Pembina River in Cavalier County five miles southwest of Walhalla reveals the Carlile Formation. The river is cut deeply into the dark gray Carlile Formation shale, which is overlain (at the top of the river escarpment) by the golden-colored Niobrara Formation shale. The upper third of the Carlile Formation is represented here, along with the lower 10th of the Niobrara Formation.

Both of these formations are Cretaceous in age. Except for the Cretaceous Greenhorn Formation shale, which is found in only a very few small river-level exposures, the Carlile is the oldest geologic formation exposed in North Dakota. JPB Photo: 9/14/2011

▲ This is a closer view of the exposure of Niobrara Formation shale shown in the previous image. The shale structure is blocky, with irregular bedding. The outer part of the exposure, which has been subject to weathering and exposure to oxidation for longer than the deeper parts, is more tan in color than the grayer parts of the exposure that were deeper. Fresh breaks on the blocks are gray and show little evidence of oxidation. JPB Photo: 7/11/2011

The yellow exposures inspire the name for Little Yellowstone State Park in the Sheyenne River Valley in Ransom and Barnes counties. Surface expressions of the Carlile and Niobrara formations are cutbanks in places where rivers are carving into them. Landforms associated with them are, exclusively, the result of ongoing river erosion. None of these three shale formations—the Greenhorn, Carlile and Niobrara—cover broad areas sufficiently extensive to allow pasture or cropland to develop on them.

Pierre Formation

The Pierre Formation (ranging from 84 million to 70 million years old) is a gray marine shale. It is widespread and found at the surface in several places in North Dakota: in the southwest in Bowman County,

▲ This river cut along Beaver Creek in Seeman Park, Linton, Emmons County, exposes Cretaceous Pierre Formation shale bedrock. Deep cuts in the Pierre shale, such as this one, are uncommon in most places that the shale occurs, as the soft sediments tend to slump, forming gentle slopes. The Pierre shale is exposed along the Sheyenne River in Barnes and Griggs counties, in many places along the James River, in some of the tributary gullies to the Pembina River in Cavalier and Pembina counties, and in the southwestern part of the state. Small, isolated exposures of Pierre shale may be seen in many places in eastern North Dakota.
JPB Photo: 7/11/2011

south-central in Emmons and Sioux counties, northeast in Walsh and Cavalier counties, and along the slopes of the Sheyenne River, through Benson, Eddy, Nelson, Griggs, Barnes and northern-most Ransom County. As it is quite soft and erodes easily, the landforms on the Pierre shale surface tend to have low relief and gentle slopes. In places where the Pierre shale covers extensive areas, the land is rolling and may have a scrubby, sometimes desolate character. The soils that develop on the Pierre surface tend to be poor, resulting in marginal farmland.

Fox Hills and Hell Creek Formation

A variety of landforms are developed on the marine Fox Hills Formation (70 million to 69 million years old) and on the continental (non-marine) Hell Creek Formation (69 million to 65 million years old). Continental materials were deposited by rivers and streams, not in seas. These two formations cover parts of Burleigh, Emmons and Sioux counties in south-central North Dakota, and Bowman and Slope

▲ Smoothly rolling topography developed on the Fox Hills Formation sandstone and shale 11 miles southwest of Hazelton in Emmons County. Soils developed on Fox Hills materials tend to be fertile, resulting in good farmland, although rolling buttes occur in places. Most of the buttes, like the one shown, are the dwindling remains of buttes that were once capped by Hell Creek Formation sandstone.

JPB Photo: 5/17/2010

counties in the southwest, as well as in a few scattered locations where the covering of glacial sediments is too thin to completely cover the underlying bedrock. The Fox Hills Formation consists of beds of sand and shale, which in most places weather to smooth and gentle slopes, with a few hard sandstone beds. Good farmland is developed on the Fox Hills Formation, but some exceptions are found in extreme south-central and southwestern parts of the state where acid soils are developed on the Fox Hills Formation. Such areas are most commonly used as rangeland.

The Hell Creek Formation is made up largely of sandstone, with a few beds of lignite. In places where it is exposed near rivers and streams, badlands topography has developed on the Hell Creek Formation beds, which are rich in dinosaur fossils in some places. Hell Creek Formation sedimentary rocks are of continental origin, meaning they were deposited by erosional processes operating on the land surface, not in seas. Farmland on the Hell Creek Formation beds is, as a whole, poor, even outside of badlands areas.

The Cretaceous Period appears to have ended with a bang when an asteroid hit the Earth approximately 65 million years ago in what is today part of Mexico. The asteroid impact may have caused sufficient change in the climate to cause the extinction of the dinosaurs. Geologists have studied evidence of the collision in places in southwestern North Dakota, where a thin layer of clay, rich in iridium, marks the contact between the Cretaceous beds and the overlying Paleocene sediments. Iridium is found in high concentrations in meteors and asteroids and was widely distributed by the impact.[5] The clay layer marks a deposit of air-borne, iridium-rich dust that settled to the ground following the asteroid impact. Although the dust probably settled over widespread areas, it was preserved in only a few places. Southwestern North Dakota is one of those places.

TERTIARY FORMATIONS

The surface of much of western North Dakota is covered by several formations of Tertiary age, of continental origin, ranging from 65 million to 3 million years old. Tertiary time has been subdivided into several epochs: Paleocene (65 million to 56 million years), Eocene (56 million to 34 million years), Oligocene (34 million to 23 million years),

▲ **Badlands topography developed in the Cretaceous Hell Creek Formation three miles northeast of Marmarth in Slope County. This view, to the west, shows a smooth, white to tan bed of bentonitic clay lying on top of gray, vertically rilled sandstone. Numerous concretions serve as caps on the pedestals in the gully on the left.**

The hills in the distance are badlands topography carved from Hell Creek beds. When I took the photo, a thunderstorm was approaching from the west (the dark sky on the horizon to the right). JPB Photo: 7/28/2011

Miocene (23 million to 5 million years) and Pliocene (5 million to 3 million years). Formations of Tertiary age consist of sandstone, silt and shale, interbedded with lignite, especially in the Paleocene formations.

The geologic history of North Dakota during the Tertiary began in Paleocene time, as the Cretaceous ended, with a climate that was subtropical to tropical and resulted in forested swampland. The dinosaurs were gone; mammals and birds were evolving. The Cannonball Sea flooded most of the eastern two-thirds of North Dakota, while the western part of the state was swampland, ponds and lakes—areas now covered by the Ludlow Formation. Volcanic ash, blown eastward from eruptions of volcanoes in the developing Rocky Mountains, settled in the lakes and swamps. The Ludlow Formation sediments are similar to those of the older Hell Creek, except they lack dinosaur remains. The Cannonball Formation consists of marine sediments, dark-colored sandstone and mudstone. The Ludlow and Cannonball formations may be the same age, or nearly so, ranging from 65 million to 60 million years old. Like the earlier Hell Creek Formation deposits, Ludlow beds commonly erode to badlands topography. Cannonball landscapes, on the other hand, are smooth and, where it is extensively exposed (Morton, Burleigh, Grant and Bowman counties), good cropland may have developed on the Cannonball beds.

The remaining Paleocene formations overlying the Ludlow and Cannonball were deposited in swampy environments represented today by layers of sandstone, siltstone, mudstone and claystone. They include the Slope, Bullion Creek, Sentinel Butte and Golden Valley formations, which represent a succession of multicolored rock units, rich in lignite. These formations, although widespread in western North Dakota, are seen to their best advantage in areas of badlands.

Temperatures initially became warmer as the Paleocene gave way to the Eocene Epoch. Dense, tropical forests covered the land. Much of western North Dakota was still a warm, humid, densely forested, swampy lowland. Late in the Eocene, worldwide temperatures began to fall and the climate in North Dakota changed from subtropical to drier and more temperate by Oligocene time, although it remained warmer than our modern conditions. North Dakota during the Oligocene was an open plain—a scrubland and savanna consisting of grasses and shrubs, few trees and some ponds.

During late Eocene and through Oligocene time, sediments deposited on flood plains eventually hardened into the rocks of the Golden Valley, Chadron, Brule and Arikaree formations. Most of these rocks have since been eroded away, but remnants still cap many of the buttes in western North Dakota. By the time the Miocene Epoch began, 23 million years ago, North Dakota's climate was cooling and becoming increasingly arid. The state was likely grassland, but not a lot is known about climatic conditions. An extended period of erosion that began in Miocene time eventually carved the Tertiary formations into the badlands, buttes and widespread rolling prairies we see today over much of western North Dakota.

I will describe the landforms developed on each of the various Tertiary formations in more detail later. Except where the land is too hilly, in areas of badlands and buttes, the exposed surfaces of most of the Tertiary materials provide good to excellent farm and rangeland.

Erosion of the Land

AMOUNT OF EROSION

How much erosion has taken place in North Dakota since the current erosion cycle began, 15 million years ago? A lot, but it's not really possible to know how much material has been removed from an area where the original amount is unknown. We can make some estimates by considering some of the highest elevation features, such as the Killdeer Mountains, Chalky Buttes and the tops of various o ther buttes.

Both the Killdeer Mountains and Chalky Buttes are capped by layers of resistant limestone, material deposited in lakes a little more than 20 million years ago. Since that time, a complete reversal of topography has taken place. What were lowlands then are highlands now. While they existed, during Miocene time, lakes were situated in some of the lowlands. Surrounding areas nearby were at higher elevations than the lakes themselves (that is, the lakes were surrounded by uplands). The same is true of the materials on top of many other western North Dakota buttes; they began as riverbeds or lakes, existing in settings lower than their surroundings. These same materials, once lowlands, are now found on uplands rising as much as 800 to 1,000 feet above nearby areas. Might we infer, then, that since Miocene time, at least 800 to 1,000 feet of material (the height of the buttes) has been removed from areas surrounding the uplands?

Yes, we might infer it, but that does not mean we would necessarily be correct in doing so. The amount of erosion that has taken place cannot be determined simply by subtracting the elevation of the lower area surrounding an upland from the elevation at the top of the upland and then assuming that the difference equals the amount of erosion (the tops of the Killdeer Mountains are at 3,100 feet above sea level, and the surrounding area is at 2,500 feet, so 600 feet of material must have been removed by erosion, etc.). It is not that simple.

At the time materials that now form the caprock on top of the Killdeer Mountains were being deposited in lakes, we do not know *how much* higher the nearby uplands were. It may indeed be that 600 feet of sediment, or perhaps more, were removed in the immediate area of the Killdeer Mountains, but that figure cannot be projected away from there for any great distance. At the same time erosion was tak-

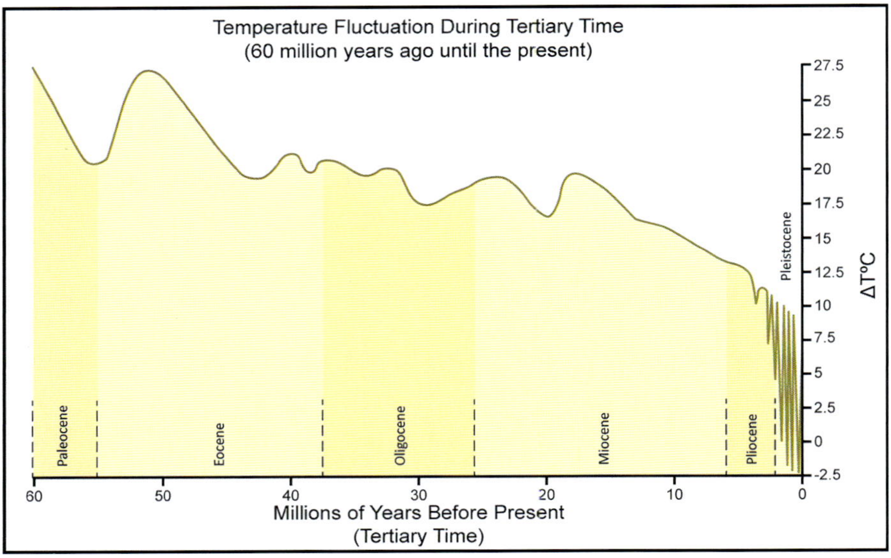

▲ This diagram shows temperature fluctuation (mean annual temperatures) in central Europe during Tertiary time, the past 60 million years. The change in overall temperatures should be similar to changes that occurred in North Dakota during the same period of time.

Temperatures peaked at 27 degrees Celsius during Eocene time, and again in Miocene time at 20 degrees Celsius. Overall, temperatures decreased gradually throughout the Tertiary. Temperatures dropped through Oligocene time, as the Antarctic continental plate moved over the South Polar Region, again in middle Miocene time, and finally in late Pliocene/early Pleistocene time, as the Ice Age intensified. The modern condition in North Dakota is approximately +4 to +5 degrees Celsius.

During Pleistocene time (far right edge of diagram), during which North Dakota has experienced numerous glaciations, temperatures have fluctuated widely, ranging from full glacial to interglacial conditions. I have shown a temperature range of about 10 degrees, but it is likely the actual range was greater. I cannot accurately show the number of changes during this relatively short period of time due to 1) the generalized scale used to depict overall geologic history and 2) uncertainty as to how many times glaciations have occurred.

I used information from two references to compile this diagram:

1. Bluemle, J.P., Sabel, J.M., and Karlén, Wibjörn, 1999, *Rate and Magnitude of Past Global Climate Changes,* Environmental Geosciences, vol. 6, No. 2, pp. 63-65
2. Zachus, James, Pagani, M., Sloan, L., Thomas, E., and Billups, K., 2001, *Trends, Rhythms, and Aberrations in Global Climate, 65 Mya to Present,* Science, vol. 292, p. 686-693.

ing place, the Earth's crust in that area was rising in response to the weight of the sediment being removed. Just as the crust sank under the weight of the more than three miles of sediment that had earlier been deposited in the Williston Basin, the land rose when some of that sediment was eroded away (more on the concept of *isostatic rebound* in my discussion of crustal rebound in chapter seven on the Red River Valley and glacial Lake Agassiz).

I attempted a variety of techniques, trying to determine how much the amount of erosion in western North Dakota might have been. The results ranged from an estimated average of 300 feet to more than 1,000 feet of material removed from the area.[6] The various methods I used gave such differing results that I prefer to simply say that extensive erosion occurred over western North Dakota during the past 15 million years.[7]

Less erosion took place in the eastern part of the state than in the west. The Red River Valley, for example, saw perhaps 50 to 100 feet of material removed, but much of that material was later replaced by glacial and lake sediment, deposited during the Ice Age. The Ice Age ended the 15 million-year erosion event, at least in places reached by the glaciers (erosion continues, uninterrupted, in much of the western part of the state).The glaciation that began in North Dakota, perhaps as long ago as 3 million years ago, provided its own version of erosion; erosion by glacial ice rather than running water. Over parts of eastern North Dakota, the glaciers probably brought in more material than had been eroded away during the millions of years preceding their arrival.

What was the North Dakota landscape like during the 15 million years of erosion? The western part of the state was probably a broad, level surface covered by gravel, much like the landscape in eastern Montana today. Buttes and valleys formed, were worn down and eroded away, a sequence that was repeated several times, and continues today.

The eastern part of the state during the 15 million-year erosion event was a level to gently rolling plain, less hilly than the west. I do not know when the Red River Valley first formed, but it has existed as a lowland for several million years, and a river has flowed north through it ever since it was formed. In places, the river carved a valley through granite and limestone, now buried beneath 100 to 500 feet of glacial deposits. Before it was glaciated, relief in the area that is now the Red River Valley was likely greater than in the rest of eastern North Dakota. Intricate valleys were carved from a variety of much older materials, ranging from Precambrian granite to layers of limestone and sandstone. All of this relief is today buried beneath several hundred feet of glacial and lake deposits, under the flat floor of the Red River Valley.

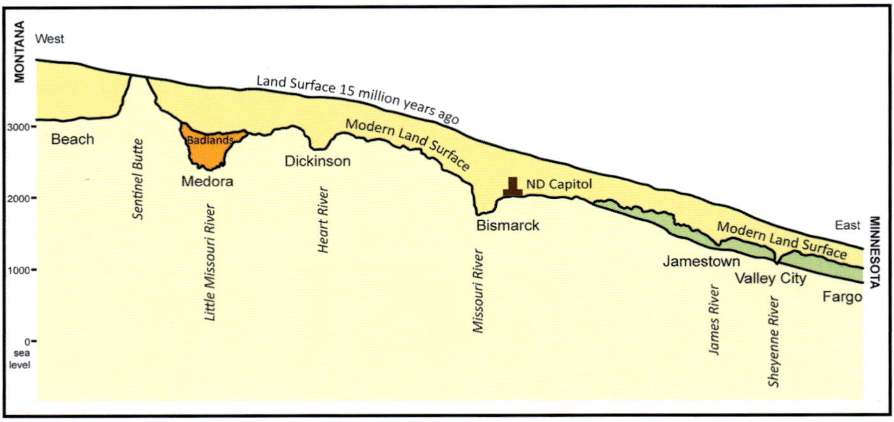

▲ This is a greatly simplified profile/cross section (a vertical slice) that approximately follows Interstate Highway 94 across North Dakota, from west to east. The vertical scale of the cross section is shown along the left margin in 1,000-foot increments. The length of the cross section is about 350 miles. Vertical exaggeration is approximately 180 times.

The top-most east-west line represents the land surface 15 million years ago. Erosion has removed the materials (shown in gold) beneath that ancient surface down to the modern land surface. The North Dakota Capitol building is about half as high as is the average thickness of the covering of material that was eroded away in the past 15 million years. The amount of material eroded from the area of the badlands (reddish orange) was greater than elsewhere. This erosion was accomplished more recently and quickly than was the erosion of areas outside of the badlands.

Features shown along the modern land surface include Sentinel Butte, the Little Missouri Badlands, Heart River Valley at Dickinson, Missouri River Valley at Bismarck, James River Valley at Jamestown, and Sheyenne River Valley at Valley City.

East of Bismarck, beginning half way between Bismarck and Jamestown, the erosion removed materials to depths greater than the current land surface. However, glaciers deposited material on the surface (shown in green), partially compensating for materials that were removed by erosion during the 15 million-year erosion event.

NORTH DAKOTA'S MODERN LANDSCAPE

The modern landscape began forming during Miocene time, probably between 15 million and 20 million years ago, when a prolonged period of erosion began.[8] Since then, streams and rivers have carried large amounts of sediment from the state. Rivers flowed northward and transported sediment they had picked up in North Dakota into the Labrador Sea in the North Atlantic Ocean, between southernmost Greenland and Labrador.

A single major river system drained most of Canada throughout much of Oligocene, all of Miocene and most of Pliocene time, until it was destroyed by Pleistocene glaciation. While it existed, for more than 20 million years, the river—the Bell River—and its tributaries in Canada and part of the United States, including all of the rivers in North Dakota, carried sediment eroded by those tributaries, northeastward to the Labrador Sea.

Beginning 5 million years ago, parts of western North Dakota were covered by coarse gravel, deposited by rivers flowing northeastward from the eroding Rocky Mountains. During a several-million-year erosion event, gravel beds, 10 or more feet thick, were deposited over widespread areas, extending eastward across eastern Montana, into western North Dakota. Much of eastern Montana is still covered by these late Pliocene gravel beds, sometimes referred to as the Flaxville and Wiota gravels. The gravel beds are overlain in places by till, but they never overlie till, so they are preglacial in age. Both the Flaxville and Wiota gravels consist of well-rounded, often brownish quartzitic pebbles (mainly chert), up to 2 inches in diameter. Agates are common in the gravels. The Flaxville gravels occur on uplands and high terraces above the Yellowstone River. The Wiota gravels are younger, likely consisting in part of Flaxville gravels that were re-worked by local streams.

In North Dakota, erosion has removed most of the gravel deposits, but remnants of the gravel surface remain in parts of McKenzie and Williams counties, as well as in the western-most parts of Slope, Bowman and Golden Valley counties. The easiest places to see the gravels are along Interstate Highway 94 in eastern Montana.

The final event in the long erosion story began as the Pliocene came to a close, between 3 million and 3.5 million years ago, when glaciers formed in Canada and advanced into the state. The glaciers de-

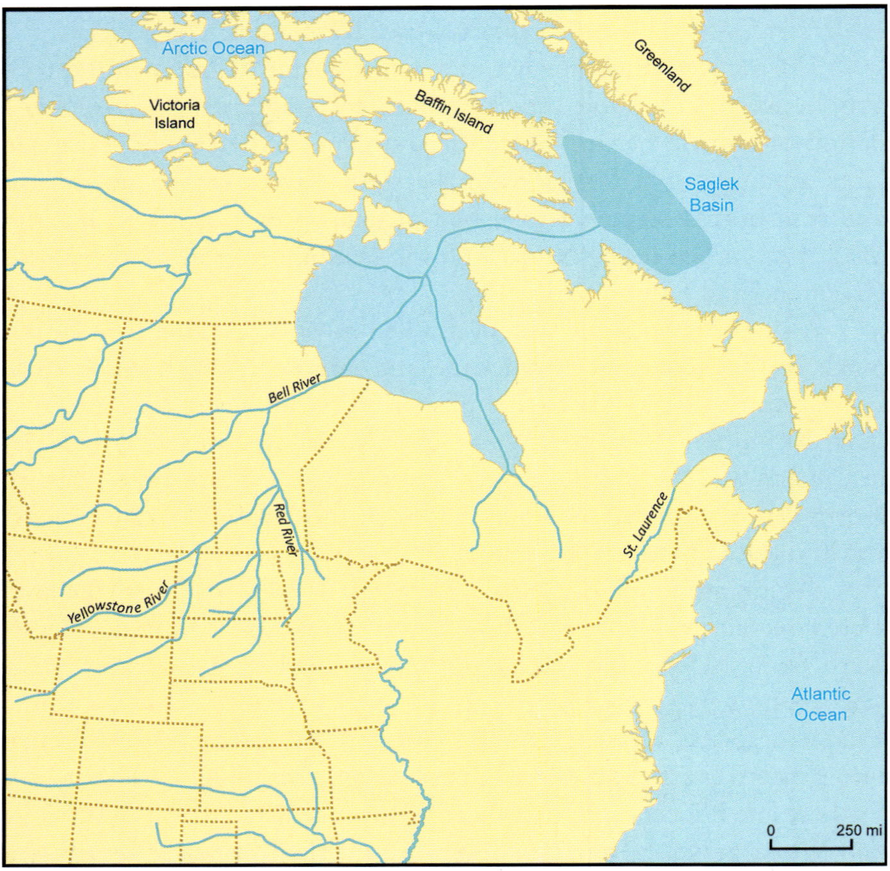

▲ This map shows the preglacial drainage pattern in North America east of the Rocky Mountain continental divide and west of the Mississippi River. During much of the interval beginning about 30 million years ago, in Oligocene time, and lasting until the beginning of the major series of glaciations at the end of the Pliocene/beginning of the Pleistocene 3 million years ago, much of Canada and the north-central U.S. was drained by a large Canadian river known as the Bell River. The Bell River and its tributaries in North and South Dakota, eastern Montana and northern Wyoming carried eroded sediment to the northeast. Much of the sediment was deposited in a large delta (the Saglek Basin) in the Labrador Sea, between Baffin Island and Greenland.

I have not shown drainage east of the Mississippi River because events there did not affect the area that is North Dakota.

For more detailed information about the drainage pattern in North Dakota, please refer to chapter three. In compiling this map, I used information from:

1. Sears, J.W., 2013, *Late Oligocene-early Miocene Grand Canyon: A Canadian connection:* GSA Today, vol. 23, No. 11, p. 4-10
2. McMillan, J.N., 1973, *Shelves of the Labrador Sea and Baffin Bay, Canada:* Canadian Society of Petroleum Geologists Memoir 1, p. 473-515.

stroyed virtually the entire drainage system, the Bell River, all its tributaries in Canada, and most of them in North Dakota. Glaciers picked up some of the materials they flowed over, carried them southward, and deposited them over northern and eastern parts of the state. The result was an entirely different kind of topography than North Dakota had known up to then. The glacial landscape consisted of small, closely spaced landforms, in sharp contrast to the wide-open style landscape of western North Dakota. In places where the glacial deposits form level to rolling surfaces, they serve as excellent farmland.

To summarize, North Dakota's modern landscape is the product of two different processes: 1) erosion by rivers and streams, which removed large amounts of sediment, resulting in the landscape we see today in the western part of the state; and 2) glaciation, which deposited sediments of various kinds on the surface, forming a complex suite of glacial landforms in the northern and eastern parts of the state. One way to contrast the two landscapes, western and eastern, is to visualize the west as an area *sculpted* from already-existing materials, and the eastern landscape as an area *built* of materials, newly delivered by glaciers.

Summary

I have provided the briefest possible sketch of North Dakota's geologic history, recorded during the past 90 million years in the various rock formations exposed at the surface throughout the state. During the first third of that time, most of the state was submerged in shallow, tropical seas. Then, as conditions became cooler and drier, a long period of erosion beginning between 15 million and 20 million years ago stripped away much of the sediment that had been deposited. Continued cooling culminated in glaciation and a series of ice ages.

The timelessness of the Ice Age is difficult to imagine. The comings and goings of the glaciers and everything that accompanied them, tundra, rivers of ice, running water and cold winds, occurred many times. In fact, we are probably still living in the most recent Ice Age. The work of the glaciers, and especially the record of their departure, is inscribed on the land. It can be seen in the hills, lakes and ponds of the Missouri Coteau, in the broad, flat expanse of the Red River Valley, in the route of the Missouri River through North Dakota, and in the rolling plains between. Even the badlands along the Little Missouri River ultimately owe their origin to glaciation. The boulders scattered across the land testify: The glacier was here! Perhaps the most important heritage of the Ice Age is our farmland, the nutrient-rich soil that makes our crops grow so well. Once you are aware of the fact of the Ice Age, its reality cannot be ignored.

Geography and Climate

North Dakota is located at the geographic center of North America and ranks 19th in size, compared with the other 49 states. It covers 70,762 square miles, of which 68,994 are land. North Dakota is bounded on the north by Canada (Manitoba and Saskatchewan), on the west by Montana, on the south by South Dakota and on the east by Minnesota. David Laskin writes:

> An evil genius could not have devised a more perfect battleground for clashing weather fronts than the prairies of North America. When conditions are right, which they frequently are, vigorous fronts unleash the worst weather in the world over this region—super-cell thunderstorms spawning tornadoes in late spring, huge globes of hail falling from anvil topped cumulonimbus clouds in summer, blizzards in winter.

North Dakota's location at the geographic center of North America results in climatic extremes, a pronounced continental environment featuring hot summers and cold winters. The weather can be extremely variable during short periods of time because the region is the confluence area for competing air masses that sweep across the midcontinent, bringing heat or cold, moisture or drought. Winter cold spells result when polar air descends from the Canadian Arctic, dominating the area for a while. Gulf winds blowing up the Mississippi River Valley bring warm, moist air. Winds from the west produce dry, so-called Pacific air masses, bringing summer droughts or unusually warm winter weather—even chinooks, which may affect the western part of the state. North Dakota experiences large annual, daily and seasonal changes in temperature. Light to moderate precipitation tends to be irregular, both in time and coverage. Most of the time, we have low relative humidity, plentiful sunshine and practically continuous air movement.

In places where the winds blow onshore from the ocean, into Oregon and Washington, for example, the climate tends to be temperate, with cool summers and mild winters, because ocean water has a high capacity to store heat energy. However, before such Pacific air masses reach North Dakota, they are modified by mountains, changing both their temperature and moisture characteristics. The western slopes of the mountains interrupt most of the otherwise moderating influences of the Pacific air, thus amplifying the continental characteristics of North Dakota's climate.

North and south are a different story. No mountain ranges impede air masses originating in the polar areas or the Gulf of Mexico, so air masses originating in these regions easily overflow the state, sometimes with only minor changes in their original characteristics. Those air masses flowing southward remain frigid; those flowing northward retain much of the heat and moisture with which they started. The usually rapid progression of air masses (weather systems) over North Dakota from the different source regions causes frequent, and sometimes extreme, weather changes and usually prevents our weather from becoming monotonous.

Day length and solar zenith angle at North Dakota's northerly latitude are important factors determining the climate. Daylight ranges from less than nine hours in December to more than 16 hours in June; noon sun angles are much higher in summer than they are in winter. These combined factors produce much more radiational energy in summer than in winter, re-enforcing the large seasonal temperature changes and the general north-south temperature gradient across the state.

The average annual temperature in North Dakota ranges from 37°F in the northeast to 43°F along the southern border of the state. January is the coldest month, with average temperatures ranging from 2°F in the northeast to 17°F in the southwest. July is the warmest month, with temperatures averaging 67°F in the northeast to 73°F in parts of the south. The range of normal average monthly temperatures between the coldest and warmest months is 54°F in the southwest and 65° F in the northeast. These large annual ranges attest to the continental nature of North Dakota's climate. The National Oceanic and Atmospheric Administration (NOAA) states that North Dakota has the highest average number of heating degree-days of the lower 48 states (the number of degrees that the daily average temperature falls below 65° Fahrenheit determines heating degree-days; for example, if the day's high temperature is 60 and the low is 40, the average temperature is 50 degrees; 65 minus 50 equals 15 heating degree-days).

North Dakota's winters are noted for their blizzards (I will not bother to define them here). The word "blizzard" appears to be peculiarly North American. European newcomers to the prairies had never experienced a blizzard. One account says that early German settlers referred to these sudden winter storms as *blitzartig*, meaning "the storm comes lightning like." Blitzartig eventually became blizzard.

The highest temperature ever recorded in North Dakota was 121°F at Steele on July 6, 1936, and the lowest temperature measured was minus 60°F at Parshall on February 15, 1936—a 181-degree range within a five-month period. The 181-degree difference is the second-greatest separation between the record low and the record high temperature within any of the 50 states (Montana's 187-degree difference is the greatest), but North Dakota's 181-degree difference is easily the most extreme change occurring within a five-month period.

Weather influences our natural and cultural life. Floods, droughts, dust storms, tornadoes, cold spells, heat waves and blizzards illustrate weather's influence on human affairs. We tend to remember extremes in the weather such as those I just mentioned. Many people still alive recall the drought of the 1930s, which began in earnest in 1931 and lasted 10 years. During those years, rains virtually disappeared, temperatures soared, dust clouds darkened skies and choked every living thing. Many lakes dried up and some rivers slowed to a trickle. Crops withered and died across the parched land. The drought could rightly be blamed on nature, but the dust storms resulted largely from overzealous plowing of fields, which exposed millions of acres of topsoil in the western Dakotas and elsewhere in the West. During the 1930s, the jet stream weakened, and any moisture it did pick up fell farther south than usual. Droughts are naturally recurring events, part of the overall climatic variability of the Plains. Tree ring and other evidence show serious 19th century droughts in the midsection of the continent in 1800, 1808, 1809, 1819, 1823 and 1852. From the 1890s to the 1970s, droughts occurred in the Great Plains every 20 years.

Droughts tend to occur when the zonal (westerly) flow of Pacific air prevails because it redirects the Gulf moisture eastward into the Great Lakes region. Persistent droughts in the Great Plains are usually associated with La Niña activity. Frontal activity decreases during regional droughts, but localized convective thunderstorms tend to increase in frequency. They are, however, unpredictable in their movement and provide limited moisture to only localized areas.

Recent research helps us understand why the 1930s were so dry. When climate researchers at NASA studied weather records of the tropical Pacific and Atlantic oceans through the 1930s, they found that sea-surface temperatures played a major role in causing the drought. Severe drought conditions involved cooler than normal Pacific Ocean temperatures, coupled with warmer than normal Atlantic Ocean temperatures. This combination changed the course of the low-level jet stream. The jet stream normally flows westward over the Gulf of Mexico, picks up moisture from the Gulf, and then turns northward, bringing rain to the Plains.

Though North Dakota's relief is less pronounced than in states to the west, it still produces important variations in temperature and precipitation. Topography affects temperature patterns because air temperature usually decreases with elevation at a rate of 2°F to 4°F per 1,000 feet. Air in the lower portion of the atmosphere warms at a rate up to 6°F per 1,000 feet in descending a slope. Topography can affect precipitation in several ways. It can increase precipitation by providing the lift necessary to start the process. When moist air flows uphill and the upper layers of moist air reach condensation level, precipitation will begin. When moisture in the lower layers is plentiful, and the air aloft is cold and unstable, a lifting of only a few hundred feet may be enough to cause precipitation. On the other hand, precipitation is inhibited when air flows downhill. Convection can be enhanced over higher terrain due to large temperature contrasts during strong daytime heating of the highlands. Evaporation loss is less because raindrops fall a shorter distance from clouds over higher ground.

I have stressed temperature and precipitation in this brief discussion of North Dakota's climate. However, climate cannot be so simply described. It includes factors such as soil temperature, humidity, wind velocity and direction, atmospheric pressure, solar and terrestrial radiation, and snow cover. Solid and liquid particulate matter and gases in the atmosphere are important. Besides oxygen, nitrogen and water vapor, a list of the more important atmospheric constituents includes carbon dioxide, ozone, dust, smoke, salt particles, and sulphur and nitrogen oxides. Individual measurements of some of these constituents are difficult, as they may occur in very small amounts but, taken together, they help determine our climate.

Climate is strongly influenced by volcanic eruptions. In fact, the Little Ice Age, which began around 1350 and ended in the mid-19th century, was caused by several large volcanic eruptions in the second half of the 13th century. The series of eruptions cooled the Earth, and it was not until the recent transition to a warming climate, beginning in the 19th century, that temperatures began returning to "normal" conditions. The Earth's—and North Dakota's—climate are always considerably cooler when volcanic activity is vigorous. The climate tends to be warmer when volcanic activity is at a minimum.

Volcanic events, and their effects on climate, may be recorded in glacial ice. An ice-core taken in Greenland revealed a thin layer of ash, deposited when Mount Tambora in Indonesia blew its top in April 1815. Tambora's eruption blasted 12 cubic miles of debris and volcanic ash dust into the atmosphere (estimates of the amount of material vary widely). Some of the dust settled on the Greenland glacier and was later covered by more ice; the thin layer of dust acts as a precise marker of the volcanic event. A dust cloud circled the globe for at least a year following the eruption, dimming the sun and cooling temperatures across the Northern Hemisphere. The finest ash particles remained in the atmosphere for several years, blocked sunlight and disrupted atmospheric circulation patterns. Snow fell in the summer and crops failed. The year 1816 is sometimes referred to as the "year without a summer." The ruined crops that resulted led to the "year of the beggar" in 1817. Natural climate change can be rapid, extreme and socially disrupting.

Soils in North Dakota

*Note: This discussion of soils is based, in part, on information included in a soon-to-be-published book (*The Northeastern Plains*) by Dr. C. Thomas Shay, as well as comments and publications by Dr. Eric Brevik.*

North Dakota's soils are developed on an array of materials that were deposited in different ways, at different times, in different parts of the state. Soils are developed on Tertiary-age bedrock that is millions of years old in western North Dakota, and they are developed on glacial deposits only a few thousand years old in the east. The materials on which our soils are developed—the parent materials—have widely varying origins: marine seas, glacial and river deposits, freshwater and lake sediments, and wind-blown materials. Regardless of the kinds of materials beneath them, the soil forms a kind of living mantle that develops during long periods as climate interacts with organisms upon a parent material.

Knowledge, or at least awareness, of soils dates to the earliest known practice of agriculture 11,000 years ago, when implements for harvesting and tilling were first used. Ancient Greek scientists had a clear understanding of soils, which included recognition of soil properties and the differences among soils from one place to another. They realized that the soil provides nutrition to plants, and they developed a concept of the soil profile. Plato and Aristotle linked soil to the giving of life by comparing it to a woman or mother.

The process of soil formation was still only poorly understood in the 19th century, when most observers still believed soils were inert substrates, a kind of loose, crusty material on the surface of the ground. In 1837, Charles Darwin presented a scientific paper explaining how earthworms help form soil. He demonstrated their importance in affecting the rate of weathering of mineral materials in the soil, humus formation and differentiation of the soil profile. Darwin was the first author of a scientific publication on the biological functions of soil, the first person to recognize the importance of animals in soil formation.

Like geology, modern soil science evolved as a result of the efforts of several individuals. The Russian scientist Vasily Vasil'evich Dokuchaev (1846–1903) was the primary figure responsible for establishing the idea of genetic soil science at the end of the 19th century.

Dokuchaev recognized five factors that combine to create the properties of any soil. They are: 1) parent material (material from which a soil is formed); 2) topography (scope, aspect and landscape position of a given soil); 3) climate (precipitation and temperature); 4) organisms (plants and animals, both large and small, in the soil); and 5) time (how long a soil has had to form). Different combinations of these five factors lead to the creation of soils with different properties. The factors are not completely independent of one another. For example, climate at any given location will influence the organisms that we find there, and topography and parent material influence the organisms. The interactions among the soil-forming factors can be very complex.

In the early 20th century, another Russian, Constantin Constantinovich Nikiforoff, influenced regional soil studies. Nikiforoff, who came to Minnesota from Russia as a refugee in the 1920s, became involved in mapping soils in the Red River Valley. Since Nikiforoff's pioneering efforts in Minnesota and North Dakota, many scientists have mapped and classified soils in the Plains states and provinces.

Most soils in the glaciated part of North Dakota formed on flat to rolling prairie landscapes, built upon beds of sand, silt and clay deposited during the Ice Age. These soils are geologically young and still retain an abundance of their original nutrients, especially calcium. Calcium stabilizes organic matter in the soil, making higher concentrations of nutrients available. Fine textures enhance fertility, as silts and clays hold more water and nutrients than coarse sands. Our dark prairie soils, which we call Mollisols, appear similar across the region because rainfall tends to be scarce and decomposition slow, allowing considerable organic matter to build up.

If you stop to inspect any road cut or freshly plowed field in North Dakota, you may see a thin layer of partially decomposed leaves, stems and other litter just below the plants growing on the surface. This earthy-smelling material teems with life—microscopic bacteria, fungi, worms, millipedes, beetles, mites, rodents and a host of other organisms—the front-line troops of soil development. In fact, far more species live below the prairie sod than above it. Soil microbes play a major role in producing organic matter. An array of creatures helps break down roots, leaves and stems into tiny particles and, ultimately, into the chemicals used by plants. The next layer down is a dark gray to black organic-rich layer; it is 6 to 12 inches deep, dense with roots, called the "A" horizon. Below the A is the "B" horizon, a brown or grayish-brown layer, usually thicker than the A horizon, and with fewer roots. Some prairie plants have roots that penetrate even deeper. A few may reach 15 feet, ensuring that they can survive everything but the most severe drought. The "C" horizon, which lies beneath the B, is the parent material, unaltered bedrock, gravel or glacial sediment, for example.

The soil-forming process is slow. Through time, physical and chemical weathering breaks down mineral particles so that their nutrients can be taken up by growing plants. Eventually, these nutrients are leached downward, beyond the reach of most plant roots.

Consider a single blade of grass as it decomposes. As it stops growing, carbohydrates and amino acids leave the blade and move to parts of the plant that are still living. In turn, waste products form in the drying grass blade. The blade may stay on the plant, but once it touches the ground, its fate is sealed. It becomes broken into small fragments as its water-soluble compounds dissolve and pass into the soil. A host of invertebrates and microbes, bacteria and fungi feed on the fragments as breakdown proceeds. Initially rapid, the process slows down through time, but after the actions of hundreds of different species during a few months, nothing is left of the original grass blade.

The process I have described is essentially similar throughout the state. However, the materials in a given area, such as on the glacial lake clay and silt in the Red River Valley, or the sand, silt or lignite of the Missouri Plateau, may behave somewhat differently from one another, depending upon differences in their texture and chemical composition. The climate differs considerably from east to west across the state, only slightly from north to south, in both temperature and moisture characteristics. The geology and topography are important; better soils tend to develop on more level areas and on certain geologic formations. Erosion may remove soil on steep slopes as fast as it forms. The type of vegetation (grassland, woodland, etc.) growing over an area influences soil formation (and, in turn, is influenced *by* the soil).

Landform Areas

Introduction

North Dakota's topography has been described as a three-step plain, "stepping" upward from northeast to southwest. From east to west are three distinct levels: 1) the Red River Valley (elevations from 750 to 1,100 feet); 2) the Glaciated Plains (1,100 to 2,000 feet); and 3) the Missouri Plateau/Missouri Coteau (2,300 to 3,500 feet above sea level).

North Dakota's lowest elevation is 750 feet, at the point where the Red River crosses into Manitoba, and the highest is at White Butte, 3,506 feet above sea level near Amidon. The three levels have all been modified—by relentless water and wind erosion, which created the buttes and badlands of the southwestern part of the state; glaciers, which formed a variety of characteristic landforms; and floods and lakes of glacial meltwater, which carved valleys and flooded broad areas. The landscape ranges from exceptionally flat plains formed by glacial lakes to ruggedly eroded buttes and badlands.

Physiographers subdivide and classify the land in a variety of ways. My physiographic map of North Dakota shows an "official" way (Great Plains west of the boundary between the Coteau Slope and Missouri Coteau, Central Lowland east of the boundary). Most people, though, mentally place the dividing line at the Missouri River, and that makes sense, too.

The Missouri River in North Dakota is a natural and definitive boundary. Eastward from the river valley's eastern bank lie green and golden fields of grain, growing on a hummocky landscape at first, but finally smoothing out to the Red River Valley. To the west, cutbanks of eroded sandstone and clay rise to blue-tinged buttes and vast expanses of prairie. Herds of cattle replace many of the fields of grain. Roughly carved ravines along rivers like the Knife, Heart, Cannonball and others hint at the transition to the badlands farther west, sculpted by the Missouri River's junior namesake, the Little Missouri River.

John Steinbeck, in his book, *Travels with Charley*, wrote the following, often-quoted passage:

> Someone must have told me about the Missouri River at Bismarck, North Dakota, or I must have read about it. In either

> case, I hadn't paid attention. I came on it in amazement. Here is where the map should fold. Here is the boundary between east and west. On the Bismarck side it is eastern landscape, eastern grass, with the look and smell of eastern America. Across the Missouri on the Mandan side, it is pure west, with brown grass and water scorings and small outcrops. The two sides of the river might well be a thousand miles apart."[9]

Physical Regions

North Dakota is part of the Interior Plains Region, an area stretching from the Rocky Mountains to the Appalachians. Within the state, the Missouri River Valley marks the boundary between two fundamentally different areas. To the north and east of the Missouri are the Central Lowlands, an intricately varied, glacially modified landscape. To the west are the rolling plains, buttes and badlands of the Great Plains, a landscape resulting from erosion by running water.

THE GREAT PLAINS

Artists and authors, photographers and poets: These and others have tried to capture the essence of the Great Plains. All have come up short. The state of Montana calls itself The Big Sky Country, and that's close. My own attempts, both in choosing prose to describe the vastness and in providing representative photos of geologic features, have been frustrating. Thomas Hart Benton, in 1937, wrote:

> In the West proper, there are no limits. The world goes on indefinitely. The horizon is not seen as the end of a scene. It carries you on beyond itself into farther and farther spaces. Cozy-minded people hate the brute magnitude of the plains country. For me the Great Plains have a releasing effect.

The effect of space on the Great Plains can be more disconcerting than Benton suggests. A student from Tokyo, along on a field trip I was leading through North Dakota, said she felt like she was going to fall off the face of the Earth. Her feeling is called "agoraphobia" or fear of wide-open places; the word is Greek and translates as "fear of the marketplace." To a lesser extent, some of us who have spent a lifetime on the prairies and plains may be uncomfortable in large cities. Even though I cannot describe the Great Plains as well as I'd like, I will try to explain their geology and physiography, and include some photos showing some of their features.

In North Dakota, the Great Plains include the Missouri Plateau, a broad plain interrupted by buttes, and the Little Missouri Badlands, with their intricately carved landscape. The valleys, hills and buttes of the Missouri Plateau were formed by erosion of flat-lying beds of sandstone, siltstone, claystone and lignite. The Little Missouri Badlands, carved into the Missouri Plateau, border the Little Missouri River from Bowman County northward to its confluence with the Missouri River in McKenzie County. Both the Missouri Plateau and the Little Missouri Badlands owe their shapes mainly to differences in resistance to erosion among flat-lying bedrock layers. Periodically, seams of lignite burned beneath the ground, baking the materials above it and hard-

▲ The Great Plains can be seen in the distance to the northwest from the top of Cow Butte (a similar view could be had from any of dozens of western North Dakota buttes). In the text, I note the difficulty in trying to show the scope of the Missouri Plateau and the Great Plains. The land is low to rolling if no buttes happen to be present. The landscape is sometimes referred to as "featureless," but that doesn't really do it justice. The vastness is difficult to show in a photo.
JPB Photo: 7/29/2011

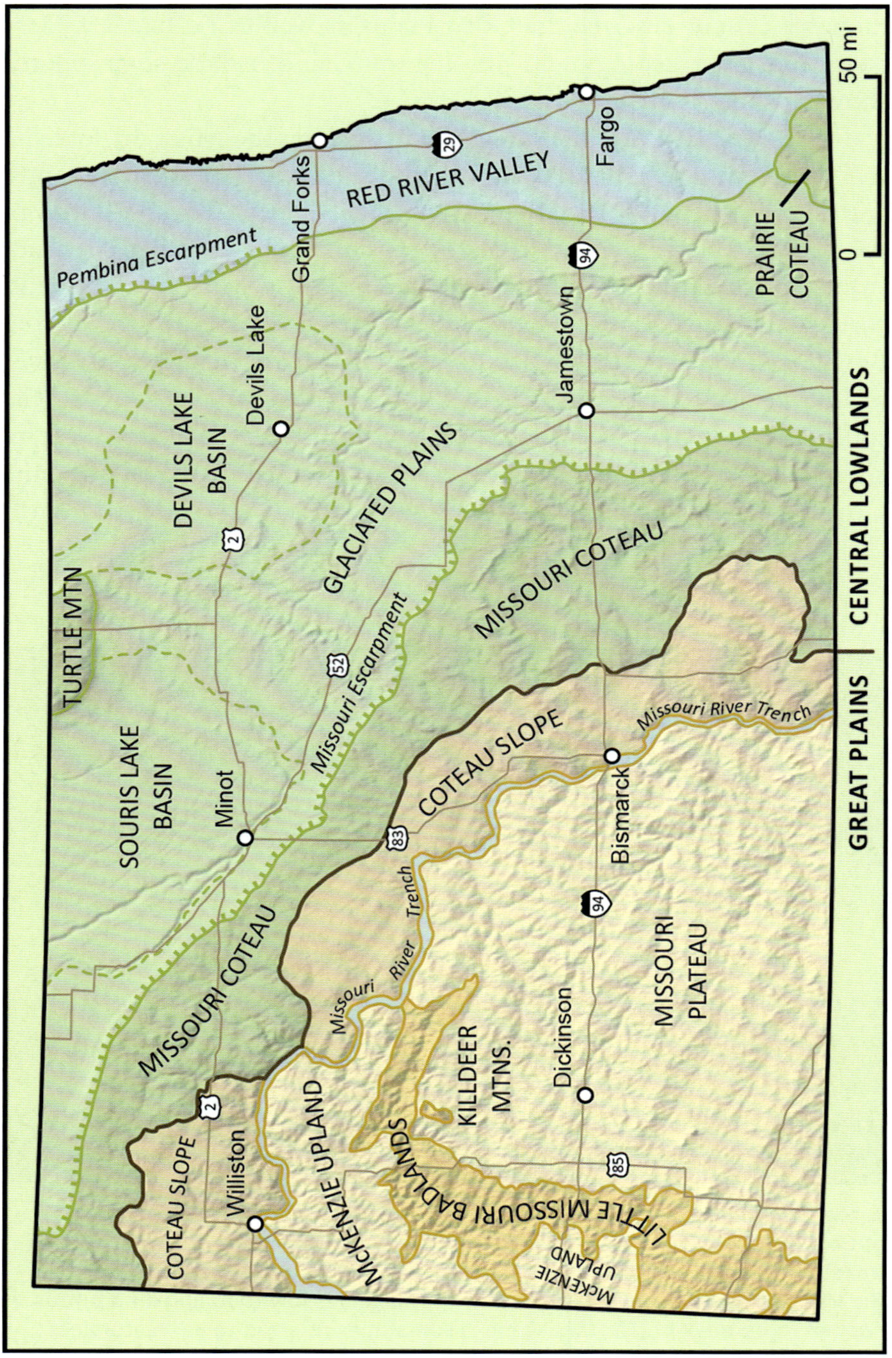
RED RIVER VALLEY
Pembina Escarpment
Grand Forks
Fargo
29
94
PRAIRIE COTEAU
0
50 mi
Devils Lake
DEVILS LAKE BASIN
Jamestown
GLACIATED PLAINS
2
TURTLE MTN
MISSOURI COTEAU
Missouri Escarpment
52
CENTRAL LOWLANDS
GREAT PLAINS
SOURIS LAKE BASIN
Minot
COTEAU SLOPE
Missouri River Trench
Bismarck
83
MISSOURI COTEAU
Missouri River Trench
94
MISSOURI PLATEAU
Dickinson
KILLDEER MTNS.
McKENZIE UPLAND
BADLANDS
LITTLE MISSOURI BADLANDS
McKENZIE UPLAND
85
COTEAU SLOPE
Williston

◀

INTERIOR PLAINS: GREAT PLAINS — Non-glacial Landforms

Missouri Plateau: Rolling to hilly plains except in badlands areas and near prominent buttes; the Missouri Slope Upland surface is 200 to 400 feet higher than the McKenzie Upland and slopes upward to the west; it has mainly gentle slopes and local relief that may exceed 500 feet.

McKenzie Upland: Rolling to hilly plains except near badlands and prominent buttes; evidence of glaciation near the Missouri River; local relief ranges from 300 to 500 feet; slopes are gentle in 50 to 80 percent of the area.

Little Missouri Badlands: Rugged, deeply eroded area along the Little Missouri River; local relief is commonly more than 500 feet with steep slopes in 50 to 80 percent of the area.

Missouri River Trench: Lowland flood plain along the Missouri River, which serves as a nominal boundary between glaciated and unglaciated land.

Coteau Slope: Rolling to hilly plains east of the Missouri River with both erosional and glacial landforms; local relief ranges from 300 to 500 feet and slopes are gentle in 50 to 80 percent of the area.

CENTRAL LOWLANDS — Glacial Landforms

Missouri Coteau, Prairie Coteau and Turtle Mountain: Hummocky, dead-ice moraine resulting from collapse of super-glacial sediment as glacial ice was melting; local relief is from 300 to 500 feet.

Missouri Escarpment: Steep, glacially modified escarpment marking the boundary between the lower Glaciated Plains and the higher Missouri Coteau; elevations may rise as much as 650 feet in a mile onto the Missouri Coteau.

Pembina Escarpment: Steep, pre-glacial (but glacially modified) escarpment marking the boundary between the Red River Valley and the Glaciated Plains in northeastern North Dakota.

Glaciated Plains: Rolling, glaciated landscape with a great variety of glacial, fluvial and eolian landforms; mainly gentle slopes with local relief less than 100 feet, but ranging up to 300 feet in places.

Devils Lake Basin: Closed basin within the Glaciated Plains with drainage to Devils Lake; rolling, landscape including the plains of the glacial lakes Minnewaukan and Cando.

Souris Lake Basin: Area drained by the Souris River within the Glaciated Plains; (part of the area was flooded by glacial Lake Souris); flat to gently sloping plain; wind-blown dunes in places.

Red River Valley: Flat plain resulting from sedimentation on the floor of glacial Lake Agassiz.

Note: This map depicts physiographic regions generally recognized in North Dakota. Considerable variations characterize the landforms that occur within each region, but overall similarities allow generalization about the geomorphic processes that operated to form each region. "Local Relief" is defined as the maximum difference in elevation within any township-size area. "Gentle slopes" refer to areas with slopes of less than 8 percent (4° 35').

▲ In northeastern Golden Valley County, numerous small, isolated buttes, all close to the same height, stand on the otherwise level Missouri Plateau along the Belle Isle Road. The plain is developed on the Tertiary Bullion Creek Formation, and the buttes are all capped by a clinker bed at the base of the Sentinel Butte Formation. The resistant clinker cap accounts for the accordant tops on the buttes. JPB Photo: 9/15/2009

ening them to clinker. The land surface overlying the area where the lignite burned was lowered by an amount approximating the thickness of the original lignite seam.

The contrast between the two areas, Missouri Plateau and Little Missouri Badlands, is the result of the great difference in amount of erosion; more intense close to the Little Missouri River; less so farther away from it. Even though part of the Missouri Plateau south and west of the Missouri River was glaciated during the Ice Age, it happened so long ago that, in most places, the only visible evidence is an occasional boulder or thin patch of glacial sediment. Everything else relating to the early glaciations was eroded away hundreds of thousands of years ago.

THE CENTRAL LOWLANDS

Cross the Missouri River and you will immediately notice that the land to the east is strikingly different than west of the river. The Coteau

Slope, a rolling to hilly area with both erosional and glacial landforms, borders the river on its northeast side (the word "coteau" is Canadian/French and means "little hill"; a second definition is "hilly upland").[10] It is an area that may be thought of as transitional, with characteristics of both the Great Plains and the Central Lowlands. Parts of Emmons, Burleigh and McLean counties bordering the Missouri River offer good examples of the Coteau Slope landscape.

Continuing to the east or northeast, we come to the intensely glaciated landscape of the Missouri Coteau. The Missouri Coteau (aka Coteau du Missouri or Coteau des Missouri) is a narrow (20- to 70-mile-wide) northwest-trending upland that extends for 825 miles from south-central South Dakota to west-central Saskatchewan along the eastern margin of the Missouri Plateau. The signature features of the Missouri Coteau are the innumerable prairie potholes with their lakes and sloughs. Indeed, the area is often referred to as North Dakota's waterfowl factory. The Missouri Coteau in North Dakota includes much of Logan, McIntosh, Stutsman, Sheridan, McLean, Ward, Mountrail, Burke and Divide counties.

The hummocky topography of the Missouri Coteau results in thousands of ponds and sloughs, but few streams flow any distance through the area. The Prairie Coteau, in southeastern North Dakota, and Turtle Mountain in the north, are areas with topography essentially identical to the Missouri Coteau, but Turtle Mountain is wooded. All three of these areas have rolling and hummocky topography, countless potholes and wetlands, and verdant mixed-grass prairie. The Coteau still has more than half of its original native prairie.

When General Sully traveled across the Missouri Coteau in 1865, from the Cannonball River to Devils Lake and back, Eugene Marshall, a member of Sully's expedition, painted a dismal picture of the area:

> The distinguishing feature of the soil in the Coteau is sterility. The aspect of the country is the very picture of barrenness and desolation . . . The country is sparsely covered with the short . . . buffalo grass which is almost the only valuable vegetable product in the country.

Marshall continued at length, describing how barren and unproductive the area was that he was crossing. Conditions were dry at the time, but he was traveling in late summer. I do not know if the dryness

▲ **This dead-ice moraine landscape is on the Missouri Coteau three miles southeast of Ruso near Otis Lake in McLean County. Dead-ice moraine is hummocky. Streams do not flow through the area.** JPB Photo: 6/20 2009

was due to long-term drought conditions, or if the weather was just seasonally dry, as it often is at that time of year.

The most extensive part of the Central Lowlands in North Dakota is the rolling, glaciated landscape known as the Glaciated Plains (sometimes referred to as the "drift prairie"). Much of the region has gently sloping, rounded hills, less rugged than areas to the west where slopes are steeper and the landforms are larger. The Glaciated Plains extend from Langdon to Ellendale, north to south, from Finley to Crosby, east to west. They include a variety of features: shorelines and floors of ancient glacial lakes; deep channels carved by cataclysmic floods of glacial meltwater; sinuous ridges of sand and gravel several miles long; straight-as-an arrow ridges shaped by flowing glaciers; massive blocks of material shoved into place by glacial ice. North Dakota's Glaciated Plains boast some of the most remarkable glacial features to be found anywhere. All date to the most recent glaciation, the Wisconsinan. Earlier glaciations during the past 2½ million years shaped earlier landscapes, but the changes they wrought are no longer apparent. Some were removed by erosion; others are concealed beneath more recent glacial deposits.

THE RED RIVER VALLEY

The exceptionally flat plain that is the Red River Valley marks the former floor of glacial Lake Agassiz, once the largest freshwater lake in North America. In North Dakota, the Lake Agassiz plain consists of a strip of land 30 to 40 miles wide along the Red River, but in northern Minnesota and Canada, the lake covered a much broader area. The North Dakota portion of the Red River Valley is covered by flat-lying silt and clay beds, deposited in still water at the bottom of the lake. The otherwise smooth lake plain is marked by shallow grooves scratched into its surface by icebergs. Rivers flowing into the lake deposited sand, which the wind later blew into dunes. Along the lake margin, ancient beaches and wave-eroded scarps mark former shorelines of the lake.

Summary

The boundaries between North Dakota's different landform types separate distinctive assemblages of geologic features. The Missouri Plateau has badlands, buttes and wide-open spaces. The Glaciated Plains, Missouri Coteau, Turtle Mountain and Prairie Coteau (in the southeastern corner of the state) are all intensely glaciated landscapes with intricate, fresh topography that includes a variety of unique glacial landforms. The remarkably level landscape of the Red River Valley mirrors the floor of glacial Lake Agassiz. Beaches and wave-cut scarps mark the lake's former shorelines and separate the Red River Valley from the Glaciated Plains to the west. Recognizing how the state's landforms are distributed is a first step toward understanding North Dakota's geology.

CHAPTER TWO

THE MISSOURI PLATEAU

Surface Materials

The Missouri Plateau is a broad expanse of rolling to rugged plains that rise westward toward the Rocky Mountains. In western North Dakota, the Missouri Plateau encompasses a roughly circular area, 100 to 150 miles in diameter, centered on Dickinson. Much of the area is a smooth, even mundane landscape, but, in places, flat-lying beds of clay, silt, sand or sandstone are carved into gullies and ravines. Buttes of all sizes rise above fields of grass and cropland. Hay and pasture predominate, but we encounter fields of wheat, flax and canola, too, especially near the eastern margin of the Missouri Plateau, where precipitation is more reliable for growing crops. A few erratic boulders are scattered on the surface near the northern and eastern edges of the Missouri Plateau, along with some glacial deposits of clay or gravel. Glacial materials are minimal, though, and they are not responsible for the character of the land. Rather, erosion during millions of years, coupled with the varied resistance of the surface materials to that erosion, has resulted in a landscape that ranges from smooth and rolling plains to rugged badlands and buttes.

The Bullion Creek and Sentinel Butte formations (please refer to the stratigraphic column on page 6) are at the surface in most places on the Missouri Plateau. On the fringes of the Missouri Plateau, in Sioux and Bowman counties, for example, erosion has been more vigorous and carved deeper, exposing the older Pierre, Fox Hills, Hell Creek and Cannonball formations. At higher elevations over the central part of the Missouri Plateau, small areas of younger formations, Eocene to Miocene in age, are found near and on the tops of the larger buttes.

Erosion has modified the various geologic formations of the Missouri Plateau in a variety of ways. Formations like the Pierre, Fox Hills and Cannonball tend to form smooth slopes as they erode, resulting in only a few low buttes or badlands areas. In contrast, the Hell Creek, Ludlow, Slope, Bullion Creek and Sentinel Butte formations contain numerous resistant beds so that, as they have eroded, they have formed steep slopes, resulting in prominent buttes and badlands.

The varied landscape of the Missouri Plateau owes much of its character to the many different materials found on the surface, whether beds of sandstone, lignite, clinker, petrified wood and concretions are in any given place. Coupled with the type of erosion that has occurred, and is occurring today, the character of these materials determines the kinds of landforms that have developed. Buttes exist today where the surface materials are harder and more resistant to the erosion processes, especially the forces of running water. The badlands along the Little Missouri River began to form when an early glacier diverted the river from its northerly route into Canada, forcing it to flow toward the east instead. When its route was changed, the river flowed over a steeper gradient and began to erode more intensely, cutting deeply into the formations over which it was flowing.

In this chapter, I will describe how the geologic materials found on the Missouri Plateau, coupled with geologic events such as erosion, glaciation, drought and other climatic changes, have shaped, and continue to shape, the Missouri Plateau.

Lignite and Petrified Wood

Late in the Cretaceous Period, beginning 70 million years ago and continuing through the Paleocene Epoch until 56 million years ago, western North Dakota's climate was subtropical. Trees up to 12 feet in diameter and more than 100 feet tall grew in a setting similar to today's Dismal Swamp in Virginia, Georgia's Okefenokee Swamp or the Florida Everglades, with meandering rivers, swamps and vast forested flood plains. Modern evidence for this fossil forest includes widespread seams of lignite coal, fossil tree leaves, pollen, and logs and stumps of petrified wood.

Lignite is a soft coal that underlies much of the western two-thirds of North Dakota. It began as an accumulation of dead plant material in tropical or semitropical basins: swamps, lagoons and marshes. As the basins filled with stagnant water, the plant debris became submerged

▲ Lignite is exposed six miles southeast of Medora in Billings County. The lignite bed marks the contact between the Bullion Creek Formation, beneath, and the Sentinel Butte Formation, which overlies the lignite. In this exposure, the Bullion Creek beds are light gray and the Sentinel Butte beds are darker in color.

JPB Photo: 9/9/2009

so that atmospheric oxygen could not reach it. When the plants died and fell into the water, they began to decay, but before all the plant debris could decompose, the bacterial action causing the decay stopped. Most of the bacteria committed suicide by filling the stagnant swamp water with their own toxins to such an extent that they died. The only bacteria that remained were ones that did not need oxygen for respiration. However, these anaerobic (the word means "living without air") bacteria are less efficient at decomposition. As a result, large amounts of submerged organic materials did not decompose, and thick beds of peat accumulated.

Streams meandering through western North Dakota during Paleocene time changed course frequently and, when they did so, they sometimes deposited sand and silt on top of the partially decomposed vegetation (peat). After the layers of peat were buried beneath thick layers

▲ **The tree trunk (to the right of the pickax) is buried by silt and sand, which were deposited quickly, thus preserving the tree trunk. The vertical part of the tree grades upward from lignite at the base to a combination of lignite and petrified wood. I took this photo in 1979 in the Glenharold Mine (now closed and reclaimed) in Mercer County. The pickax is about 4 feet long.** JPB Photo: scan of 1979 slide of sediment, the weight of the overlying beds gradually compressed the peat to lignite. Layers of swamp vegetation, some of them more than 50 feet thick, were eventually transformed into beds of lignite coal only a few feet thick.

Seams of lignite, horizontal black bands, can be seen eroding out of hillsides today. They range from a few feet to as much as 40 feet thick in Slope County and even thicker in Wyoming and Montana. If a peat bog happened to be buried by river sediments before the decay process had progressed very far, and trees were still growing in the swamps, some lignite may have formed, but some of the trees were instead changed into petrified wood. Occasionally, a petrified tree stump, rooted in a lignite bed, can be seen.

Petrified wood formed when minerals gradually replaced the buried plant material. The petrification process requires rapid burial of the wood to prevent decay. This sometimes happened when rivers shifted course or overflowed their banks, burying a forest floor under a layer of sand and silt. Other times, forests were partially covered by volcanic ash, blown to the area from volcanoes in the rising Rocky Mountains. After burial, ground water seeped through the ash and wood, coating cell walls and filling the intercellular cavities with minerals. Usually, the cellular structure of the wood was destroyed, leaving only a rough cast of the original log, but sometimes growth rings, bark, knots and even the shapes of the wood's tiny cells are preserved with remarkable fidelity. This more detailed preservation was possible because some molecules, such as silica and other inorganic materials, are much smaller than organic molecules so, rather than molecule for molecule replacement, the organic molecules were coated and surrounded with silica. Cavities in petrified wood may be encrusted with quartz crystals.

Petrified wood ranges from solid, well-silicified specimens to splintery, or "coalified," wood that tends to disintegrate when it is exposed to weathering, or it may simply fall apart when you pick it up. The degree of petrification can vary, even within a single specimen; individual stumps or logs may contain both well-silicified parts and other parts that are splintery coal. Most of North Dakota's petrified wood is brown or tan on weathered surfaces and dark brown where it is freshly broken, but colors can range from white to gray, with streaks of black. Traces of minerals add color to the fossilized wood: yellow, brown and

▲ **The wall of this bandstand in the Elgin City Park in Grant County is constructed of pieces of petrified wood collected in the area.** JPB Photo: 6/23/2010

▲ **This petrified stump is in the South Unit of the Theodore Roosevelt National Park in Billings County. The badlands topography in the background is carved from the Bullion Creek Formation.** JPB Photo: 9/9/2009

red may indicate iron; black and purple hues suggest carbon or manganese mineralization.

Petrified wood may occur as entire logs or stumps, some standing upright where they once grew, or it may consist of scattered limbs and fragments, strewn over the land surface. A fallen log was probably cylindrical when it fell down, but the petrified logs we find today often have oval cross sections. This is because, after they were buried, they became compressed and flattened by the weight of overlying sediments. Most of North Dakota's fossil wood is Paleocene in age, but petrified wood can be found in smaller amounts in the older Hell Creek Formation as well as in some of the younger bedrock units.

Fossil leaves, commonly found along with petrified wood, help us identify the species of trees that grew in and near the swamps where petrified wood is found. Many specimens belong to the plant genus *Metasequoia*, the dawn redwood. Fossils of dawn redwood were first discovered in 1941, and the tree was thought to be extinct, but living specimens were discovered in south-central China in 1945. Today, the dawn redwood is widely used as an ornamental tree in warmer climates.

During the Paleocene, while *Metasequoia* trees were growing in North Dakota, a variety of other kinds of vegetation were present, too. We know them primarily through studies of fossil pollen and the delicate imprints of leaves in mudstone, siltstone and carbonaceous shale. Along with the leaf fossils, we find remarkably preserved petrified cones of *Sequoia dakotensis* (giant evergreen trees), the leaves of tree ferns and various kinds of petrified wood. Cypress, black spruce, palm, tamarack, cedar and other plants grew in dense, subtropical forests. A large number of fossils of deciduous plants are found as well. They include poplar, ginkgo, magnolia, hazelnut, oak, walnut, sycamore, box elder, elm, maple, birch, alder, hickory, dogwood, buckthorn, viburnum, witch hazel, horse chestnut and bittersweet. Evidence for all of these still-living trees, and others, have been found in western North Dakota's Paleocene formations.

So much fossil wood is strewn over the surface in some places that such areas are referred to as petrified forests. North Dakota's best-known petrified forest is in the South Unit of Theodore Roosevelt National Park, where large numbers of tree stumps have eroded out of the

Sentinel Butte Formation. Some stumps are still upright, in the positions in which they grew 60 million years ago. They were preserved when the forest floor was flooded, burying the bases of the trees. The unburied parts of the trunks and branches decayed and disappeared. Petrified stumps may be anchored in a lignite bed or a buried soil horizon, which may mark a former forest or swamp floor.

Petrified wood is often used in landscaping; many western North Dakota driveways and flower beds are decorated with fine specimens. An outstanding example of a petrified stump, collected in McKenzie County, may be seen in the Long-X Visitor Center in Watford City. The stump, probably bald cypress, is 9 feet in diameter and weighs approximately 8 tons. Perhaps the most elaborate use of petrified wood

▲ Petrified wood is located in Lemmon, South Dakota, a half mile south of the North Dakota-South Dakota state line. The park covers a city block in Lemmon. The petrified wood used to construct the pillars and other monuments was collected in the area. The sign reads: "IN MEMORY OF OLE S. QUAMMEN, Father of Mrs. Harry C. Olson, the creator and donor of this, the world's largest Petrified Wood Park of its kind. Mr. Quammen's picture is displayed in the museum."

JPB Photo: 6/24/2010

in an ornamental sense is in the Petrified Wood Park in Lemmon, South Dakota. In this park, completed in 1932, O. S. Quammen constructed hundreds of pillars and intricate structures of petrified wood, much of it from North Dakota.

In 1990, the level of Lake Sakakawea was low, revealing several petrified logs weathering out of the Sentinel Butte Formation along the lake shore in Mercer County. Pieces of an 80-foot-long petrified log, collected from the area, along with two stumps from the Amidon area, are now displayed on the North Dakota State Capitol grounds. Still another large petrified log was uncovered during construction of Interstate Highway 94 west of Dickinson, This 120-foot-long, 6-foot diameter log was offered to nearby towns as a tourist attraction, but it was reburied when no one wanted it.

▲ **This 80-foot-long log was collected along the shore of Lake Sakakawea in Mercer County when the reservoir level was low. The cross sectional shape of the log is oval rather than round because, even though the log was round when it fell, the weight of the overlying sediment that buried it also flattened it into an oval shape. Since I took this photo, the log has been moved to the east side of the Heritage Center.** JPB Photo: 9/26/2009

Clinker

As you travel through western North Dakota, notice the multicolored layers and brick- or glasslike masses of baked and fused clay, shale and sandstone. These baked materials, known as clinker, but often referred to locally as scoria, formed in areas where seams of lignite coal burned, baking the nearby sediments to a natural brick. Clinker beds range in thickness from a few feet to more than 50 feet in North Dakota, with even thicker beds in Wyoming and Montana.

The first recorded reference to clinker that I know of was by William Clark, who made the following entry in his journal while wintering at Fort Mandan (March 21, 1805):

> Saw an emence quantity of Pumice Stone on the sides & feet of the hills and emence beds of Pumice Stone near the Tops of them, with evident marks of the hills having once been on fire. I Collecte Somne of the different sots i.e. Stone Pumice & a hard earth, and put them into a funace, the hard earth melted and glazed the others two and the hard Clay became a pumice Stone glazed.

When Lewis and Clark arrived at Beulah Bay, 20 miles west of present-day Riverdale, on April 10, 1805, Lewis noticed a seam of lignite burning along the face of an outcrop. He commented: "the bluff is now on fire and throws out considerable quantities of smoke which has a strong sulphurious smell."

On April 16, 1805, Meriwether Lewis wrote the following:

> I believe it to be the strata of coal seen in those hills which causes the fire and birnt appearances frequently met with in this quarter. where those birnt appearances are to be seen in the face of the river bluffs, the coal is seldom seen, and when you meet with it in the neaghbourhood of the stratas of birnt earth, the coal appears to be presisely at the same hight, and is nearly of the same thickness, togeter with the sand and a sulphurious substance which usually accompanys it.

Following Lewis and Clark, numerous explorers mentioned seeing clinker as they traveled through the region. They included Larocque (1805), Maximilian (1833), Nicollet and Fremont (1839), and Audubon (1843). Some of these explorers believed the clinker beds had a volcanic origin, but Lewis and Clark were correct in their appraisal of how clinker formed.

Several early explorers reported seeing coal fires in the northern Great Plains. Through the years, range fires have ignited lignite beds many times. At Buck Hill, in the South Unit of Theodore Roosevelt Na-

▲ The land surface has collapsed over a burning lignite vein in the South Unit of the Theodore Roosevelt National Park in Billings County. The overburden on the burning lignite is falling into the space where the lignite has burned. Notice the red-hot embers of burning lignite at depth. The materials overlying the burning lignite form cracks as they break up (shown by the cracks in the land surface). Oxygen reaches the burning lignite through the cracks, allowing the burning to continue. This lignite vein burned for 26 years, from 1951 until 1977. While it was burning, fumes from the burning lignite caused juniper trees in the vicinity to grow in a columnar configuration. Since the fire went out, the trees have reverted to their normal growth habit. I took this photo in 1972. JPB Photo: Scan of 1979 slide tional Park, a lignite seam burned from 1951 until 1977. During early October 1976, prairie fires burned over large areas in the southwestern part of the state, igniting underground lignite seams in at least 30 locations over a 7,000-acre area near Amidon. Most of the fires were extinguished before the following spring, but some of them burned for several months. Again, in July 1988, several lignite seams were ignited by widespread fires in the badlands. Juniper tree roots burning downward from the surface, into the coal, ignited some of the fires.[11]

Lignite is a low-grade and volatile-rich coal. It burns easily when it is ignited by spontaneous combustion or wildfires. Because so many coal seams have burned, over such a broad area and in such a variety of situations, it seems likely that spontaneous combustion was responsible for igniting many of the fires that formed clinker. Lignite contains considerable sulfur and can ignite spontaneously. The presence of the iron-sulfide minerals pyrite and marcasite, coupled with moisture, can result in the production of heat. This chemical reaction takes place easily, and it can heat the lignite sufficiently to ignite it. Indeed, the term "pyrite" is derived from a Greek word meaning "stone that strikes fire," and both pyrite and marcasite were used in ancient times as sources of sparks for starting fires. Pyrite and marcasite are often found in large quantities in and near lignite seams known to have ignited spontaneously.

Burning lignite is limited to depths where adequate air circulates from the surface. The level of the water table may control this depth. While veins of coal are burning, fumes from the smoldering coal can alter the growth habits of nearby vegetation, causing it to grow in unusual shapes. After the fires go out, the vegetation reverts to its normal shape, common elsewhere in the badlands. Near Amidon, a stand of junipers grew as columnar-shaped trees for many years while a nearby lignite seam burned, producing ethylene gas, which altered the growth habit of the trees. Since the fire went out, the trees have resumed their normal, more bush-like shapes.

Heat from burning lignite beds hardens, melts or sinters the overlying and surrounding rocks into brick or glass (sintering is a process that fuses material into a hard mass without melting it, much like bricks are baked in a kiln). When lignite burns, it may be transformed to an ash bed that takes up only a fraction of the space the lignite did before it burned. Thin layers of white ash, mostly potash, lime and other inorganic, noncombustible minerals, can sometimes be found at the base of clinker beds.

The baking process oxidizes iron-rich minerals, mainly to red shades, but black, gray, purple, yellow and other hues are common. The hue and intensity of the colors depends upon the mineral composition, the grain size of the material that was baked and how hot a temperature was reached during the baking process. The brick-red color, which is most common, is due primarily to the presence of the mineral hematite (iron oxide: the same as common rust). Following a rain shower,

▲ This clinker plug is a mile north of Sentinel Butte in Golden Valley County. This fused, glassy mass of exceptionally hard and weather-resistant clinker formed when a mass of sediment fell into the hottest part of a burning coal vein and melted to a glass. The plug stands above the surrounding area because it is more resistant to erosion than were the surrounding materials (now eroded away).
JPB Photo: 8/28/2010

▲ **Wave-cut exposure of clinker and lignite is visible along Lake Sakakawea's Nishu Bay in McLean County. A partially burned lignite vein is exposed here by wave action along the lake. Trees grow above the area underlain by lignite, but not over the clinker, because more moisture is available from the lignite. On either side of the unburned lignite, beneath the reddish clinker, is a white bed of ash, all that remains of the lignite that has burned. I took this photo in 1968.**
JPB Photo: scan of 1968 slide

wet clinker beds are much brighter in color.

Clinker resists erosion because it is harder than unbaked rocks and because heating and subsidence during the baking process produce fractures that allow water to infiltrate, minimizing surface runoff. Erosion often leaves clinker as a cap on red-topped knobs, ridges and buttes standing above the nearby, more subdued topography developed on less-resistant, unbaked materials. Some widespread areas of clinker are particularly scenic; good examples can be seen along the Red Hills Road south of Sentinel Butte and on parts of the Scenic Loop Drive in the South Unit of Theodore Roosevelt National Park.

By the time the materials overlying a burning lignite bed cool and collapse, they are hard, and usually partially fused by baking. As they slump, falling into the burned-out space, the baked, melted and sin-

▲ **This clinker exposure is in the South Unit of the Theodore Roosevelt National Park in Billings County. The columns are aligned along the paths the heat took as it escaped from the hottest part of the "furnace," moving from hotter areas toward cooler areas. The same phenomenon can be seen at Devils Tower, Wyoming. Devils Tower consists of vertical columns, indicating that the heat escaping from the cooling basalt was traveling upward, toward the surface of the ground at that time.** JPB Photo: 8/28/2010

tered materials may hold together, resulting in a mass that can be as much as 75 percent air space. After the clinker cools, the empty spaces provide convenient living places for small animals, such as rattlesnakes.

Several prominent clinker zones are found throughout the Little Missouri Badlands. The most extensive is the HT Butte clinker,[12] which marks the contact between the Bullion Creek and Sentinel Butte formations in places where the HT Butte lignite bed has burned. The clinker forms a cap on many hills and ridges over extensive areas.

▲ Red clinker overlies a white ash bed at Sentinel Butte in Golden Valley County. A 3- to 6-inch-thick, gray to white ash, all that remains of several feet of burned lignite, lies beneath a thick clinker bed. The ash bed is underlain by unbaked clay because the heat from a burning lignite seam rises, baking mainly the overlying materials. For this reason, the rock layers beneath the lignite may be essentially unaffected or they may be baked less intensely than overlying beds.

JPB Photo: 8/28/2010

Nodules and Concretions

North Dakota has a great variety of nodules (the term is taken from the Latin *nodulus*, or *nodus*, meaning knot) and concretions (from the Middle English *concret*, a word derived from the Latin *concrescere*, meaning "to grow together or harden"). Conspicuous because they differ in appearance from the materials in which they are embedded, concretions and nodules range in size and shape from spheres the size of peas or golf balls to massive, log-like structures. Black, irregularly-shaped forms may weather out of the surrounding sediments, forming an armor on eroded badlands slopes. Some nearly transparent nodules of the mineral selenite or bronze-colored marcasite contain

▲ The spherical concretions in this exposure are eroding out of Sentinel Butte Formation sandstone. This 2½-foot-diameter sandstone concretion is in the North Unit of Theodore Roosevelt National Park, McKenzie County. Cannonball-shaped concretions may be as small as an inch or less in diameter, to in excess of 10 feet.

In addition to the cannonball concretions, notice the various styles of erosion in the sandstone. They include rilled slopes in the foreground and, in the background, vertical (castellate) structures capped by a horizontal layer of concretionary material. JPB Photo 7/27/2011 fossil plants or animals at their centers. Irregularly shaped disks or lenses may be perched as caps on pedestals of softer sediment, shielding them from erosion. "Cannonballs," for which the Cannonball River is named, may be perfectly spherical balls, commonly 2 or 3 feet in diameter, but occasionally greater than 10 feet.

Concretions, discrete bodies of rock contained within softer enclosing sediments of the same composition, form when dissolved minerals precipitate out of ground water. The spherical cannonball shape of some concretions is a result of mineral growth outward from a center point, such as a grain of sand or a piece of organic material.

Common cementing materials in concretions are calcium carbonate, the mineral that forms limestone, and siderite (iron carbonate or "ironstone"). As these and other minerals precipitate from mineral-rich ground water, they fill the pore spaces between grains of sediment,

▲ **These plant casts are preserved by the deposition of iron-oxide concretionary material on the organic plant stems. These hollow plant casts are being exhumed by the wind, which has eroded the surrounding sandstone of the Bullion Creek Formation sandstone in Adams County. The larger cast is 6 inches high. I took this photo in 1972.** JPB Photo: Scan of 1972 slide cementing them together. Concretions can range from masses without structure to features in which the internal sedimentary structures are preserved in great detail.

Nodules are hard bodies of rock, similar to concretions. Unlike concretions, their composition often differs from the sediments that surround them. Nodules form by the selective precipitation of dis-

▲ The horizontal ledges on the left are concretion bands in the Bullion Creek Formation sandstone a mile north of Tracy Mountain in Billings County. The bands are formed of material that precipitated along bedding planes in the sand. The clinker capping the hill in the background marks the top of the Bullion Creek Formation; the Sentinel Butte Formation has been eroded away and is absent in this area. JPB Photo 9/9/2009

► The differences in resistance to erosion between the cemented layers ofsandstone and the thicker, uncemented layers of sand are apparent. JPB Photo: 9/9/2009

▲ **Spherical iron-oxide nodules are weathering out of a channel sand in the Bullion Creek Formation sandstone in Adams County. The nodules average between 1 and 2 inches in diameter. I took this photo in 1972.** JPE Photo: Scan of 1972 slide

solved minerals that completely replace the original sediment, rather than simply cementing the original materials together.

Concretions and nodules are found in all of the geologic formations in southwestern North Dakota and, less commonly, in some of the glacial sediments and bedrock formations in the eastern part of the state. They are conspicuous in badlands areas, concentrated at the bases of eroding slopes where they have weathered out of the surrounding materials.

Sandstone "logs" are large sand bodies, usually cemented by calcium carbonate. From a distance, they could be mistaken for logs of petrified wood, but they were never wood. Their elongate forms result from preferred growth parallel to the direction of ancient ground water flow. Such logs, common in channel sands, may be from 5 to 10 feet in diameter, and 15 to 50 feet long. As a result of being buried beneath a heavy covering of sediment for a long time, many have become flattened, with oval cross sections. Bedding and other sedimentary structures are commonly well preserved in log concretions. As the "logs" have grown, some have coalesced into broad, tabular sandstone bodies capping small buttes in western North Dakota.

Siderite (the word is from the Greek *sideros*, meaning iron), or ironstone, is an important ore of iron in some parts of the world. It occurs as heavy, dark-brownish to black nodules ranging from small, isolated spheres to complex, rounded shapes that may be contained in beds tens of feet long. Nodules form an erosion-resistant armor on

▲ **This log concretion protrudes from a hillside at Initial Rock Historic Site in Billings County (a mile north of Tracy Mountain). The tubular concretion is 2 feet in diameter and unsupported for 6 feet; it will likely collapse soon, when more of the underlying sandstone erodes away. Even though such concretions are referred to as log concretions, they are not petrified wood; the word "log" refers only to the shape.** JPB Photo: 9/9/2009 badlands slopes, where their black or maroon colors contrast with the softer gray to tan sediments from which they weather. Siderite nodules may be encased in clay sediments, concentrated along continuous horizons. They likely formed at the bottoms of shallow ponds or lagoons, perhaps as gels, where they incorporated sand and fossil fragments as wave currents rolled them, back and forth, over the floors of the lagoons.

The outer rind on siderite nodules is often weathered to dark brown limonite, a catch-all term for several soft, chemically complex mixtures of iron oxides and hydroxides. Beneath their weathered rinds, most nodules are dark purplish black due to their manganese content. Nodules may have a pattern of polygonal ridges, or raised welts, sometimes referred to as alligator skin.

▲ A log concretion in Golden Valley County shows internal bedding detail. The surface of the rock is covered by orange lichens – slow growing, symbiotic growths of fungus and algae, or fungus and cyanobacteria. Because the diameter of a lichen circle continues to expand at a constant rate, studies of lichens (lichenometry) help geologists determine how long a rock has been exposed.

JPB Photo: Scan of slide – date unknown

Badlands

Introduction

When asked what he or she knows about North Dakota's geology, an average resident will likely mention the badlands first. That's true, too, of visitors, many of whom come to the state to see our best-known natural feature, the scenic badlands along the Little Missouri River.

The badlands landscape is a rugged and hilly one, best viewed from above, looking down on the hills, not up at them, as we usually view buttes. From the rim of the "breaks," the point where we descend into the badlands, an intricately eroded landscape of sparsely wooded ridges, bluffs, buttes and pinnacles lies before us. Behind us stretch rolling plains, interrupted only by occasional buttes. Black veins of lignite coal may be seen, eroding out of steep badlands slopes. Reddish bands of clinker add vivid colors to the area. Pieces of petrified wood, as well

▲ This upstream view of the Little Missouri River is in Billings and Golden Valley counties three miles north of Bullion Butte. JPB Photo: 10/23/2009

▲ This is a panoramic view of badlands carved from the Tertiary Sentinel Butte Formation in the North Unit of Theodore Roosevelt National Park. The level bedding is apparent in this scene. Notice that certain beds can be followed across the entire vista, although they are discontinuous (eroded away in places). An example is the bluish-gray layer that forms the surface of many table-like pedestals. This layer, sometimes referred to as the Big Blue bed, is bentonitic clay, a weathered volcanic ash.

The layers shown here in the North Unit of TR Park are slightly younger than those exposed in the South Unit of the park. The layers exposed in the South Unit are buried beneath the bottomlands here. The total relief here, from valley floor to upland surface, is 500 feet. JPB Photo: 7/27/2011

▲ The Bullion Creek Formation is exposed along the West River Road four miles north of Bullion Butte in Golden Valley County. Castellated sandstone structures, resulting in towering or battlement shapes, can be seen at the top of the butte. Such structures are an example of one kind of badlands erosion. JPB Photo: 7/7/2011 as fossil stumps and logs, litter the surface.

The American Indians, who inhabited the area when the European settlers arrived, referred to badlands as *mako sica* ("land bad"). Early French explorers translated and added to this, referring to *les mauvais terrers a' traverser* ("bad land to travel across"). General Alfred Sully, preparing to cross the badlands in August of 1864, described them as "hell with the fires burned out." Theodore Roosevelt, who lived for a while in the Little Missouri Badlands in the 1880s, described them as "fantastically beautiful." I prefer TR's description.

Age of the Badlands Materials

Badlands topography is found in several places on the plains of the U.S. and Canada. The best-known badlands in the United States are the extensive Big Badlands, along the White River in western

▲ **This is a winter scene of the Little Missouri badlands in Billings County. The view is to the north, along the East River Road five miles north of the South Unit, Theodore Roosevelt National Park. The snow helps show erosion patterns in the south-facing bluffs in the distance beyond the river, which is at the base of the bluffs.** JPB Photo: 1/15/2010

South Dakota. Near Dickinson, we have the South Heart Badlands (known, too, as the Little Badlands), where we find layers of sedimentary rock, equivalent in part to those in South Dakota's Big Badlands. The South Heart Badlands are an erosional remnant of what was once (perhaps 10 million years ago) a large butte or group of buttes. They are carved from strata of Eocene and Oligocene age, ranging between 55 million and 25 million years old. The youngest beds belong to the Miocene Arikaree Formation sandstone (22 million years old), which caps some badlands buttes.

North Dakota's Little Missouri Badlands extend from near the Little Missouri River's headwaters in Wyoming at Devils Tower to near the point where the river joins the Missouri River in western North

▲ These dark gray, mound-like hills are examples of erosional topography in the South Heart Member of the Eocene Chadron Formation in the South Heart Badlands in Stark County. The dark gray mounds are composed largely of montmorillonitic (or similar) clays, which increase greatly in volume when they absorb water. When wet, the clay is sticky and slippery and, when dry, it forms a popcorn-like surface. This kind of claystone is the result of weathering of volcanic ash. JPB Photo: 9/24/2009

Dakota. The eroded beds in these, our most extensive area of badlands, are older than those in the South Heart Badlands.

The oldest badlands beds are in the southwestern corner of the state, near Marmarth, where Cretaceous-age Hell Creek Formation beds (65 million years old) have been carved into badlands. The dark and somber, gray and purple beds of the Hell Creek Formation contain dinosaur fossils. Farther north, the beds that have been eroded into badlands are too young for dinosaur fossils; the dinosaurs were already extinct when they were deposited. Small patches of Hell Creek badlands can be seen along State Highway 1806 between Huff and Fort Rice in Morton County. Apart from the southwestern-most corner of the state, where the Hell Creek Formation beds occur, the Little Missouri Badlands are carved largely from the Paleocene Bullion Creek and Sentinel Butte formations, which were deposited between 58 million and 56 million years ago.

Between 70 million and 40 million years ago, a major mountain-building event known as the Laramide Orogeny (orogeny = mountain forming) formed the Rocky Mountains in Montana and Wyoming. As the mountains rose, they were attacked by intense erosion, providing sediment to eastward-flowing rivers and streams. The rivers delivered the eroded sediment to western North Dakota's coastal plain, an area that could be likened to today's Mississippi River Delta (central North Dakota was an inland sea at that time). Sediment from the eroding mountains accumulated into thick layers of soft, poorly lithified siltstone, claystone and sandstone: materials that were deposited on river flood plains and in swamps in what is now western North Dakota. These are the sediments we see exposed today in the Little Missouri Badlands.

Clouds of volcanic ash, blown eastward from the rising Rocky Mountains during the Laramide Orogeny, collected in clay layers that were later weathered to bentonite. When wet, the clay absorbs water and swells. Bentonite can become slippery when wet, so do not try walking or driving on it. When the beds dry, they assume a surface similar in appearance and consistency to popcorn, with colors ranging from white to bluish-gray or black.

▲ This bentonitic clay—"popcorn"—surface is in the Bullion Creek Formation in Billings County along Franks Creek Road. Consisting mainly of weathered volcanic ash, the clay can absorb large amounts of water, increasing its volume as much as eight times. It feels greasy and soap-like when wet. In places where bentonite is common, getting around can be difficult, either by car or walking, as the wet clay is very slippery. It also sticks to and builds up on car tires or shoes. Bentonite is named for Fort Benton in eastern Wyoming. When it dries, the texture becomes loose, resembling popcorn. Bentonite is most often light-colored, but it can be almost any color. Several bedrock formations in western North Dakota contain localized bentonite beds. JPB Photo: 9/16/2009

Why the Badlands Formed

Even though the layers of sedimentary rock exposed in North Dakota's Little Missouri Badlands range from Cretaceous through Eocene in age (65 million to 50 million years old), the badlands themselves—the hills and valleys we see today—are not that old. Before a glacier diverted it, the Little Missouri River flowed northward through a broad, smooth valley, joining the early Yellowstone River in northern Williams County. The Little Missouri and Yellowstone rivers joined near Alamo (30 miles north of Williston) in a place now buried beneath 400 feet of glacial deposits. From there, the combined Yellowstone-

Little Missouri River flowed northeastward into Canada.

The diversion of the Little Missouri River happened sometime prior to the deposition of a volcanic ash bed on the glacial sediment blocking the channel (the ash was deposited as a result of a volcanic eruption in the area of Yellowstone Park 640,000 years ago). However, a glacier might have diverted the river much earlier than that—the 640,000-year figure is a minimum. In fact, erosion of the badlands could have begun as early as sometime between 2.5 million and 3.5 million years ago.

Since it was diverted by glacial ice, the Little Missouri River has flowed over a shorter and steeper route than it did prior to its diversion. The part of the river's route today, downstream from the point where it makes its sharp turn toward the east in the area of the North Unit of Theodore Roosevelt National Park, is east rather than north, as it had been before a glacier diverted it. When the river assumed its new route toward the Gulf of Mexico, it began a vigorous erosion cycle, cutting down more rapidly and deeply, and sculpting badlands topography. The badlands, then, are an indirect product of glacial activity even though the only conspicuous direct evidence of glaciation remaining in the area is an occasional glacial erratic on the upland in McKenzie County north of the river.

Erosion in the Badlands

Rain and melting snow, wind, frost and other forces of erosion have carved the badlands into intricate shapes. Since the Little Missouri River began to form the badlands, it has carried an enormous amount of sediment away. In the southern part of the badlands area, near the river's headwaters and close to Devils Tower in northeastern Wyoming and adjacent Montana, the river has cut down 80 feet below the level it had been flowing at before it was diverted by a glacier. Near Medora, the valley floor is 250 feet lower than the pre-diversion level. Still farther downstream, in the North Unit of Theodore Roosevelt National Park and near the confluence of the Missouri and Little Missouri rivers, the east-trending portion of the Little Missouri River flows in a valley that is 650 feet deeper today than it was before glaciers diverted the river.

The average rates of erosion in the badlands, assuming they started to form 640,000 years ago, can be calculated as follows:

> Headwaters area in Wyoming: 0.15-inch/100 years;
> Medora area: 0.5-inch/100 years;
> Confluence area near Mandaree-Missouri and Little Missouri rivers: 1.25 inch/100 years.

These rates may seem tiny, but through time, erosion has removed a huge amount of sediment. Approximately 40 cubic miles of sediment have been eroded and carried away by the Little Missouri River from the area that is now the badlands. Most of that sediment now lies beneath the water of the Gulf of Mexico.

The rates of erosion I have noted are long-term averages, but erosion goes on at highly irregular rates. Locally, during the past few hundred years, the badlands have undergone four separate periods of

▲ This scene shows badlands erosion in the Sentinel Butte Formation in the North Unit of Theodore Roosevelt National Park in McKenzie County. Some beds have a "rilled" microtopography, with vertical grooves (center of photo), while other beds retain their horizontal layering. A lag covering of reddish nodules covers the surface in places. JPB Photo: 7/27/2011 erosion and three periods of deposition. Since 1936, new gullies have been cut to their present depths. It may seem a paradox that, although running water is the main agent of erosion, badlands formation can be most intense when water is in short supply. Why? Because erosion tends to be more vigorous during times of drought when the vegetative cover is too sparse to protect the soil from the occasional rain storm or spring snowmelt. When precipitation is sufficient for the growth of heavy vegetation, the soil is better protected from severe erosion.

Streams and rivers carry sediment away from the area of the badlands, but most of the actual on-the-spot erosion is a result of slopewash. In places where vegetation is sparse, the soil and rock materials are easily weathered, forming loose surfaces that slide downslope easily, slumping and sliding during showers or when the snow cover melts.

The Badlands Landscape

The shapes, sizes and configurations of the hills, buttes, valleys and other landforms in the badlands are not entirely happenstance. Differences in hardness of the materials result in differences in resistance to erosion. Nodules and concretions help shape a landscape ranging from beautiful, to desolate—even grotesque. Hard beds of sandstone or clinker cap many of the small buttes. Variations in permeability (permeability is a measure of the ease with which water can move through porous rock) have similar effects; rain and melted snow soak into the more open and permeable sands, resulting in only minimal erosion. When water flows over the surface of tighter, less permeable sediment, such as clay, it abrades and erodes the material, carrying some of it away. The presence or absence and character of the vegetation play important roles in governing the rate of erosion. Grass usually helps control erosion more effectively than does forest vegetation.

The irregular placement of hard nodules and concretions may result in the development of rock-capped pillars, known as "hoodoos," mushroom-like shapes perched on stalks of clay. In places, slopes are covered by nodules of siderite (iron carbonate). As they weather out of the surrounding materials, becoming concentrated on the surface, the nodules form an erosion-resistant armor, which temporarily slows the rate of erosion. Clinker beds are much more resistant to erosion than are the softer surrounding beds. We commonly see badlands buttes

▲ **This rather desolate landscape is developed on the Paleocene Slope Formation topography near Williams Creek, southwestern Golden Valley County. The surface is strewn with nodules and concretions. The Slope Formation is similar to the Cretaceous Hell Creek Formation but without the dinosaur fossils. Both formations tend to form badlands.** JPB Photo: 10/23/2009

capped by red clinker beds.

Erosional "pipes" sometimes form in gullies and ravines where surface runoff is focused. Piping results where runoff can flow downward into small cracks and joints. Drought can play a role by shrinking and cracking the soil. Pipes are common in places where surface runoff erodes cavities vertically downward through the soft rock. With time, the initial pathways may widen at depth into caves the size of small rooms. The average depth of vertical pipes is from 10 to 15 feet, but some are much deeper. The tops of pipes may be partially concealed, making hiking treacherous. I have seen the bones of animals, such as rabbits and deer, at the bottoms of pipes. The animals fell into the holes and could not get out.

▲ Concretion pedestals (hoodoos) are visible in the badlands topography. The concretions act as caprock and keep the underlying softer sediments from eroding, resulting in table-like configurations. Hoodoos can form in any of several geologic formations. These examples are in the Paleocene Bullion Creek Formation in the South Unit, Theodore Roosevelt National Park, Billings County. JPB Photo: 9/10/2009

▲ This pedestal is located a half mile south of Lake Sakakawea in northern McKenzie County. Pods of such freshwater limestone are common in several of the Tertiary formations. They may occur sporadically or as semi-connected layers, and they often form small caprocks, such as this one. JPB Photo: 7/23/2010

▲ This 75-foot-high exposure cuts through two "pipes," vertical cavities in badlands topography in the Tertiary Bullion Creek Formation in northeastern McKenzie County just south of Lake Sakakawea. The pipes are 15 feet deep, top to bottom. A cross-sectional view of a pipe, as shown here, is not common. More often, they are concealed, with the only opening at the top. Notice that the tops of the pipes are partially sealed by a flat concretion. If an animal (or person) falls into such a pipe, it can be difficult to impossible to get out. I've seen small animal remains in some pipes where I was able to see to the bottom. JPB Photo: 7/23/2010

Summary

The geology is only part of the badlands story. The weather and climate, vegetation, animals, birds, insects, sounds and aromas—all of these, along with the human history and the ranching heritage, work together to complete the story of the badlands.

I think the North Dakota badlands are particularly beautiful because of their parklands: wooded areas that occur in draws and on north-facing slopes. Heavy vegetation in the badlands in places like Little Missouri State Park adds to the scenery. Evergreens, such as the Rocky Mountain juniper, ponderosa and creeping juniper, are interspersed with quaking aspen, cottonwood and poplar. Limber pines are found in the badlands in the southwest corner of the state, near Marmarth.

I have hiked and camped in the badlands many times. Evening summer showers accentuate the colors, and the clinker beds assume intense shades of red and orange. The fresh, pungent aroma of wet sage and cedar enhance the experience. At night, the stark, intricately eroded pinnacles can seem unreal. In the moonlight or in a night lightning storm, it is easy to imagine the strange shapes as ruins of a magical city, rather than structures of mere sand and clay. Blend in the sound of coyotes conversing, and the badlands environment is complete.

Buttes of Western North Dakota

Introduction

North Dakota's buttes figure prominently in our history and lore. Early explorers relied on them as landmarks in their travels, as did early settlers traveling west in wagon trains through the state. More than any other landforms, buttes bring to mind our western heritage.

One definition of a butte is "a conspicuous, isolated hill with steep, sometimes almost vertical sides, and a small, flat top." The word "butte" comes from a French word meaning "knoll" or "mound." Mesas (the word is Spanish for tables) have flat tops, too, but their surfaces cover larger areas, sometimes several square miles. A general rule for differentiating buttes and mesas is that mesas have tops wider than their heights, while butte tops are smaller. Given those criteria, North Dakota has several mesas, but the term is seldom used here.

General Geology

Buttes and mesas are found in areas of broad, flat-lying geologic formations. Most have caprocks of erosion-resistant material such as layers of sandstone, clinker, quartzite, limestone or chert. Some rise several hundred feet, but others are simply pointed hills, any flat top that might once have existed, eroded away long ago. Some buttes, capped by cemented gravel deposits, are aligned into long ridges or strings of hills that coincide with the routes of ancient rivers in which the gravel was deposited. Such ridges are usually discontinuous because the gravel cap has not been preserved everywhere.

The recipe for making a butte involves several stages, each of which may be separated from the other stages by millions of years. The first step is the deposition of a potential caprock material. This may happen in any of a variety of ways, but the material deposited needs to be something with the potential to eventually harden into an erosion-resistant caprock. It might, for example, be a layer of sand that, millions of years later, becomes cemented into sandstone.

The second step involves the burial of the potential caprock beneath one or more layers of some other kind of sediment. These new layers may then remain as a covering until a new cycle of erosion begins. During the time the potential caprock is buried (it might be millions of years), ground water may seep through it, slowly cementing the loose materials, depositing minerals such as calcium carbonate, silica or iron oxide that transform it into a much-harder rock.

A variety of additional events could occur during the butte-forming process, but the final step is a long period of erosion, during which the softer, overlying and surrounding sediments are removed, leaving the more-resistant materials standing above the nearby countryside. Running water is the most important erosional force that shapes buttes, but wind can play an important role. Modern buttes in North Dakota are currently undergoing active erosion. Their shapes are constantly changing, although at a geologic rate. The change is so slow that we are unlikely to notice any change during our lifetimes. Events such as landslides can instantaneously alter the shape of a butte, but from our human perspective, the overall pace of erosion is slow.

The current cycle of erosion in western North Dakota, which began 15 million years ago, was greatly accelerated when glaciers first advanced into the state, perhaps between 2 million and 3 million years ago. In addition to depositing thick layers of sediment in northern and eastern parts of the state, where they built a landscape dominated by glacial features, the early glaciers diverted all the rivers from their northeasterly routes, forcing them to flow eastward and southward, toward the Gulf of Mexico. The downstream parts of rivers like the Cannonball, Heart, Knife and Little Missouri were cut off, their earlier routes diverted to steeper routes. As a result, their gradients increased, causing rates of erosion to accelerate, deepening valleys and carving buttes.

Buttes are most numerous in places where two or more flat-lying geologic units of differing character occur, stacked one above the other. If several such units are present, one or more are likely to be much more resistant to erosion than the others. For example, a widespread layer of thick, soft silt might be overlain by a thinner, much harder layer of sandstone, and another layer of softer sediment might overlie that. While buttes were being sculpted into their modern shapes, lower surrounding areas were being eroded, too, and great amounts of material were carried away to other places.

The differences in resistance to erosion allow some layers to remain while others are removed, resulting in buttes, mesas, hills, valleys, gullies and ridges—virtually any kind of landscape irregularity imaginable. Some modern butte surfaces preserve small parts of older landscapes. The flat upland surfaces on Sentinel Butte, Bullion Butte, the Killdeer Mountains and other buttes have existed for millions of years, although not necessarily in their current roles as the tops of buttes. No land surfaces in North Dakota are older than the tops of our buttes.[13]

The relative amount of upland land surfaces increases proportionally as we travel westward. In eastern Montana, the land surface that corresponds to the tops of our buttes covers most of the area because much of the land surface there has not yet been carved into buttes. In North Dakota, where we are at a greater distance from the mountains, most of the area that was once uplands has been eroded away, leaving only the scattered buttes we see today. To be sure, the Montana land surface is incised by rivers and streams, but the amount of incised area is much smaller relative to the uplands than in North Dakota. For example, Adams County, North Dakota, has several small buttes ranging from 2,500 to 2,900 feet in elevation. The upland surfaces on these buttes constitute only 1 or 2 percent of the total area of the county; most of Adams County stands at elevations several hundred feet lower than the butte tops. In contrast, 150 miles to the west, in Custer County, Montana, most of the land is at the upland surface level, and that upland amounts to approximately 80 percent of the area, with most of the remaining, lower area restricted to land along the Powder River.

North Dakota has dozens of named buttes and hundreds of unnamed ones, but I will discuss only a sampling, grouping them according to their differing caprock materials.

Limestone and Sandstone Caprock

Some of our larger buttes are capped by deposits of fresh-water carbonate (limestone) or sandstone, which may include beds of water-lain volcanic ash (tuff). Notable examples of such buttes are the Killdeer Mountains and Chalky Buttes. Some other major buttes as well, including Bullion Butte, Blue Butte and Sentinel Butte, are capped by massive sandstone layers.

Killdeer Mountains—the name of the Killdeer Mountains is said to be a translation of a Lakota name, *Tahkahokuty* or *tah-kah-o-*

▲ **This is a view from the west of North Killdeer Mountain in Dunn County. The 400-foot-thick caprock near the top of the slope is the Miocene-age Arikaree Formation. The caprock is breaking up and falling down the grassy slopes of the butte (on the right).** JPB Photo: 7/29/2011

kuty, meaning "The Hunting Ground Where They Killed the Deer."[14] The Killdeer Mountains consist of two mesas in Dunn County that rise 800 feet above the nearby countryside. The entirety of the Killdeer Mountains extends over 115 square miles, but the level surface on top covers 4,800 acres (7.5 square miles). A 300- to 400-foot-thick layer of resistant rock on top consists of hard beds of carbonate, sandstone and siltstone, along with beds of volcanic ash. The main hard caprock on the Killdeer Mountains is early Miocene in age, 22 million years old.

The area that is now the Killdeer Mountains was once a lake basin. Nearby sediments washed into a lake, or lakes, along with ash that was blown in from volcanoes erupting west of North Dakota. The ash, and other sediments, carried first by air, and then by streams to the lakes, accumulated to thicknesses of hundreds of feet. After the lakes became filled with sediment and eventually dried up, the ash-rich sediment lithified into solid rock, forming hard, tuffaceous (tuffaceous = rich in volcanic ash) limestone beds.

By the time the Miocene lakes had dried up, 15 million years ago, erosion had already begun shaping the Killdeer Mountains. Since then, streams have continued to carve downward for millions of years, removing large amounts of sediment. Then, sometime prior to 640,000 years ago, a glacier diverted the Little Missouri River, forcing it to flow eastward over a steeper route than it had previously. In addition to carving badlands, the vigorous cycle of erosion that has resulted from the river's diversion has steepened the slopes of the Killdeer Mountains to the rugged shapes we see today.

The hard limestone beds, which had been deposited in the Miocene lakes, withstand erosion better than do the surrounding sediments. Therefore, they remain standing as a caprock on the Killdeer Mountains, high above nearby areas, whereas the surrounding softer sediments have been eroded and carried away by streams and rivers. With continued erosion, pieces of the main caprock break away and roll down the side of the hill, forming a boulder-strewn slope.

The higher elevations near the center of the Killdeer Mountains are covered in most places by the tuffaceous rock that had collected in the Miocene lakes. This layer of resistant rock is known as the Arikaree Formation, which is referred to informally as the Wormy Marker Bed because it was riddled by burrowing organisms (actually clams, not worms) before it eventually hardened.

▲ **Boulders of Arikaree caprock are strewn on the south slope of South Killdeer Mountain in Dunn County. The boulders, pieces of the Miocene Arikaree Formation ("Wormy Marker Bed"), average between 10 and 15 feet in diameter. The Arikaree Formation, the main resistant bed on top of the Killdeer Mountains, acts as the "caprock" on the butte.** JPB Photo: 9/25/2009

▲ **Before it hardened, teredo worms ("shipworms"—not worms at all, but clams) burrowed in the limey mud at the bottom of a lake The limestone, and the burrows in it, formed in early Miocene time, 22 million years ago. This photo shows part of one of the boulders that has rolled down from the top of the butte.**
JPB Photo: 9/25/2009

A cave, Medicine Hole, is located near the southern edge of South Killdeer Mountain. The 2-foot-wide cave entrance is located 100 feet below the top of the mesa. It can be reached by walking up a trail leading from a small park on the southeast corner of South Killdeer Mountain (but the area is restricted, so please do not enter it without permission). Although Medicine Hole is contained in both sandstone and carbonate rocks of the Arikaree Formation, it was not formed by water seeping through the rocks and dissolving material as are most large caves (it has no stalactites and stalagmites, typical of solution caves). Rather, Medicine Hole is essentially an east-west trending fissure formed when a block of Arikaree caprock broke loose. The block is now in the process of falling away from the main caprock on Killdeer Mountain. The fissure extends at least 70 feet down into the ground and is 100 feet long at the surface. Air issuing from the cave's entrance at the top of the plateau suggests that the fissure has one or more openings somewhere farther down the slope.

Bullion Butte—A large, horseshoe-shaped butte in Billings

▲ The entrance to Medicine Hole is on top of the Killdeer Mountains. Medicine Hole is an opening to a crack, which extends east-west for 100 feet. The opening is 2 feet wide and opens into a fissure formed where a massive block of Arikaree rock is breaking away from the main part of South Killdeer Mountain. I took this photo in the 1970s. JPB Photo: Scan of 1970s slide

County, Bullion Butte rises 700 feet above the surrounding area, to an elevation of 3,336 feet. Bullion Butte's flat, uppermost surface covers 200 acres. The 90-foot-thick sandstone that forms the caprock on Bullion Butte is similar to the Golden Valley Formation sandstone bed that caps several large buttes in western North Dakota (Sentinel Butte, Square Butte and others). As much as 50 feet of brightly colored claystone of the Oligocene South Heart Member of the Chadron Formation (ranging in age from 34 million to 30 million years old) overlie the Golden Valley caprock on Bullion Butte, but these beds are not at all as resistant to erosion as is the Golden Valley sandstone, and they are not responsible for most of the butte's elevation. The slopes surrounding

▲ **This view is from the north toward Bullion Butte in Billings County. The main caprock on Bullion Butte is a 90-foot-thick bed of Eocene Golden Valley Formation sandstone. The sandstone was deposited between 49 million and 56 million years ago; the part of the formation on Bullion Butte is probably 55 million years old. This sandstone is overlain by 30 feet of brightly colored claystone beds, part of the South Heart Member of the Oligocene Chadron Formation, perhaps 32 million years old. Bullion Butte rises to an elevation of 3,336 feet.** JPB Photo: 10/23/2009

Bullion Butte are mainly sandstone.

Lion's Cave, on the side of Bullion Butte, was named for a pair of mountain lions that made their den there and harassed cattle during the late 1890s. The caprock has many overhanging ledges along the north and northeast edge of the butte that might serve as caves or entrances to caves.

Chalky Buttes—The Chalky Buttes, in Slope County, include White Butte and Rattlesnake Butte. Originally named the Chalk Buttes because of their white color, these buttes rise 400 to 500 feet above the surrounding landscape and their upper surface covers 3,300 acres. White Butte, at 3,506 feet above sea level, is the highest elevation in North Dakota. The Chalky Buttes have a sandstone caprock of Oligocene-age White River and Miocene-age Arikaree rocks.

Several prominent buttes, all capped by resistant sandstone beds, lie immediately west of the Chalky Buttes. They include Black Butte, Tepee Butte and Slide Butte.[15] Black Butte has a flat, grassy surface on top, two miles long and a quarter to three-quarters of a mile wide. The sandstone cap on Black Butte forms a cliff along the rim on top. Black Butte rises to the second-highest elevation in North Dakota, at 3,465 feet.

Sentinel Butte—A prominent landmark five miles south of Interstate Highway 94, in western Golden Valley County, Sentinel Butte is probably the best-known butte in North Dakota. It rises to an elevation of 3,430 feet and stands 600 feet above the surrounding area. Sentinel Butte is actually two buttes separated by a small saddle. Its upper surface covers 225 acres. Sentinel Butte was named for two Arikara Indian sentinels, American Black Eagle and Standing Together, who were killed there by Sioux Indians in 1864. A plaque on top of Sentinel Butte commemorates the two sentinels.

The main caprock on Sentinel Butte is a 75-foot-thick sandstone of the 55 million-year-old Eocene age Golden Valley Formation. This sandstone is overlain by the Chadron Formation, a layer of green, calcareous claystone, Oligocene in age (30 million years old). Thin beds of limestone near the top of the butte contain fish fossils. Prominent reddish clinker beds occur on the north face of Sentinel Butte. Camels Hump Butte (3,273 feet elevation), five miles north of Sentinel Butte, and Square Butte (3,345 feet elevation), seven miles east of Sentinel

▲ White Butte in Slope County, at 3,506 feet, is the highest elevation in North Dakota. The caprock on the butte is the Miocene Arikaree Formation. Below that can be seen the smooth, dark gray slopes of the South Heart Member of the Oligocene Chadron Formation. Under the Chadron, the vertical yellowish slopes at the base of the butte are the Chalky Butte Member of the Chadron. The flat upper surface of White Butte covers 3,300 acres. JPB Photo: 10/22/2009

▲ Black Butte in Slope County is the second-highest point in North Dakota at 3,465 feet. The upper surface of Black Butte covers 400 acres and stands 550 feet above the nearby countryside. Caprock chunks, strewn on the west slope of the butte, shown here, cover a surface that is largely landslide topography. Geologists have not agreed on the age of the massive sandstone caprock on Black Butte, as it is unlike any other nearby unit. It has been variously identified as part of the Chadron Formation of Oligocene age or the Paleocene Sentinel Butte Formation. No fossil evidence has been found to verify either possibility. JPB Photo: 7/28/2009

▲ This is a view from the top of Sentinel Butte. The light-colored beds in the foreground are part of a thin layer of Oligocene White River Formation on top of the butte. The Oligocene beds are rich in fish fossils and overlie massive sandstone of the Eocene Golden Valley Formation, which forms the main caprock on Sentinel Butte. This view is to the west, into Montana. JPB Photo: 8/28/2011

Butte, have caprock materials similar to those on Sentinel Butte.

Blue Buttes—A sandstone-capped butte known as Blue Butte, located 15 miles north of Mandaree in McKenzie County, rises to an elevation of 2,712 feet. Nearby, but apart from Blue Butte, is a separate group of a dozen buttes, referred to, collectively, as the Blue Buttes. This scenic group of buttes is located 10 miles northeast of Blue Butte, 20 miles east-northeast of Watford City. The Blue Buttes include several named buttes (Chimney Butte, Table Butte, Stony Butte, Saddle Butte, Aura Butte and others) as well as a number of smaller, unnamed buttes. The buttes rise to elevations from 2,640 to 2,830 feet. All of the Blue Buttes are capped by a massive, yellowish-brown sandstone, 75 to 100 feet of the Eocene-age Camels Butte Member of the Golden Valley Formation. The slopes of most of the buttes are strewn with blocks of the sandstone caprock that has fallen from the top.

▲ **These buttes are among a grouping of several in the Blue Buttes area of northeastern McKenzie County east of Keene. The sandstone caprock on the buttes is the Camel Butte Member of the Eocene Golden Valley Formation, 55 million years old. The materials beneath the caprock are Paleocene Sentinel Butte Formation. The reddish exposures in the center foreground are clinker. Chimney Butte is on the left, Table Butte on the right.** JPB Photo: 7/23/2010

▲ **This is another view of the Blue Buttes in McKenzie County. The sandstone caprock on the Blue Buttes (not well exposed in this view) is the Camel Butte Member of the Eocene Golden Valley Formation. The Blue Buttes rise to elevations of about 2,645 feet.** JPB Photo: 7/27/2011

Other Sandstone-capped Buttes

Several dozen sandstone-capped buttes occur in a 60-mile-long, 20-mile-wide zone that extends from Grant County on the southwest to western Burleigh County on the northeast. These buttes, which apparently share a common origin, include the Dogtooth Buttes near Lark in Grant County, the Square Buttes in eastern-most Oliver County and several small, isolated buttes including Keever, Solberg and several more buttes along Burnt Creek and near Baldwin in western Burleigh County. The Square Buttes and Dogtooth Buttes are especially prominent features. From a distance, the Dogtooth Buttes present a striking, jagged skyline that resembles a rugged mountain range.

▲ Cemented sandstone forms a cap on a range of jagged buttes (the Dogtooth Buttes) near Lark in eastern Grant County. The sandstone caprock, part of the Paleocene Bullion Creek Formation, was probably deposited along a shore by streams flowing into a sea. Similar sandstone beds outcrop in the Square Buttes in Morton County; about seven miles north of Bismarck along U.S. Highway 83 in Burnt Creek Valley; near Baldwin (Keever and Solberg Buttes); and near Regan, in western Burleigh County. I am unsure whether the sandstone beds that occur in all of these locations are equivalent to one another. JPB Photo: 6/11/2009

All of the buttes in this grouping are capped by a sandstone bed that occurs in the lower part of the Bullion Creek Formation. The sandstone marks the contact between the Cannonball Formation and overlying Bullion Creek Formation. It may have been deposited by streams, perhaps as a sandy delta in a sea at a time when the sea was being replaced by terrestrial conditions. In places, the sandstone contains coarser gravel deposits that include pieces of lignite and petrified wood, materials the streams picked up as they flowed over a landscape developed on (probably) the Hell Creek Formation. The sandstone caprock on the Dogtooth Buttes is more cross-bedded and gravelly than it is elsewhere and it may, in part, have been deposited in a river bed.

Fox Hills Formation Sandstone Caprock

The Linton Member sandstone of the Fox Hills Formation caps buttes in Sioux and Emmons counties in the south-central part of the state. This same 70 million-year-old sandstone was quarried in the early 1900s and used in constructing buildings in Linton and other nearby towns (see photo of Fox Hills sandstone-capped butte in chapter five).

Hell Creek Formation Sandstone Caprock

Sandstone layers of the Cretaceous Hell Creek Formation cap several buttes. Small Hell Creek sandstone-capped buttes are common in the south-central part of the state in Emmons, Sioux and Morton counties. In Slope and Bowman counties in the southwest, Hell Creek sandstone beds cap several prominent buttes and form extensive areas of badlands. Beds of lignite in the Hell Creek Formation have burned in places, resulting in clinker caprock on some buttes, the clinker consisting of baked beds within the Hell Creek Formation.

Clinker Caprock

Buttes capped by reddish clinker beds are found in parts of western North Dakota where seams of lignite have burned, baking nearby sediments to layers of natural brick. The resistant clinker results in rugged and scenic topography with groups of clinker-capped hill tops, often all at the same level. Most clinker-capped buttes are small and unnamed, but a few of the larger ones have been given formal

▲ **This is Bullion Creek Formation topography with clinker capping the tops of buttes five miles southeast of Medora in Billings County. The clinker formed when the HT lignite bed burned.** JPB Photo: 9/9/2009

▲ **Dozens of buttes, all capped with clinker, occur in the western part of McKenzie County. The clinker cap is at the same elevation on all of the buttes. The clinker marks the position of the burned HT lignite bed, which marks the boundary between the Bullion Creek Formation (beneath) and the overlying Sentinel Butte Formation (but the Sentinel Butte Formation has been removed by erosion in this area).** JPB Photo: 7/22/2010 names. Pretty Butte in Slope County is capped by a prominent clinker bed, formed when a lignite seam in the Hell Creek Formation burned, baking the nearby sediments. Determining the age of a butte capped by clinker can be difficult because the lignite that formed the clinker burned long after it was deposited.

Silcrete Caprock

Many buttes are capped by silcrete, an extremely resistant material formed as a result of a concentration of silica by flowing ground water during long periods of time (see discussion of how silcrete forms in chapter five). Some silcrete deposits mark the locations of wetlands, subsequently buried beneath layers of other sediment. When it is later exposed, after millions of years of erosion, a resistant silcrete bed may form a cap on a butte or ridge.

Cow Butte—Cow Butte, 15 miles west of Hettinger in southwestern Adams County, is a small butte, rising to an elevation of 2,880 feet. Along with several nearby buttes, Cow Butte is capped by a hard bed of silcrete. The silcrete surface on the butte has a prominent pattern of polygons. Grass grows in the silt filling the cracks between the polygons, accentuating the pattern on top of the butte. Geologists do not agree on the origin of the polygons. One explanation is that they formed as frost wedges during the Ice Age at a time when permafrost extended over southwestern North Dakota. Alternatively, the polygons may have formed as desiccation cracks when wetland sediments, rich in silica, dried out. I have seen footprint markings in the silcrete surface on Cow Butte; they look somewhat like tapir footprints.

Rocky Ridge—Some buttes and ridges, such as Rocky Ridge in Hettinger County, have coverings of silcrete that have broken up or are in the process of disintegrating. Even after erosion has reduced a butte to little more than a swell on the landscape, the silcrete that once formed a continuous cap may remain as a pile of rubble. At first glance, the silcrete pieces on the surface of Rocky Ridge look as though they might have been gathered and piled there, but the rubble accumulation is naturally occurring.

▲ Erosion-resistant silcrete covers the surface of Cow Butte seven miles south of Reeder in Adams County. The silcrete is in place ("in situ") in the position in which it was deposited, but it is fractured and forms a polygonal pattern. Grass growing in the cracks accentuates the pattern, which is visible on air photos (and even on satellite photos) on this butte and many similar buttes. The cracks have been interpreted by some geologists to have formed as permafrost polygons during the Ice Age. However, permafrost polygons I have studied involve loose material, mainly coarse gravel. The polygonal pattern on Cow Butte (and nearby butte tops) is developed in solid silcrete, which has been solid for millions of years. I think it is more likely that the polygons shown here formed as a desiccation pattern when swamp sediment, destined to become silcrete, hardened and shrank as it dried out. The same process forms modern mud cracks. JPB Photo: 6/24/2010

▲ Rocky Ridge is six miles north of Hettinger in Adams County. The ridge, which is several miles long and 20 to 25 feet high, is spectacular for the density of silcrete chunks on its surface and along its sides. The pieces of silcrete are all that remain of what was once a flat, continuous cover on a butte. I stress that the piles of silcrete are natural. They have not been gathered into piles. The part of the ridge shown here is 200 yards long. JPB Photo: 7/29/2011

▲ This sod-capped pedestal—a miniature butte—is in Golden Valley County six miles south of the Theodore Roosevelt National Park Elkhorn Ranch site. The sod cover helps retard erosion. Such sod pedestals (or sod tables) are short-lived features in a geologic sense, but they help illustrate the idea of butte formation.
JPB Photo: 9/15/2009

Sod-capped Buttes

Sod can form a kind of caprock, although usually only on small pedestals, which might be considered to be very small buttes. These pedestals, or sod tables, are short-lived, compared with a resistant caprock, but they help govern the way erosion occurs in badlands areas and near many of the larger buttes. Sod tables illustrate how a covering of grass helps prevent erosion.

Buttes Without Caprock

As erosion continues, the caprock on buttes eventually collapses. Late in the life of a butte, the caprock may break up and litter the hill slopes. The array of broken-up caprock boulders on the slopes can serve to protect a butte effectively for a while and prevent it from be-

▲ **Rocky Butte is seven miles southeast of Beach in Golden Valley County. The view is to the southeast. Rocky Butte is an example of a butte in which the caprock has broken up and no longer forms a continuous intact surface. The feature might be termed an "expiring butte." Even so, the numerous chunks of caprock, pieces of sandstone from the Eocene Golden Valley Formation, are effective in helping protect the butte from eroding quickly.** JPB Photo: 9/16/2009

ing quickly eroded away, but when the caprock has deteriorated to the point that it no longer forms an intact layer, the butte is nearing the end of its existence. North Dakota has many examples, large and small, of such expiring buttes. A scenic example is Rocky Butte in Golden Valley County. Another possible example is the area of the South Heart Badlands south of Dickinson. These badlands, in their entirety, are the remnants of what was once a large butte (or perhaps several buttes), now reduced to badlands topography.

Buttes in Name Only

The definition of a butte calls for a resistant caprock, steep slopes, a flat upper surface and the stipulation that it formed as a result of erosion. North Dakota has a variety of hills that do not fit this definition at all, yet we still refer to them as buttes. Examples are Butte St. Paul, Boundary Butte, Bear Butte and Burr Oak Butte, all in the

Turtle Mountain area; Dogden Butte near the town of Butte in McLean County; Devils Heart Butte in Benson County; and Butte de Morale in Wells County, north of Harvey. All of these features were deposited or shaped by glacial ice.

The vagaries of names such as these are not necessarily logical and cannot always be explained. Certainly, the buttes I just mentioned are no more out of place than are various North Dakota features we refer to as mountains, places like Turtle Mountain, Devils Lake Mountain and the Killdeer Mountains.[16]

Glaciated Buttes

Although most of North Dakota's buttes occur on the Missouri Plateau, a few buttes are found in other areas outside of the Missouri

▲ This view of the southern escarpment of Turtle Mountain is 11 miles east of Bottineau in Bottineau County. Turtle Mountain is located where a cluster of small buttes was overridden by glacial ice and buried beneath glacial deposits. Turtle Mountain rises approximately 600 feet above the lowland (from which I took this photo) to the hills in the distance. Of this, a minimum of 100 feet of elevation is due to the thickness of the glacial deposits on top of Turtle Mountain. Areas between the buried buttes have a thicker covering of glacial deposits. The preglacial surface, Cretaceous Hell Creek Formation in this area, consisted of a range of buttes that once rose 500 feet above the surrounding area.
JPB Photo: 8/31/2010

Plateau. Most of them occur on the Coteau Slope or, in some instances, in glaciated areas. Some of these buttes, farther east than most, were in the path of the advancing glaciers. These buttes were overridden by ice and covered by glacial deposits. The best example is Turtle Mountain,[17] a broad upland area of rugged glacial topography coinciding with an area that was once a range or grouping of perhaps a dozen buttes. When glaciers overrode the buttes, they buried them beneath 200 to 500 feet of sediment. Before they were glaciated, the 300- to 400-foot-high buttes were capped by beds of sandstone in the Fox Hills and Hell Creek formations.

Summary

Western North Dakota has hundreds of buttes. I have climbed or walked to the tops of several dozen of them, always with a sense of exhilaration and discovery. No two are the same, and I have no favorites, although some have more interesting geologic stories to tell than others. The tops and slopes of a butte constitute distinct environments, each in sharp contrast to the lower land surrounding the butte. Understanding the modern plant and animal life associated with our buttes is an important part of experiencing the overall butte environment. Many butte tops are vegetated with native prairie. Some have no vegetation at all, the surface entirely bare caprock. The surfaces of our buttes are the oldest landscapes in the state. Climb to the top and walk around on a butte. You will be roaming over an area on which long-extinct mammals once trod.

CHAPTER THREE

THE MISSOURI RIVER VALLEY

Geologic History of the Missouri River Valley

Agreement on the origin of the name of the Missouri River is impossible; too many contradictory explanations exist. The name apparently comes from a Siouan Indian word, *ouemessourita* or *emissourita*, translated by early French explorers as "those who have wooden dugout canoes," or "river of the large canoes," or "town of the large canoes," or any of several other possibilities. One source says the term was the name the Illinois Indians used for the native people who lived in the Mississippi River Valley, probably on the eastern (Illinois) side of the river; another source says they lived in what is now the state of Missouri.

The Missouri River originates near Three Forks, Montana, where the Gallatin, Jefferson and Madison rivers come together, and it flows 2,341 miles to St. Louis, where it joins the Mississippi River. This makes the combined Missouri-Mississippi River, at 3,709 miles, the fourth longest in the world, after the Nile, Amazon and Yangtze. The entire length was once riverine environment, but due to the dams that have been built along its route, approximately a third of the length is now reservoirs—lakes rather than river. Listed from upstream to downstream, the dams are: Fort Peck in Montana; Garrison in North Dakota; Oahe, Big Bend and Fort Randall in South Dakota; and Gavins Point on the South Dakota-Nebraska border.

Along with its valley, the Missouri River is a product of glaciation. Before North America was glaciated, all the drainage in North and South Dakota, eastern Montana and northern Minnesota was north and northeastward into Canada. There was no "Missouri River" carrying drainage from the northern midcontinent region to the Gulf of Mexico. My definition of the Missouri River requires that its water ultimately reach the Gulf of Mexico as it does today, and that it carry water draining from the Rocky Mountains and northern Great Plains. Prior to glaciation, no such river existed. Why is the situation today so different than it was before the Ice Age?

The modern Missouri River Valley in North Dakota consists of several discrete valley segments that differ markedly from one another. Some of the segments are broad: six to 12 miles wide from edge to edge, with gentle slopes from the adjacent upland to the valley floor. Others are narrow: less than two miles wide, with rugged valley sides, and even badlands slopes in places. Most of the wide segments trend from west to east, whereas most of the narrow segments are north-south. The Bismarck-Mandan area is one of only a few exceptions, and I will explain why shortly.

The west-east segments of the Missouri River Valley are wide because they coincide with much older valleys that existed long before the area was glaciated. Old, mature river valleys, which formed during long periods of time, tend to be broad with gentle slopes, whereas younger valleys formed quickly and are usually narrower with steeper sides. An example of a wide segment is the 40-mile-long, west-east segment of the Missouri River Valley upstream from Garrison Dam. This part of the valley, now flooded by Lake Sakakawea, was once the route of a river that flowed east to Riverdale. The river did not turn south at Riverdale, though, as it does today; rather, it continued eastward from Riverdale and on past Turtle Lake and Mercer, flowing into northeastern North Dakota. For convenience, I will refer to this ancient river as the McLean River. East of U.S. Highway 83, the route of the old

McLean River Valley is a broad, low area, partly buried beneath tens to hundreds of feet of glacial sediment. Lake Audubon, Turtle Lake, Lake Brekken, Lake Holmes, Lake Williams, Lake Petersons, Pelican Lake, Blue Lake, Brush Lake and other smaller lakes mark the former route of the McLean River through eastern McLean County. However, continuing farther east, the McLean River Valley becomes so deeply buried beneath glacial deposits that it is impossible to know its route from

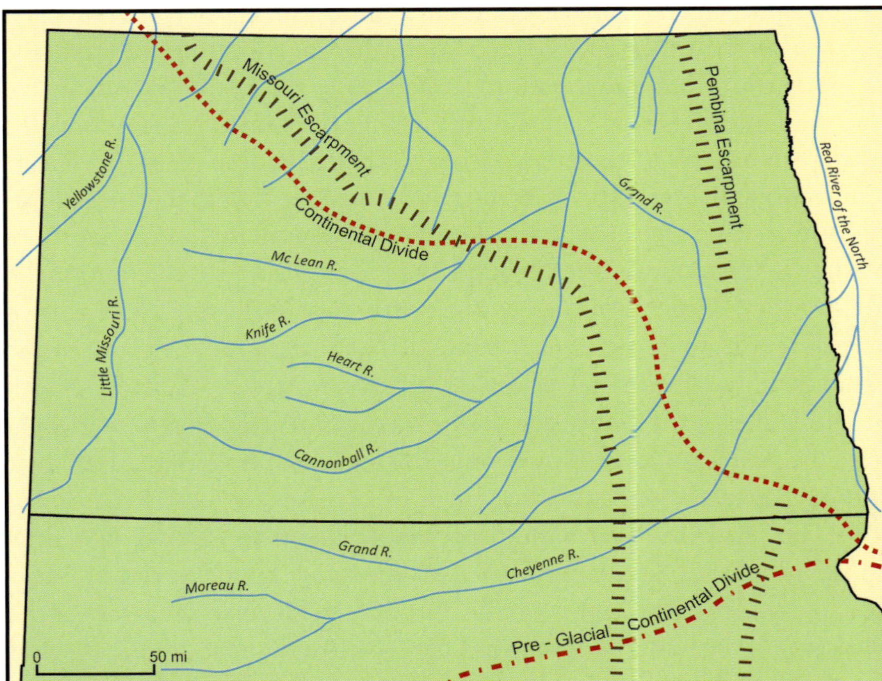

▲ This map shows the drainage pattern in North Dakota prior to glaciation. Some of the river valleys, which are now buried beneath thick layers of glacial sediment, may be much older than others. The ancient river systems in northern and eastern North Dakota are today buried beneath glacial sediments that date back to unknown times in the Pleistocene Epoch—from a few hundred thousand years ago to as much as 3½ million years ago. The routes of most of these valleys through much of the northern and eastern parts of the state are known only through data obtained by test drilling.

The modern, north-south continental divide separating drainage to Hudson Bay from that to the Gulf of Mexico is shown by the dotted line. The preglacial continental divide was farther south, in South Dakota.

surface topography. Fortunately, hundreds of test holes were drilled during studies of the ground water resources of the glacial deposits, so we have a good idea of the route the river followed into northeastern North Dakota.

Another wide, west-east trending segment of the modern Missouri River Valley, between Stanton and Washburn, is an eastward continuation of the modern Knife River. Prior to glaciation, the Knife River flowed east in its modern valley to Stanton, but from there it continued on past Washburn. A few miles east of Washburn it turned slightly northeastward. The Knife River joined the McLean River near the town of Mercer and the combined Knife-McLean River continued northeastward to the Devils Lake area, then north along the east side of the Turtle Mountain area into Canada.

Still another wide segment of the Missouri River Valley in northwestern North Dakota extends from near the modern Missouri River/Yellowstone River confluence, northeastward to Williston. This six- to-eight-mile-wide section of the valley coincides with the preglacial route of the Yellowstone River through that area. Prior to glaciation, the Yellowstone River continued to the north, past Williston. That part of its valley is now buried, but it coincides with the modern route of the Little Muddy River as far as Zahl, 30 miles north of Williston. North of Zahl, the old Yellowstone River Valley into Canada is so deeply buried that its route is known only through drill hole data. The river entered Saskatchewan six miles north of Crosby.

The Missouri River Valley between Williston and New Town, now flooded by Lake Sakakawea, follows the same route as did an east-flowing, mid-Ice Age, but probably not preglacial river. This part of the Missouri River Valley is somewhat narrower than most other east-west segments of the valley in North Dakota, and it is younger than most of them. It is a continuation of a mid-Ice Age river that flowed east from Montana. In Montana, the route of that river coincides with the modern route of the Missouri River past Wolf Point, Poplar and Culbertson. The Montana part of the mid-Ice Age river joined the north-flowing Yellowstone River at Buford.

At Bismarck-Mandan, the Missouri River Valley is two miles wide at the Interstate Highway 94 crossing, but on the south side of Bismarck the valley broadens to six miles wide. The widening southward seems contrary to my earlier comment that north-south segments of the valley tend to be narrow. There is a reason for this exception, though. The valley widens at Bismarck-Mandan because, prior to glaciation, the Heart and Little Heart rivers, which today flow into the Missouri River, joined a few miles east of Bismarck. The combined (preglacial) Heart-Little Heart River continued flowing eastward, joining the Cannonball River in southern Burleigh County, near Moffit. The old, combined Heart-Little Heart Valley still exists as a broad lowland south and southeast of Bismarck, now a wide spot in the Missouri River Valley. At the location where the Heart River flowed eastward, it deposited a thick, gravel deposit which now lies buried 100 feet beneath the Missouri River. Bismarck's new (2013) water-intake structure withdraws

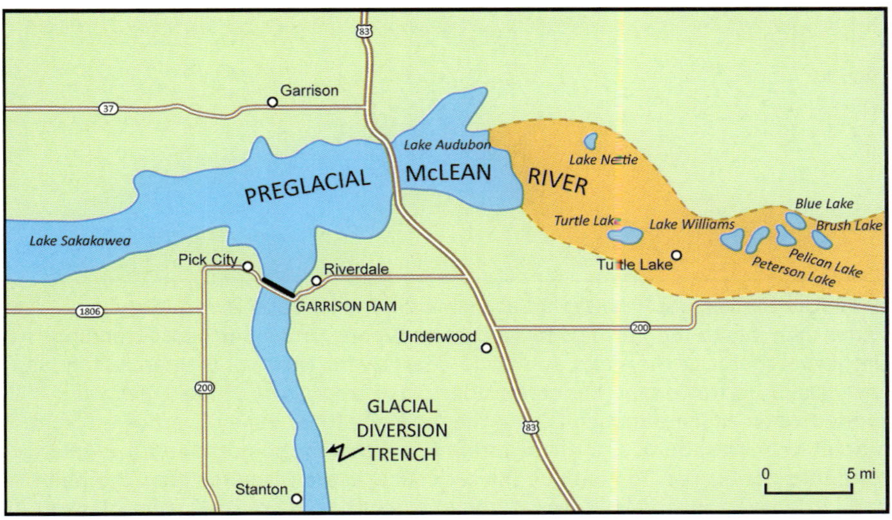

▲ The map shows a portion of the route of the preglacial McLean River, which flowed from (at least) as far away as the New Town area through a broad valley (now flooded by Lake Sakakawea), eastward between Garrison and Riverdale to the Turtle Lake area, and on into Sheridan County. When the McLean River was blocked by a glacier (east of the area shown here), a proglacial lake formed in the valley ahead (to the west) of the glacier. When the lake rose high enough, it overflowed southward from a point near Riverdale, at the site of the modern Garrison Dam. The south-flowing river eroded a narrow diversion trench from Riverdale past Stanton. The modern Missouri River flows through this diversion trench.

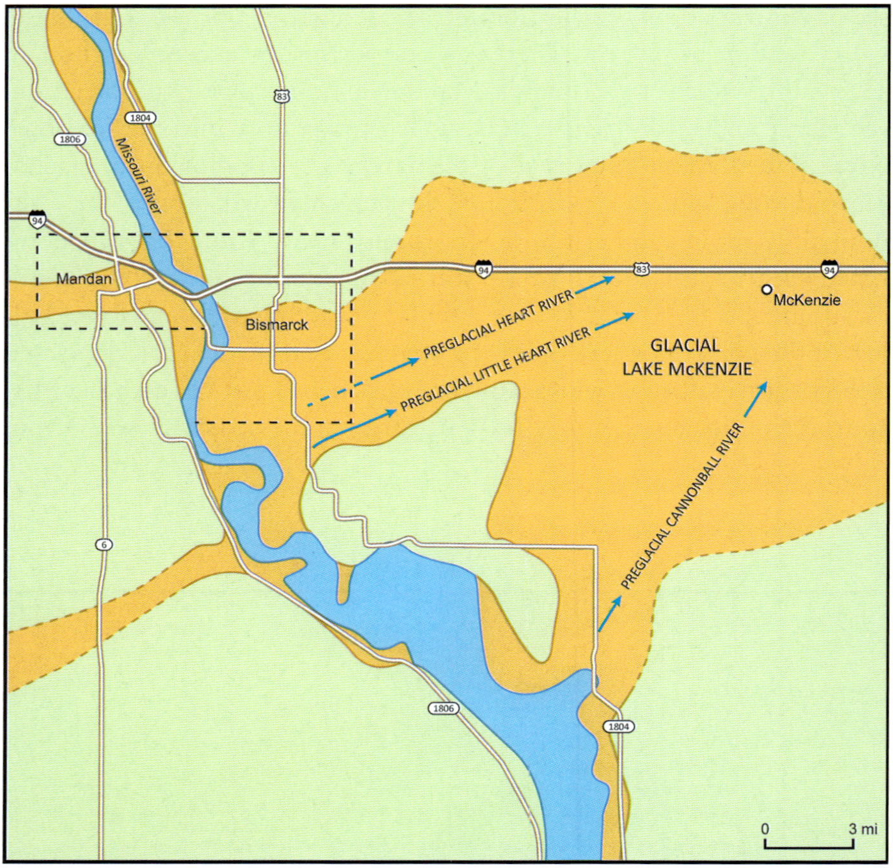

▲ South of Interstate Highway 94 at Bismarck, the Missouri River Valley is wider than it is to the north because it corresponds to the northeast-trending preglacial routes of the Heart and Little Heart rivers. North of Bismarck-Mandan, the valley is narrower and has steep sides. The narrow part of the valley was eroded when an ice-dammed lake in the old Knife River Valley overflowed from a point near Washburn, and the resulting flood carved the north-south segment to the Bismarck area. To continue the sequence, in the Bismarck area, the Heart and Little Heart river valleys were flooded when those two rivers were blocked by glacial ice to the east. The dammed water resulted in the formation of glacial Lake McKenzie, which, in turn, overflowed, carving yet another narrow valley south from Huff. That valley is now flooded by Lake Oahe.

ground water from this old Heart River gravel deposit.

Most of the narrow, north-south segments of the modern Missouri River Valley correspond to places where glaciers diverted then-existing rivers southward. Glaciers in the central part of the state diverted northeast-flowing rivers, like the Knife, Heart and Little Heart, forcing them to flow southward from the points of diversion, forming the north-south segments of the modern Missouri River. Glaciers advancing into northwestern North Dakota diverted north-flowing rivers, like the Yellowstone and Little Missouri, away from their routes into Canada, forcing them to flow to the east and south.

When the McLean River Valley was blocked by a glacier in the Riverdale area midway through the Ice Age, a large proglacial lake formed ahead (to the west) of the ice in the valley. This lake might be considered to be the original Lake Sakakawea: an early ice-dammed lake that predated the Corps of Engineers version of Lake Sakakawea by thousands of years. When the lake overflowed, near where Gar-

▲ **This is a view of the Missouri River from Fort Lincoln State Park. The river is shown in flood stage in August of 2011. View is to the southeast.** JPB Photo: 8/21/2011

rison Dam is today, the resulting flood quickly carved a narrow spillway trench south to the Stanton area. Similarly, the Knife River, which flowed past Stanton and on to the Washburn area, was dammed by glacial ice near Washburn. The Knife River Valley was flooded west of Washburn. The resulting lake overflowed and spilled southward into the Burnt Creek-Square Butte Creek drainage, carving a narrow trench from a few miles east of Washburn to the Bismarck-Mandan area. The modern Missouri River flows in that trench today.

The youngest and narrowest segment of the Missouri River Valley in North Dakota is at New Town, between Four Bears Bridge and Van Hook Bay. As recently as 13,000 years ago, a glacier blocked the river from its route around the north and east side of New Town. The old river route (prior to 13,000 years ago) is now a broad valley, known as

▲ **This view of Lake Sakakawea is from Crow-Flies-High Hill at New Town in Mountrail County. McKenzie County is in the distance on the right. Lake Sakakawea floods a narrow, north-south valley that was eroded when the Missouri River was diverted southward from its earlier route around the north side of New Town more than 13,000 years ago.** JPB Photo: 7/16/2010

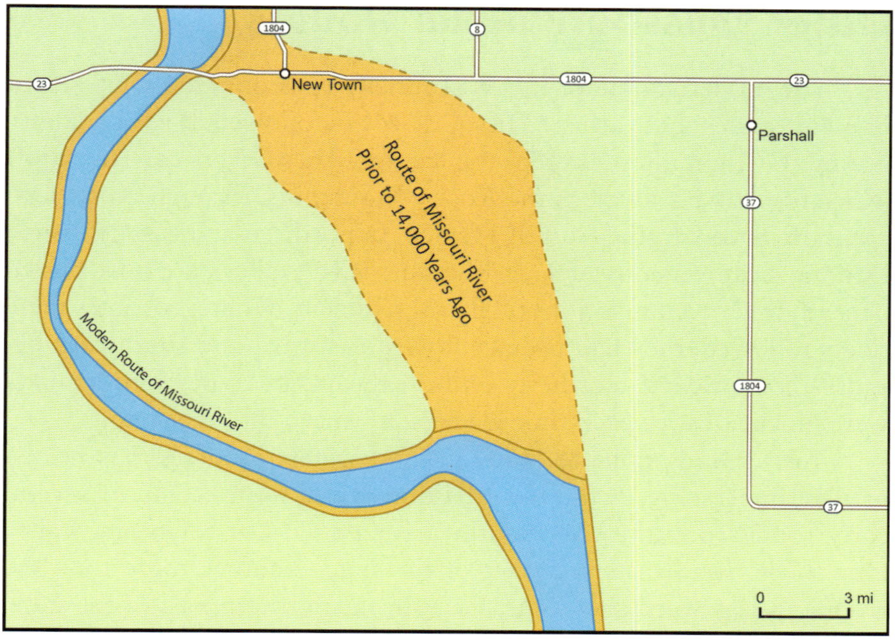

▲ This map shows the old route of the Missouri River at New Town (within the dashed lines southeast of New Town) and the more recent route, formed when the glacier diverted the river farther southwest (within the solid lines). This diverted loop of the Missouri River is the youngest portion of its valley in North Dakota.

the Van Hook Arm, flooded by Lake Sakakawea. The glacier dammed the valley, causing a lake to form upstream (to the west) of the point of blockage. Thick layers of lake sediment, known as the Crow-Flies-High silt, were deposited in the ice-dammed Crow-Flies-High Lake. Crow-Flies-High Lake extended westward from the New Town area to near Williston. In many places between these two cities, exposures of the bedded silt deposits occur at elevations as high as 70 feet above the modern, maximum reservoir level (1,850 feet) of Lake Sakakawea. The lake rose until it overflowed southward, cutting the channel now spanned by the Four Bears Bridge.

Other "Missouri" River Routes

The modern route of the Missouri River is only the most recent of many routes that earlier "Missouri" rivers followed at various times during the 2½ million years of the Ice Age. These rivers carried runoff water from as far away as the Rocky Mountains, through North Dakota, on its way to the Gulf of Mexico. In northern and eastern North Dakota, early versions of such Missouri River valleys are now buried beneath thick accumulations of glacial sediment. Whatever routes these rivers followed, they had to have flowed eastward and southward because their original, northerly and northeasterly routes into Canada were blocked by ice each time glaciers advanced into the state.

Test drilling, done to study ground water resources, has helped us determine some of the old Missouri River routes. One of several early routes took the river southward past Cooperstown and Valley City to the southeastern corner of the state. Another route took the river southeastward past Jamestown. In the northern part of the state, rivers like the Yellowstone were diverted from their northerly routes to easterly and southeasterly routes, past places like Columbus, Kenmare and Minot. The array of buried river valleys is really amazing – and so complicated, with so many possible routes, that we cannot be sure which rivers used which routes, and when they used them. All of these valleys are now buried beneath hundreds of feet of glacier sediment.

Not all of the early Missouri River routes through North Dakota are deeply buried. In the western part of the state, a version of a Missouri River formed when an early glacier advanced at least as far southwest as the Hebron area. The margin of that glacier coincided with what is now a prominent, broad valley known as the Killdeer-Shields channel. The channel extends southeastward from the Killdeer Mountains, past Hebron and Glen Ullin, to the Fort Yates area, crossing the modern Missouri River Valley, and continuing through southwestern Emmons County into South Dakota. No river flows through the Killdeer-Shields channel today, but an early Missouri River flowed in it, perhaps for a longer period of time than the current Missouri River has flowed in its modern route. Interstate Highway 94 crosses the valley half way between Dickinson and Mandan. Good views of the Killdeer-Shields channel can be seen just north of Richardton and between Hebron and Glen Ullin where Highway 94 crosses State Highway 49. Highway 49 and the Burlington Northern Santa Fe Railroad follow the old channel from Hebron to Glen Ullin.

Summary

I realize that my description of the changes in the routes the various Missouri Rivers have followed since the Ice Age began is complicated. Even so, it does not begin to account for the evolution of all of the changes in the vast array of routes that rivers followed during the Ice Age in North Dakota.

The modern Missouri River Valley is a composite valley consisting of older, wide preglacial segments, formed during long periods of time prior to the Ice Age, and younger, narrow segments that were cut quickly at various times during the Ice Age. The parts of the Missouri River Valley that extend from west to east are wider and older than are the narrower segments that extend from north to south. Many of the early Missouri River routes followed for varying periods of time during the Ice Age in northern and eastern North Dakota were later buried beneath thick deposits of glacial sediment.

The current route of the modern Missouri River Valley is only the latest in a continuing series. After the next glacier has come and gone, a new version of the Missouri River will likely follow a different route than does the river today.

CHAPTER FOUR

THE GLACIATED PLAINS AND MISSOURI COTEAU

The Ice Age

North Dakota's climate during much of the past 15 million years of erosion I discussed earlier was probably warm and semiarid, resulting in grasslands. However, global cooling had already caused an ice cap to build over the Antarctic polar region by at least 30 million years ago. Beginning between 3 million and 4 million years ago, glaciers built and spread over much of the northern United States and most of Canada; James Bay in Ontario was glaciated 3.5 million years ago; Greenland has had an ice cap for at least 2.7 million years. Vast ice sheets covered parts of North America as many as 20 times during the past 3.5 million years, and they occur in some mountainous regions today. The relative warmth we have enjoyed for the past 10,000 years is part of an interglacial period. Sometime in the future, the world will likely once again cool and glaciers will again begin to push south from the Arctic.

North Dakota has been glaciated several times already during the Ice Age, and it will be glaciated again, but the most recent glaciation, which we refer to as the Wisconsinan, is the one we know best,[18] as it left behind the array of landforms we see throughout the eastern and northern parts of North Dakota, in all but the southwest part of the state. Glacial features can be seen elsewhere in North America, but in many places, they are obscured by forests, mountains or lakes, preventing them from being recognized, studied and understood as easily as they can in North Dakota.

The onset of the Ice Age was gradual, preceded by a long period of cooling through Tertiary time. When the Tertiary began, 65 million years ago, warm and moist conditions prevailed and North Dakota was tropical or subtropical. Then, during the Oligocene, the Antarctic continental plate moved into its position over the south polar area, and worldwide cooling began in earnest.[19] Glaciers likely formed over Antarctica in Oligocene time, and possibly earlier. By Miocene time, 23 million years ago, the northern polar Arctic Ocean may have been frozen over at times, as it is today. Recent studies indicate that during much of Pleistocene time, an ice cap existed in an area northwest of Alaska in what is now the Arctic Ocean. As long as Antarctica remains in place over the South Pole, we will continue to have recurrent episodes of continental glaciation.

The reasons for glaciation and what has caused it at various times in the geologic past are not completely understood. I have discussed some of the theories relating to this topic in more detail in the appendix (*The Holoarctic—Antarctic Glacial Period*). In addition to the location of a continental-size land mass being positioned over a polar region, a variety of additional circumstances are likely involved. For example, it is known that the Panamanian Seaway closed 4 million years ago. This event likely diverted equatorial water in the Atlantic north, resulting in more moisture for snow and ice accumulation, and more intense zonal westerly winds. These circumstances may have contributed to North American glaciation. Another event was the rising of the Himalayan Mountains and the Tibetan Plateau through Cenozoic time. This resulted in more arid conditions in central Asia and an increase in dust accumulating in the atmosphere. The increased dust cut down on the amount of heat the Earth received from the sun.

The first extensive North American glaciers formed during late Pliocene time, perhaps 3.5 million years ago or even earlier. Since then, North Dakota has been buried beneath glaciers several times. Approximately 70 percent of North Dakota's area, including most of the land-

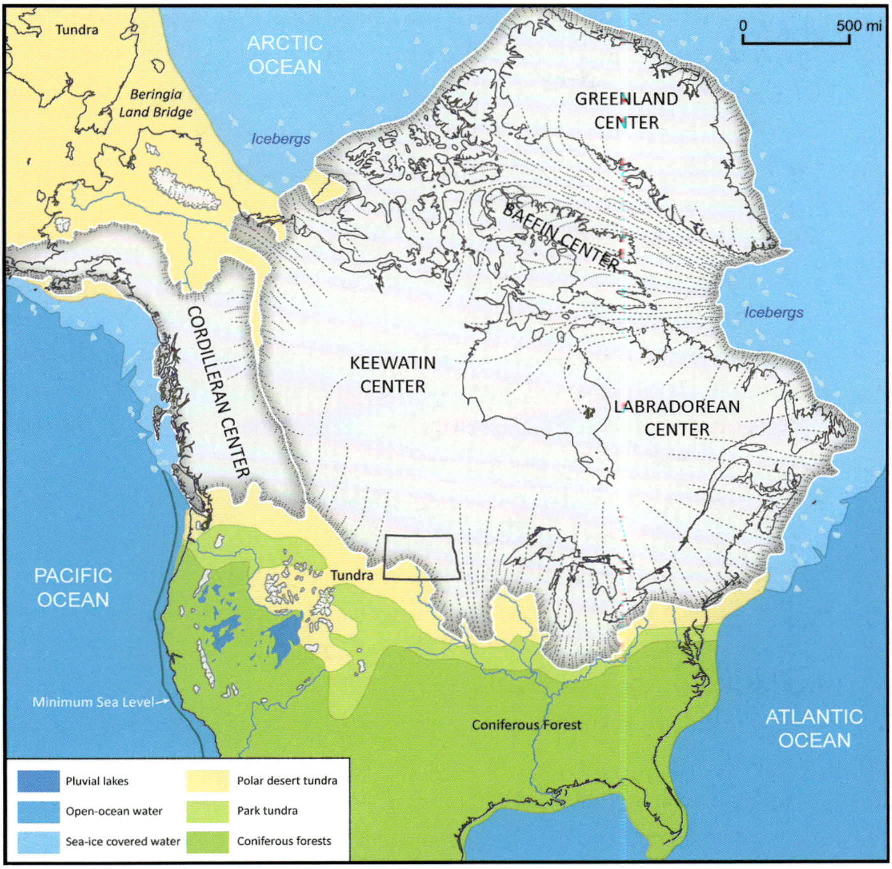

▲ This map shows one of many interpretations depicting the extent of glaciation in North America. It also shows the general distribution of the main vegetation zones beyond the ice sheet. During the Wisconsinan glaciation, glaciers flowed southward from the Keewatin Center, west of Hudson Bay. Ice covered all but the southwest part of North Dakota. The several ice caps flowed into one another, resulting in an uninterrupted expanse of ice that covered North America, from the Pacific to the Atlantic oceans. I have outlined the location of North Dakota.

This map is one I modified from one that was originally included in the following publication:

- Anderson, B.G., and Borns, Jr., H.W., 1997, *The ice age world: An introduction to Quaternary history and research with emphasis on North America and Europe during the last 25 million years,* Oslo, Norway, Scandinavian University Press.

scape north and east of the Missouri River, was glaciated during the most recent part of the Ice Age. We know from test drilling near Devils Lake that at least six glaciations have occurred in that area. Some of the earliest—late Pliocene and early Pleistocene glaciations—may have reached the southwestern corner of the state, but if this happened, their deposits have since been completely eroded away.[20]

The most recent glacial advance, the Late Wisconsinan Glaciation, is the one responsible for most of the glacial landforms we see today in North Dakota. Layer upon layer of snow, year after year, eventually built up to create a dome of ice centered in the Keewatin district of Nunavut, west of Hudson Bay. The increasing weight at the center of the mass of ice pushed the developing glacier west and south into an ever larger, ever thickening pool of ice.

North Dakota's landscape when the Ice Age began was benign, less rugged than today. Rivers like the Red, Yellowstone and Little Missouri flowed through broad valleys, toward the north and northeast. The badlands had not yet been carved and the Missouri River did not exist. The Pembina, Missouri and Turtle Mountain escarpments were not the abrupt features we see today.

Glaciation came on slowly, spreading southward. It was inexorable, a long winter one year, a shorter and cooler summer the next until, finally, unremitting winter reigned for thousands of years. Snow piled on snow near Hudson Bay, compressing the bottom layers into ice. Then, somewhat like thick molasses or tar, mobile ice flowed out from under the incredible weight of year after year of snow, pulverizing the rock and soils beneath, so that the Earth's crust sagged under the burden. Unable to withstand the downward pressure, the whole mass—ice, soil, and rock—flowed southward out of Canada onto the North Dakota tundra. Once the glacier began to move, it moved quickly, perhaps advancing more than a mile a year. Ice eventually covered all except the southwestern corner of the state, but in that area, the ground was frozen to a depth of several hundred feet.

Only a few hardy living things could endure the eternal winters. Generation after generation of plants and animals were forced southward to more temperate climates. Keep in mind that this entire scenario was repeated at least six or seven times. By the time the most recent glacier reached its maximum extent, 18 thousand years ago, glacial ice covered all except the southwestern part of North Dakota and extended southward as far as central Iowa. Earlier glaciers had advanced even farther, into Kansas and northern Missouri.

The Ice Age has been a time of dramatic climate change. With the current concern about the importance of climate change to human society, the relevance of understanding the climate during the geologic past should be apparent. Detailed studies of North Dakota's glacial deposits provide a glimpse of climate change during the past several thousand years, long before instrumental records can be applied. In addition to studies of the glacial deposits, geologists are studying the glaciers themselves in places like Greenland, where recent studies show that the melting of the glacial ice has not been a steady wasting. Rather, the Greenland ice sheet was smaller between 3,000 and 5,000 years ago than it is now, a response to warmer ocean temperatures at that time than currently; temperatures on land peaked earlier than that, between 9,000 and 5,000 years ago.

Cores drilled in the Antarctic ice provide a record of climate change for more than 700,000 years. All of these geological climate records, now being studied in detail, contain lessons about the impact of past climates on the North Dakota landscape, soils, lakes and life. The lessons to be learned are critical to understanding modern climate change.

Glaciers and Glacial Deposits

Glaciers are giant bodies of ice formed from snow that survive from year to year. Accumulations of snowfall from past years compact into a substance called *firn*, a recrystallized residue of snow left over from past seasons. Each year, new layers of snow bury and compress the previous layers. Compression forces the snow to re-crystallize so that it forms grains similar in size and shape to grains of sugar. The grains gradually grow larger and the air pockets between the grains get smaller, causing the snow to slowly compact and increase in density. After two winters, when the snow has turned into ***firn***, it is two-thirds as dense as water. Through time, the larger ice crystals become so compressed that any air pockets between them are tiny. In old glacier ice, crystals can reach several inches in length. For most glaciers, this process takes at least 25 to 100 years or more.

During the summers, when it is warm enough for rain instead of snow, the rainfall adds to the mass of a glacier, eventually freezing and becoming part of the glacier. With time and additional snow cover, the whole mass gradually solidifies into hard ice: a glacier. Ice formation may go on almost indefinitely. Until the shrinkage identified in 2013, the Greenland ice cap grew for more than 100,000 years to a thickness of two miles (although, during those 100,000 years, there were many times when the ice cap was shrinking, as it is today).

The color of pure glacial ice, if it is clear and lacking the various rocks and sediments often found in glaciers, is ice-blue. Any air trapped in the ice, thousands of years ago, and pressurized for thousands of years by the great overlying weight of the glacier, may escape with force from the ice when it melts. Drop a piece of glacial ice into a glass of warm water and it may literally explode. By analyzing these trapped pockets of air, scientists can learn what our atmosphere consisted of in the past.

When an ice mass becomes thick enough and heavy enough to flow, it officially becomes a true glacier. A glacier flows slowly—at a glacial pace; by comparison, a snail can travel much faster. Glaciers flow away from the places where they are thickest. They may flow at a few feet a year, although in some circumstances, glacial ice may move by sliding on a layer of water that percolates downward from the ice surface during summer melt. With water as a lubricant, a glacier can slide hundreds of feet per day, rather than a few feet per year. In the Northern Hemisphere, glaciers expand southward, away from polar regions because the temperatures to the south are warmer than to the north. A glacier tends to flow most easily when it is warmer and less brittle. It can move much faster at 30° F than at minus 30° F.

A glacier does not glide placidly over the land. It scratches and grinds the underlying bedrock surface, picking up pieces of rock and soil, dragging them along and using them as tools to scour the ground beneath it. Glaciers do not normally flow uphill, but they do fill lowlands and overtop them, much like floodwater.

Glaciers in mountainous areas, unlike broad ice sheets in places like Greenland and Antarctica, come with a sense of scale. The moun-

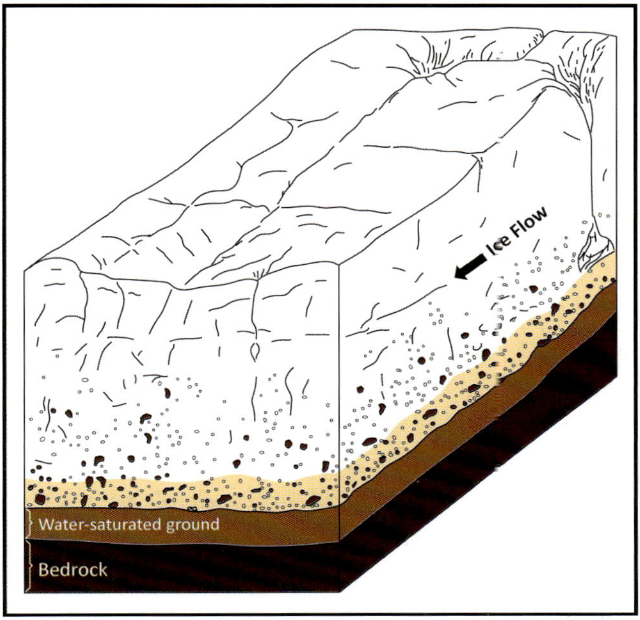

▲ This cross-sectional view of part of a glacier shows how material over which a glacier is flowing is incorporated into the moving body of ice. The heaviest concentration of sediment is near the base of the glacier, with the ice becoming relatively free of debris upward. The materials at the base of the glacier were thawed, resulting in a muddy/slushy surface over which the glacier flowed. The entire mass—ice, along with its entrained debris—moves forward as a single mass.

tains in the distance and the valley walls that hold the glacier help you orient yourself. But suppose you are standing on a snow-covered, continental-size glacier (dressed in a heavy parka). You see nothing but frozen wasteland, nothing but whiteness. No buildings, no fences, no trees, no landmarks. Only icy emptiness. No sound but the wind. On a cloudy day, sky and ice blend, making it practically impossible to distinguish the horizon marking their boundary.

The most recent major glacial episode in North Dakota is referred to as the Wisconsinan glaciation. It began approximately 90,000 years ago and ended 11,500 years ago, but glacial conditions were not continuous during that entire time. An initial pulse of glaciation (the Early Wisconsinan, 90,000 to 70,000 years ago) was followed by withdrawal of the ice, which was probably complete by 65,000 years ago. Then, between 65,000 and 35,000 years ago, North Dakota's climate alternated between combinations of warmer, wetter, cooler and drier periods, much as it does today. A second major pulse of glaciation, the Late Wisconsinan, began 30,000 years ago. The Late Wisconsinan glacier reached its maximum extent between 18,000 and 16,000 years ago, when it covered all but the southwestern part of North Dakota. The position of the Missouri River approximates the maximum extent of the Late Wisconsinan glacier. Active glaciers melted completely from the state by 11,500 years ago.

By the time each glacier advanced, the land ahead of it may have become deeply frozen tundra (permafrost). The ice moved over the frozen rock and soil, picking up chunks of these materials and incorporating them into the glacier itself. To the north, west of Hudson Bay, the materials beneath the thickening ice were crystalline igneous and metamorphic rocks such as granite and gneiss. Farther south, the glacier flowed over layers of sedimentary rock, but whatever the glacier flowed over, it carried some of it along.

A moving glacier may be likened to a huge excavation and grading machine that does its job of eroding by plucking and abrasion. Plucking, the more important of the two, is based on a freeze-thaw cycle. The cycle begins when downward pressure melts the ice at the base of a glacier. Water seeps into cracks in the rock beneath the glacier. When the water freezes, the expanding ice breaks rock fragments loose and incorporates them into the debris near the base of the glacial ice.

After a glacier covers an area for a while, and a considerable thickness of ice lies on the land, the materials beneath the glacier gradually thaw, a combined result of the pressure of the overlying ice and the natural upward flow of heat from the Earth's interior. The ground surface beneath the North Dakota glaciers was not frozen most of the time, and the base of ice may have been a muddy mass, which contributed to even more sediment being incorporated into the base of the moving glacier.

In some places, ground water in the saturated sediments beneath the heavy weight of the glacier built up great pressures due to the weight of the overlying ice. Besides transforming materials beneath the ice into a mud-like mixture, the water, because it was pressurized, tended to force—squeeze—the subglacial sediment slurry upward into the base of the moving glacier. Sediments beneath the ice were smeared out as they were carried along with the advancing mass of ice.

A glacier seldom behaves like a bulldozer, pushing debris ahead of it. More often, it incorporates debris as it moves by freezing it onto its base. The boulders, gravel, sand, silt and clay beneath a glacier become part of the moving glacier mass, both ice and sediment flowing forward, and the farther the glacier travels, the more material it accumulates. Each time glaciers advanced over North Dakota, when they melted, they dropped their entire load of rock and sediment – the materials gathered from places previously overridden. We find rocks in North Dakota that came from Alberta, Saskatchewan, Manitoba and Ontario. We find chunks of shale that came from only a few dozen feet away. The sediment a glacier was carrying finally came to rest when the last ice melted.

During the active life of a glacier, every crystal of ice, every boulder, sand grain and fragment of rock within the ice is moving, slowly making its way away from the center of snow and ice accumulation. In North Dakota, the movement was southward, away from the Keewatin center of ice accumulation west of Hudson Bay. Apart from its overall southward progress, variations in the topography over which the glacier advanced locally affected its direction of flow.

When a glacier advanced southward into more temperate regions, melting took place on its surface and along its margin, especially at times when temperatures were warm. Some melting occurred during each summer season—even more when the climate warmed for periods of several years at a time. The position of the edge of an ice sheet at any given time was determined by the balance between melting and the rate the glacier was flowing. While the climate was cold, the edge of a glacier advanced. When it warmed a little, the glacier margin melted back as fast as new ice could be supplied so that equilibrium was struck. Even though the glacier was moving, its edge neither advanced nor receded. Given still warmer conditions, the surface of a glacier melted rapidly; the ice thinned and its edge melted back faster than new ice was being supplied. The area the ice had covered gradually became deglaciated.

As melting occurred along the margin of the glacier, debris that had been frozen into the ice many miles to the north was freed and deposited on the ground. This glacial sediment consisted of a blended sampling of the various kinds of rock and sediment over which the glacier had flowed. Ice advancing into North Dakota from the northeast deposited sandy, granite-rich materials it had picked up as it flowed over the Precambrian rocks of northwest Ontario. Ice advancing from the northwest brought a mixture of sandy and clayey sediment it had accumulated as it flowed over broad expanses of Cretaceous shale and Tertiary sands of southern Alberta and Saskatchewan. Ice advancing straight southward from Manitoba, up the Red River Valley, deposited carbonate-rich sediment it had picked up north of Winnipeg where Paleozoic limestone and dolostone are exposed today. Studying the composition of the glacial sediment is one way that geologists can determine the direction from which the ice came, and the kind of land it flowed over.

Eventually, as each glacier melted, the land gradually became free of ice. I emphasize that glacier retreat refers to the melting of the ice; no reversal of ice flow is involved. Three different kinds of glacial ice wasting occurred at different times and places in North Dakota. The first occurred when the glacier margin may have been far to the south, in Iowa and South Dakota. The result of melting, perhaps during a period of warmth of thousands of years, resulted in the loss of glacier mass off its surface. The view from above would not necessarily have changed much—everything was still all ice—but the thickness of glacial ice covering the land was diminished in thickness by hundreds or thousands of feet, even before the glacier margin receded appreciably.

The second way a glacier wasted was when the ice margin was nearby, and wasting involved change in the position of the active glacier margin. As a glacier melted, and after it had thinned, its active margin gradually receded because the volume of ice arriving was not sufficient to replace the ice lost at the edge due to melting. This kind of retreat caused the ice margin to melt back, sometimes in a step-like fashion, the flow of ice pausing long enough at times for the forward movement of the glacier to deliver piles of sediment to the receding ice margin. A year of glacial activity might involve the margin moving forward a short distance during a winter, then receding a slightly greater distance the following summer.

The third way glaciers wasted involved large-scale stoppage of ice movement, leaving large parts of the glacier stranded, sometimes over broad, upland areas, detached from the main body of still-active ice. This kind of wasting was especially important in North Dakota, as it resulted in areas of stagnant, or "dead," ice, which then continued to melt slowly. Landforms resulting from the melting of such stagnant ice are distinctive and much different from those constructed during the step-wise retreat of active glacial ice I described earlier.

Much of North Dakota's modern landscape reflects its latest encounter with glaciers during the Ice Age. As glaciers flowed into the

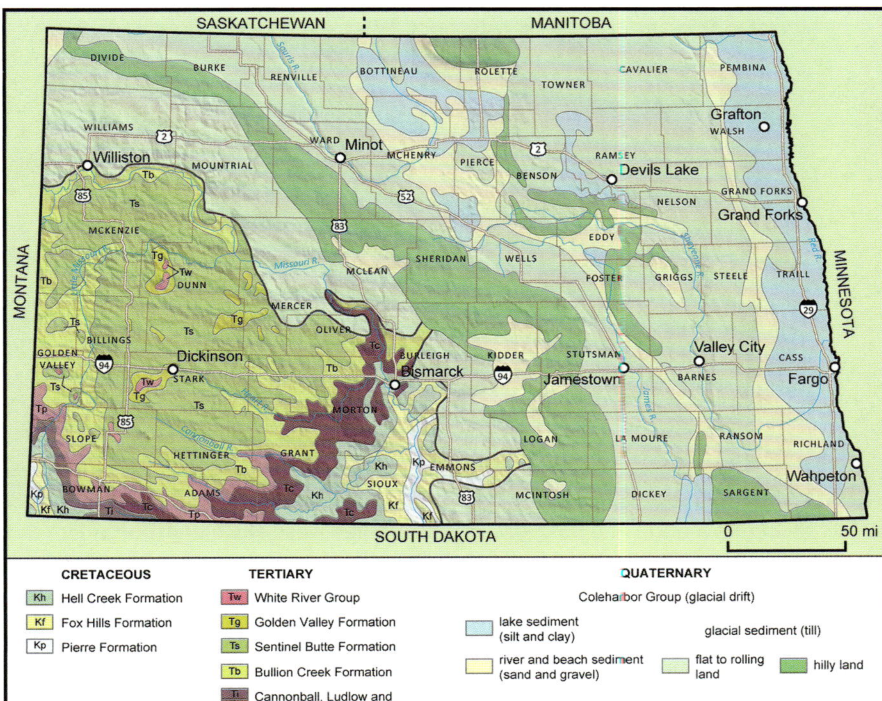

▲ This geologic map of North Dakota shows the preglacial bedrock formations in the southwestern part of the state and the glacial landforms in the north and east. The map is generalized; scattered patches of glacial material that occur in places overlying bedrock in western North Dakota are not shown, nor are small outcroppings of bedrock in places where the glacial cover is thin.

state, carrying the pulverized rock and soil debris they had picked up along their routes, they sheared off old bedrock landforms, smeared on new layers of sediment and built new landforms. They filled old river valleys with sediment at the same time rivers of meltwater were flowing away from the glaciers carving new valleys. In some places, the glacial ice forced existing rivers to follow different routes; in other places, it completely obliterated rivers and valleys. Cold winds blowing over sand that had earlier been deposited on flood plains and in lakes built dunes and spread a veneer of silt (loess) over much of the state.

Most glacial sediment is loose material. It is unconsolidated and does not hold together well (you can dig it with a shovel). Earlier gla-

▲ This till cliff is along Lake Sakakawea seven miles west of Coleharbor in McLean County. The exposure of glacial sediment consists of a mixture of cobbles and boulders contained in a finer-grained matrix of sandy silt and clay. In most exposures near the ground surface, the overall color of till tends to be buff or tan, as shown here. The minerals it contains have been exposed to oxygen for a long time and have weathered so that they are oxidized. In the subsurface, below the zone of oxidation where oxygen cannot reach, till is usually dark gray to black.

The lower part of this wave-cut exposure, with the vertical columns, is either early Wisconsinan or older in age, 70,000 years or older. The till nearer the top of the exposure is probably Late Wisconsinan, perhaps 15,000 years old.

This till deposit was buried for thousands of years until wave action eroded the shore when Lake Sakakawea flooded in them in the mid-20th century.

JPB Photo: 6/20/2009

ciers deposited sediment as well, but almost all of this earlier sediment has eroded away. In the few places where we have found it exposed, or drilled into it, the older glacial sediment may be cemented. A jackhammer may be more appropriate than a shovel for digging in such cemented deposits. However, these deposits are the exception. Most glacial deposits are not cemented.

Sediments related to glaciation in North Dakota can be grouped into three main types: till, lake sediment and outwash.

▲ The till in this close-up view was baked to a form of clinker when an underlying lignite vein burned. In this Ward County exposure, the fine-grained portion of the till (silt and clay) has been baked to a reddish hue. In addition to the change in color, the fine-grained materials have been hardened by the baking process.

The texture of the till is easy to see in this exposure because the pebbles and cobbles, which are mainly limestone, were unaffected by the heat and remain their original color, whereas the finer fraction has been transformed to red due to the oxidation of the iron minerals it contains. The two light-colored rocks near the top of the photo are 6 to 8 inches in diameter.

Notice the near-vertical and intersecting cracks (joints). Joints in till sometimes indicate a very old till, much older than most till deposits found near the land surface. However, in this instance, the jointing may be the result of heating at the time the till was baked.

Baked till is uncommon, and this is the only good exposure I have seen. I took this photo in 1965. JPB Photo: Scan of 1965 slide

1. **Till** was deposited directly from the ice, mostly in the form of mudflows, which slumped or flowed into their current position as the ice melted. Till consists of silty, sandy, pebbly clay, as well as cobbles, or even large boulders.[21]

2. **Lake sediment** is layered material deposited in lakes that formed on and near the glacier. Such sediment is also known as *glaciolacustrine* sediment. It consists of layers of fine-grained silt and clay, deposited on the lake floor, along with some sand and gravel, which formed as beaches along the shores of lakes dammed by glaciers.

▲ These bedded silt deposits of glacial Lake Mose are 12 miles east-northeast of Finley in Steele County. The lake was ice-marginal, dammed between the glacier that filled the Red River Valley to the east and higher land to the west. It preceded Lake Agassiz, and although I am unsure of its extent, it extended at least 130 miles from north to south and was between 30 and 50 miles wide. Its deposits are widespread, found in many places in Nelson, Griggs, Steele, Barnes and Ransom counties. Lake Mose deposits are as thick as 40 feet in places where they are exposed and they tend to be uniform, with few pebbles. They are overlain in most places by till deposits, indicating that a glacier readvanced over them, likely ending the duration of the lake.

I took this photo in 1963. JPB Photo: Scan of 1963 slide

3. **Outwash** consists of material deposited by running water. Some outwash may be cemented into a kind of stony concrete, but most of it is loose sand and gravel that was washed out of the melting glacier (hence the name "outwash"). Outwash was deposited by

▲ This is a gravel exposure two miles north of Pickardville in Sheridan County. This glacial outwash gravel grades upward from interbedded sand and small gravel at the base of the exposure (indicating deposition by slowly running water) into coarser material at the top of the exposure. The material was deposited by more rapidly running water. Some of the larger cobbles near the top are as much as a foot in diameter, indicating a strong water flow.

The gravel exposure's reddish hue is due to oxidized iron minerals, which help to partially cement the gravel.

I took this photo in 1979. JPB Photo: Scan of 1979 slide

streams and rivers flowing in valleys carrying meltwater or as broad, often level, sheets of sand ahead of a melting glacier.

Where they are present, sediments deposited directly by glaciers, and by wind and water associated with glaciation, form a thick covering over much of the preglacial (bedrock) surface. In Sheridan County, in central North Dakota, the glacial sediment is more than 700 feet thick in places. Ten miles northwest of Tolley in northwestern Renville County, it is at least 800 feet thick, the thickest I can document in the state. The amount and thickness of glacial sediment can vary considerably across short distances, so it is likely that even thicker deposits exist in places. Over much of the glaciated part of the state, the glacial materials average 150 feet thick.

At any given location, the glacial deposits commonly consist of two or more layers of till, interbedded with lake beds, alluvial sediments or other materials. In some places, soils, which had developed on the surface of an earlier glacial, river, lake or wind-blown deposit, were buried when a new layer of glacial material was deposited. These old, buried soils (paleosols) were formed during long intervals of weathering and exposure, like the one we are enjoying now. Paleosols are among the best indicators in the geological record for multiple episodes of glaciation during the Ice Age. The characteristics of a paleosol help us understand the climatic conditions (forest or grassland, wet or dry, cool or warm, etc.) at the time it formed.[22]

Drill-hole data in eastern and southeastern North Dakota provide evidence that at least a half dozen glacial advances have occurred there since the Ice Age began. Parts of southwestern North Dakota were glaciated during some of the earlier glaciations, but glacial landforms are not found there because they were eroded away long ago. Glacial lake sediments and river gravels containing glacially derived materials can be found as far southwest as Dickinson and near the Killdeer Mountains, and I have tentatively identified patches of cemented till and glacial river gravels near Bowman and Rhame, places usually considered never to have been glaciated.

Modern soils are an important link to our geologic past. Fresh glacial deposits consist of a mixture of materials, and because their sources are so varied, they provide the combination of nutrients necessary for fertile soil. In the glaciated part of the state, North Dakota's soils consist of the weathered exterior of materials left by glacial action. In the thousands of years that have elapsed since the ice sheets disappeared, constantly changing climate, physical and chemical weathering, accumulation of prairie and woodland plant litter, development of root systems and burrowing activity by organisms have all contributed to the transformation of glacial deposits into the rich soils that form the basis for much of our agricultural wealth.

Landforms Shaped by Glaciers

All of the glacial landforms in North Dakota were formed during the most recent (the Wisconsinan) glaciation. Landforms related to glaciation can be seen in many other places in North America, but forests, mountains and urban developments often prevent them from being analyzed as thoroughly as they can be on the open plains of North Dakota.

Ground Moraine

The landform known as ground moraine consists of a blanketing layer of till that was deposited by a glacier. Ground moraine is the most extensive landform in North Dakota.[23] It may include several layers that accumulated during successive glaciations so that, at any given place, the total thickness of glacial deposits in an area of ground moraine may be several hundred feet.[24]

When the main body of a glacier, the broad expanse of ice behind the terminus, melts in place, the result is usually ground moraine. Rock and sediment within the ice are gradually deposited as mudflows at the base of the glacier. Lower areas become filled in as material flows to them from nearby higher areas. The result is usually a leveling effect that leaves the land less hilly than it was before it was glaciated.

Ground moraine has gently rolling topography, and the soils developed on it tend to be fertile, built on a rich mixture of minerals. In most places, ground moraine itself is gently rolling and featureless, but ground moraine surfaces are commonly interrupted by a variety of other geologic features such as meltwater trenches, eskers, kames, drumlins and small lake plains. These are the landforms you are likely to notice when you are in an area of ground moraine, not the ground moraine itself. Places like Carrington, Kenmare, Langdon, Lakota, Leeds, Tower City or Forman are situated in areas of ground moraine: areas of low-relief and gently rolling landscapes that provide excellent farmland.

In several places in North Dakota, the ground moraine surface is marked by small, linear ridges known as washboards. Washboard ridges (or "washboard moraines") are usually too subtle to be recognized by a ground-based observer, but they are obvious on air photos. If an area of ground moraine has any noticeable relief, that relief is sometimes due to washboards.

Washboard ridges formed along the leading edge of the glacier. They may have from 5 to 10 feet of relief between the tops of the ridges and the intervening troughs. The spacing between the ridges can vary

▲ In this photo, a canola crop is growing on ground moraine in northern Towner County five miles northwest of Rock Lake. Ground moraine has flat to gently rolling, mainly featureless topography, and it is usually excellent farm land.
JPB Photo: 7/30/2009

from only a few tens of feet to several hundred feet. The ridges consist of material that was carried upward, along shear planes in the ice, near the edge of the glacier. They are not true ice thrusts (see discussion of

▲ This air photo shows partially submerged washboard ridges along the east side of McHugh Slough in northeastern Nelson County. The area in this view, six miles northeast of Lakota, is approximately three miles north to south. The image is positioned so that the top is slightly west of north. The road on the right side of the photo is north-south.

Water in the slough partially floods the area, helping accentuate the ridges. The ridges are only 1 to 2 feet high, but the water happens to be at just the right level to flood the depressions between the ridges. The washboard ridges are very slightly arcuate, with their concave sides facing north (the arcuate configuration becomes more apparent in a larger view). The margin of the glacier that deposited the washboards was receding northward when the area was becoming de-glaciated.

Washboard ridges like these formed when material at the edge of a glacier froze and the ice advanced a very short distance. The ice then receded as far as the next ridge to the north and the process was repeated, again and again. If the glacier had advanced a greater distance – during a prolonged period of colder time – the washboard ridges would have been overridden by ice and destroyed. For washboard moraine ridges to form, and be preserved, the glacier has to be in an overall stage of receding at the very end of active glaciation in that area.

Google Earth image of part of northwestern Nelson County six miles northeast of Lakota.

ice thrusting) because they consist of what was once the loose material at the ice margin, now collected into ridges. At least some washboard ridges may have formed as annual or otherwise periodic accumulations of material at the edge of the glacier while its margin was receding. Very probably, the formation of a washboard was the last action a glacier did while its margin stood in a given area; if the glacier had re-advanced over a washboard ridge, it would have destroyed it. Most washboards formed when the outermost edge of the glacier became frozen to the ground and therefore picked up the underlying material as it advanced a short distance.[25]

However they formed, washboards are always parallel to the former margin of the glacier and thus are reliable indicators of glacier flow direction, which is normal to the long axes of the washboard ridges. Washboards can be found on the ground moraine surface almost anywhere, but they are much more numerous in some places, such as northern Nelson County, east of Valley City in Barnes County and near Jamestown in Stutsman County.

The broad area of ground moraine covering most of the eastern half of North Dakota forms the basis for much of the land we commonly refer to as the "prairie" (the word is French and originally meant "a meadow grazed by cattle"). The land was largely grassland prior to the introduction of modern agriculture. Joseph Nicollet, a French geographer who immigrated to the United States in 1832, wrote the following in 1839 while traveling through the area near present-day Northwood, in what is now Grand Forks County:

> May I not be permitted, in this place, to introduce a few reflections on the magical influence of the prairies? It is difficult to express by words the varied impressions which their spectacle produces. Their sight never wearies. To look at a prairie up or down; to ascend one of its undulations; to reach a small plateau . . . moving from wave to wave over alternate swells and depressions; and finally, to reach the vast interminable low prairie that extends itself in front—be it for hours, days, or weeks, one never tires; pleasurable and exhilarating sensations are all the time felt; ennui is never experienced. Doubtless there are moments when excessive heat, a want of fresh water, and other privations, remind one that life is a toil; but

> these drawbacks are of short duration. There is almost always a breeze over them. The security one feels is knowing that there are no concealed dangers, so vast is the extent which the eye takes in; no difficulties of road; a far spreading verdure, relieved by a profusion of variously colored flowers; the azure of the sky above, or the tempest that can be seen from its beginning to its end; the beautiful modifications of the changing clouds; the curious looming of objects between earth and sky; taxing the ingenuity every moment to rectify . . . everything, is calculated to excite the perceptions, and keep alive the imagination. In the summer season, especially, everything upon the prairies is cheerful, graceful, and animated. The Indians, with herds of deer, antelope, and buffalo, give life and motion to them. It is then they should be visited; and I pity the man whose soul could remain unmoved under such a scene of excitement.

Nicollet was surrounded by ground moraine as he wrote these words, although he likely had no notion of the glacial origin of the landscape.

Drumlin-like Landforms

Drumlins are hills elongated in the direction the glacier flowed. They are common in Ireland and, in fact, the name "drumlin" is derived from an Irish Gaelic word *druim*, meaning "back" or "ridge." Drumlins are found in the eastern United States, too; they are abundant near Boston, Palmyra, New York (near Rochester) and Wausau, Wisconsin. Irish drumlins, and the drumlins in the places I just mentioned, might be 100 feet high, 500 feet long and 200 feet wide: two or three times as long as they are wide. Most drumlins have blunt noses pointing in the direction from which the ice approached and a gentler slope tapering in the other direction; they look like inverted spoons.

An array of drumlin-like ridges can be seen in the north-central part of the state, mostly in McHenry County. It is unlikely that these features, which I will refer to as drumlinoid, formed in the same way as did the drumlins near Boston or in Ireland. Most such landforms in North Dakota are found in the area between Minot and Harvey.

They are especially abundant in an area between Verendrye and Balfour, where 200 of them occur. Some of the ridges rise up to 70 feet above the surrounding landscape. The largest of the ridges is a 17-mile-long feature known as Hogback Ridge, which is 375 feet wide and 25 to 40 feet high along most of its length.[26] Hogback Ridge, along with nearby similar, smaller ridges, trends from northwest to southeast, precisely indicating the direction in which the most recent glacier flowed as it moved through the region. The ridge resembles a large railroad or highway grade and, in fact, a road once followed the crest along a part of the ridge. All of the ridges are exceptionally straight, so straight, in fact, that they might at first be mistaken for artificial features. Some of the drumlinoid ridges, including Hogback Ridge, are bordered on one or both sides by long, narrow, pond-filled depressions 3 to 5 feet deep.

The differences in the shapes of these drumlinoid landforms from place to place are due, in part, to differences in the composition and texture of the materials over which the glacier was flowing and, in some instances, to characteristics of the glacier itself. In the eastern U.S. and Ireland, the glaciers moved over sandy materials. In North Dakota, the predominant substrate beneath the ice was rich in clay. Clay is plastic and can be molded more easily than sand, allowing much longer ridges to form. The forces exerted by the glacier responsible for forming drumlins may have been similar from place to place, but the resulting landforms were quite different because of variations in the composition and textures of the materials over which the ice was flowing.

The character of the materials beneath the glacier was not the only factor that determined the formation of Hogback Ridge and nearby ridges. Shortly before the drumlinoid-forming event, the glacier had receded from the area, likely into Canada. As it receded, the first Glacial Lake Souris formed ahead of (southeast of) the glacier in the area where the drumlinoid landforms are now.

The withdrawal of the glacier, the result of a long period of warmth that began at least 16,000 years ago, might have been destined to be permanent if the climate had not turned sharply colder 12,800 years ago. This change to colder climates is known as the Younger Dryas event, and it caused the glacier to re-advance quickly.[27] The long period of warmth preceding the Younger Dryas had caused the glacier to melt to the point that its volume was only a fraction of what it had been a

▲ This is an air view of Hogback Ridge, a 17-mile-long drumlinoid ridge in McHenry County between Verendrye and Balfour. I took this air photo from a point three miles southwest of Karlsruhe in 1987. The glacier that formed the ridge was moving southeastward, toward the camera view.

Two University of North Dakota geology students and I made several excavations in Hogback Ridge and determined that the ridge formed quickly, perhaps in a single year, when a thin, surging (rapidly flowing) glacier moved over the area. Immediately prior to the formation of the ridge, the glacier thrust a small block of sediment southeastward near Verendrye. The glacier then continued to flow to the southeast and, as it moved over the ice-thrust block, a cavity formed at the base of the ice. Because a thin glacier is brittle (or at least less plastic than a thick glacier), the cavity did not close, as it would have if the ice had been thicker. As the glacier flowed over the area, it forced—squeezed—the soft, subglacial sediment into the cavity at the base of the glacier as it moved southeastward. Our 1987 excavations revealed sedimentary structures that formed when material was forced into the cavity from both sides.

JPB Photo: Scan of 1987 slide

few thousand years earlier. The thinness of the glacial ice, along with its rapid advance, allowed the cavity at its base to remain open as the glacier advanced.

The substrate beneath the glacier was likely saturated because it had recently been a lake floor (or it may be that the glacier advanced into the still-flooded lake). This allowed the glacier to advance over a lubricated surface of soft, lake-floor sediment, sliding on a layer of water. With water as lubricant, a glacier can slide tens of yards a day, instead of just a few yards a year. Were it not for the combination of circumstances I have described, and perhaps others of which I am unaware, it is unlikely the long drumlinoid landforms could have developed.

End Moraine

At times, the margin of a glacier melted at the same rate at which the ice was advancing, allowing the glacier margin to remain stationary. When this happened, greater amounts of sediment accumulated along the position of the ice margin, somewhat like material being dumped at the end of a broad, moving conveyor belt. The thicker accumulations of sediments at the positions of such stationary ice margins are marked by zones of hummocky topography that correspond to the former position of the edge of the glacier. These features are referred to as terminal moraine or end moraine. Examples of end moraine can be seen near Martin in Pierce, Sheridan and Wells counties, Edinburg in Walsh County and along the east side of the James River between Kensal and Juanita in Foster, Stutsman and Barnes counties.

Missouri Escarpment

The several escarpments in North Dakota are among the most prominent features in the state. In northeastern North Dakota, the Pembina Escarpment (known, too, as the Manitoba Escarpment) is a prominent feature that marks the western edge of the Red River Valley. The escarpment that marks the edge of the Prairie Coteau in the southeastern corner of the state and the escarpment that borders Turtle Mountain are both obvious and well-defined features. These and

▲ **This is the east face of the Missouri Coteau Escarpment seven miles west of Forbes in Dickey County. This view is to the northwest, from the North Dakota-South Dakota state line.**

The Missouri Escarpment has many short, tree-lined ravines (some shown here). In this area, the escarpment is marked by a rise in elevation of about 300 feet from the lower land on the east to the upland on the west. East of the escarpment, relatively smooth ground moraine predominates; to the west is hummocky dead-ice moraine. East of the base of the escarpment, cropped land predominates and, on the Missouri Coteau, to the west, most of the land is pasture. JPB Photo: 7/28/2011

other features existed in less pronounced forms prior to glaciation of the state, but they were re-shaped and enhanced by glacial processes. Before glaciation, the Prairie Coteau was probably an outlier of Cretaceous sandstone standing above a surrounding shale surface. Turtle Mountain was a grouping of several buttes of Cretaceous and Tertiary sandstone formations that stood above the surrounding plain.

The longest of the several escarpments in North Dakota is the Missouri Escarpment, which diagonally traverses the state from the South Dakota border, extending northward through western Dickey County to near Carrington, and from there trending northwestward to Divide County, entering Saskatchewan at the North Dakota-Montana boundary. The Missouri Escarpment marks an abrupt rise in elevation on its west- and southwest-facing side, relative to the lower area to the east and northeast. It provides a distinct boundary between the gently rolling Glaciated Plains to its east and northeast, and the hummocky topography of the Missouri Coteau to its west and southwest. Wherever you cross the Missouri Escarpment, you are likely to notice the abrupt change in topography and landforms. Depending upon location, the view from the top of the escarpment over the lower land to the north or east can be impressive. In places, the rise in elevation at the Missouri Escarpment is as much as 650 feet, occurring in a distance of one to two miles, but a less pronounced rise of 300 to 500 feet in a distance of two or three miles is more common.

The origin of the Missouri Escarpment has not been explained with certainty. A buried, but more subdued version of the Missouri Escarpment exists on the preglacial surface, in at least some places, underneath the modern escarpment. In other places, though, the surface escarpment seems to exist entirely independently of any buried preglacial topography.

In northwestern North Dakota, several small, buried and concealed faults coincide closely with the location of the Missouri Escarpment. They can be identified in the subsurface Cretaceous and Paleo-

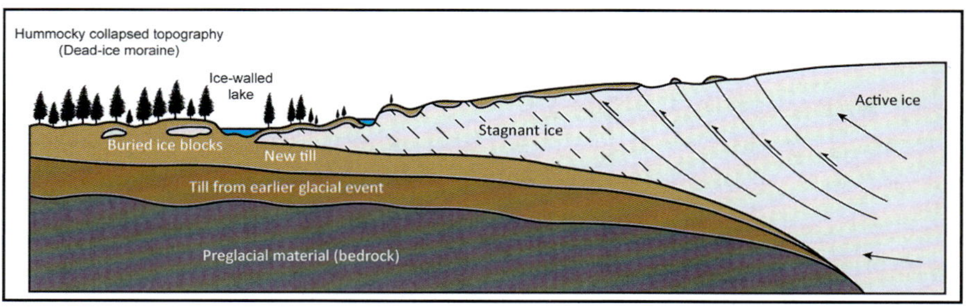

▲ This is a cross section through the edge of a glacier. It represents a situation similar to that which developed when glaciers moved over the Missouri Escarpment and onto the Missouri Coteau. As the glacial ice advanced upslope, the shearing that resulted within the ice brought enormous amounts of sediment to the surface of the glacier. The insulating effect of this thick covering of material on the ice slowed the rate of melting of the ice to the extent that plant and animal communities flourished on top of the buried glacier. The development of dead-ice moraine was the result of this lengthy melting process.

cene shale, sandstone and lignite deposits. The faults suggest that the buried preglacial escarpment, although much less pronounced than the modern one, may have resulted from upward movement along the west side of the bedrock faults. In other places, such as in Dickey and LaMoure counties, the buried preglacial escarpment, if it can be identified at all, is positioned several miles to the west of the modern escarpment. In such areas, the modern escarpment was probably sculpted entirely by glacial action, with no direct influence by the underlying topography.

The modern Missouri Escarpment in North Dakota is almost entirely a glacial feature. During the early phases of glaciation, the thick glacial ice moved up and over a preglacial escarpment and, in doing so, it sheared large amounts of subglacial debris through the ice and onto the upland, thereby building the upland higher. With time, and after repeated episodes of glaciation, this debris accumulated to a considerable thickness on the Missouri Coteau. As the Ice Age came to a close, and the Late Wisconsinan glacier became thinner, the escarpment helped divert the ice flowing along its face. As a result, the glacier, in its waning phases, refined, steepened and accentuated the face of the Missouri Escarpment.

Dead-ice Moraine

As the Ice Age ended and the climate moderated, the glaciers ceased advancing and stagnated over the uplands: their forward movement died; hence the term "dead-ice moraine." The glacial ice still existed, but it remained in a stationary position. As the "dead" glacier melted, large amounts of sediment, which had been sheared into and dispersed through the glacier, accumulated on the surface of the ice, resulting in a mantle of sediment covering the ice. This sediment, which ranged from a few feet to tens of feet thick, insulated the underlying glacial ice so effectively that it took hundreds of years for the ice to melt. Debris-covered stagnant glaciers persisted for 3,000 years over Turtle Mountain and on the Missouri Coteau, lasting until 9,000 years ago, long after actively advancing glaciers had disappeared from the rest of North Dakota.

As the stagnant ice on the uplands slowly melted, the surface of the glacier became increasingly irregular, and the soupy debris overlying the ice slumped and slid to lower areas, forming dead-ice moraine and an array of associated landforms (perhaps a better name for dead-ice moraine might be "mud flow" topography). In those places where the covering of debris was thickest, the glacier took longest to melt. A modern example of a debris-covered glacier is the Martin River glacier, 200 miles southeast of Anchorage in southeast Alaska, where dead-ice moraine is forming today.

▲ **The prairie pothole environment — North Dakota's "duck factory" — is shown in this early spring air view of dead-ice moraine topography in Stutsman County north of Woodworth. This hummocky topography formed when water-saturated sediment (essentially mud) slid off melting glacial ice. When all of the sediment finally came to rest, the result was an irregular surface marked by ridges, mounds and depressions. The low areas are ice-covered ponds, lakes and sloughs, which provide excellent waterfowl habitat.**

This topography is widespread on North Dakota's Missouri Coteau, Prairie Coteau and Turtle Mountain. JPB Photo: Scan of 1972 slide

Dead-ice moraine is prominent in such places as Hurdsfield, Max, Stanley, Woodworth, McClusky, Napoleon and Ryder on the Missouri Coteau; near Havana on the Prairie Coteau; and over the entirety of Turtle Mountain. The landforms associated with Turtle Mountain are identical to those on the Missouri Coteau and Prairie Coteau, but because Turtle Mountain has several inches more annual precipitation than the other areas, it has a woodland cover.

As I noted earlier, the elevation rises several hundred feet westward along parts of the Missouri Escarpment, which marks the eastern and northeastern edge of the Missouri Coteau. Similar prominent escarpments border the Prairie Coteau near Lidgerwood, and the slopes along the edges of Turtle Mountain. When glaciers advanced over each of these escarpments, the internal stress generated in the ice caused shearing. Large amounts of rock and sediment were carried from beneath the glacier up and into the glacier along shear plains in the ice. The debris-laden glacial ice then flowed onto the uplands, where it be-

▲ This dead-ice moraine is seven miles north of Bottineau in Bottineau County. The topography on Turtle Mountain is essentially identical to that found over the Missouri Coteau and Prairie Coteau, but the land is wooded. JPB Photo: 6/30/2010

came stranded and abandoned while the active glacier on the adjacent lowlands continued to flow around the uplands.

A variety of unique landforms resulted when stagnant glacial ice melted. Kettle holes formed where blocks of buried glacial ice melt-

▲ This treeless dead-ice moraine landscape is on the Missouri Coteau three miles northeast of Beaver Lake State Park in Logan County. Dead-ice moraine has hummocky topography with no streams flowing through the area.
JPB Photo: 5/30/2010

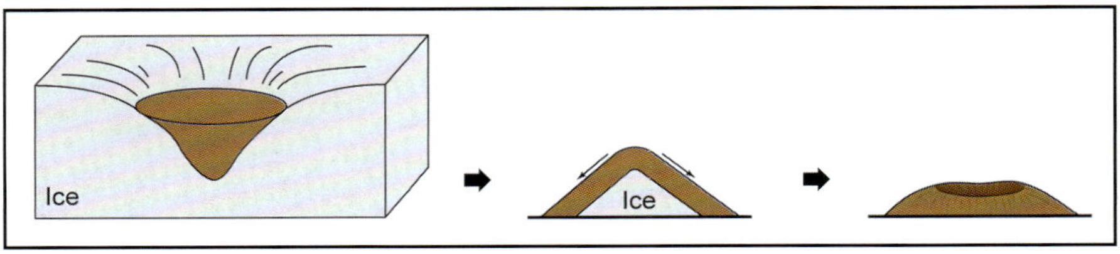

▲ This three-part diagram shows the development of a circular ridge (a "doughnut"). Left-most diagram: sedimentary material on the surface of a stagnant glacier is washed into or slides into a low area or hole in the ice. Middle diagram: when the surrounding ice melts, the ice beneath the sediments remains for a while because it is insulated by the overlying debris. However, as the ice core melts, the material on top slides off (small arrows). Right-most diagram: after all of the ice beneath the debris eventually melts, a ring-shaped depression remains in the material that had overlain the ice core.

ed, allowing the overlying materials to collapse. Today, many kettle holes are flooded by lakes, ponds and sloughs, which provide waterfowl habitat. Ridges formed where sediment slid into cracks in the ice; such ridges are referred to as disintegration ridges. Mounds of material collected in holes and depressions in the ice; ring-shaped hummocks formed if the mounds were cored by ice. When the ice cores melted, the centers of the mounds collapsed, forming circular ridges, which geologists refer to as doughnuts.

When lakes of glacial meltwater flooded areas of dead-ice moraine, flat-lying beds of silt and clay were deposited. The result, today, may be a level area where the lake once was, surrounded by rugged dead-ice moraine topography. Some lakes flooded areas where patches of glacial ice still covered the surface; other lakes formed in places where the glacial ice had already melted. If lakes flooded areas still covered by ice, the lake sediments later slid down and slumped when the ice melted, forming lake plains with irregular surfaces. An extensive area of such "collapsed lake topography" can be seen over much of eastern Ransom County near Sheldon, where, early in its history, Glacial Lake Agassiz flooded extensive areas of still-existing glacial ice.

Lakes sometimes flooded holes in the glacial ice; such lakes were on top of ice and surrounded by walls of ice. When the ice surrounding the lakes melted and the lakes drained, flat-lying beds of lake sediment remained standing as smooth, table-like areas above the surrounding landscape. Some of these perched lake plains cover areas as extensive as a square mile. Perched lake plains tend to be free of boulders and are good farmland—oases of flat topography surrounded by hummocky dead-ice moraine. Several dozen perched lake plains have been identified in North Dakota. The town of Lehr, on the Logan-McIntosh county line, is situated on such a lake plain. The town sits on a level area that stands above the surrounding dead-ice moraine. Another good example is the perched lake plain located a mile northeast of Denhoff, in Sheridan County, at the junction of State Highways 14 and 200.

The insulating blanket of debris on top of stagnant glacial ice was so thick in many places that the cold temperatures of the underlying ice had little or no effect at and near the ground surface. Forests of spruce, tamarack, birch and poplar established themselves on the debris-covered glacier. Lakes on the glacier teemed with aquatic life.

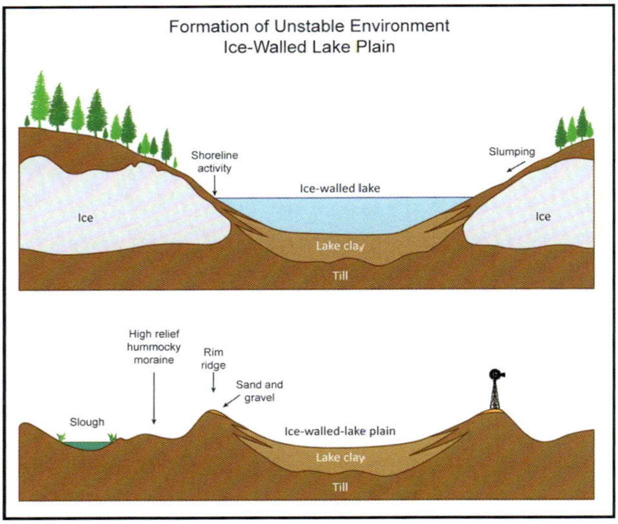

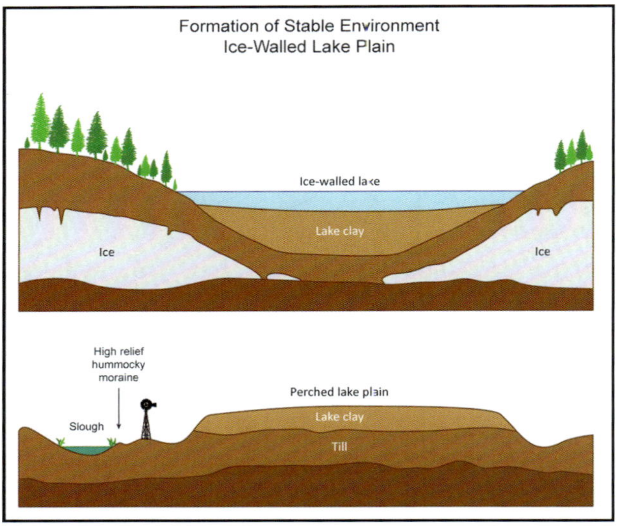

▲ Both of these kinds of lake plains are common in areas of dead-ice moraine:

1. Formation of unstable environment ice-walled lake plain. Thin sediment on the surrounding glacier ice allows it to melt rapidly. The large amounts of meltwater thus produced result in lakeward slumping of the confining walls and the deposition of sediment into the lake itself. Such lake plains often have rims at their edges.
2. Formation of a perched, ice-walled lake plain. A thick, insulating cover of sediment on the surrounding ice slows its rate of melting, resulting in a quiet-water environment dominated by fine-grained sedimentation. These lakes tend to be long-lived and show a general absence of the mass movement of water-saturated till. Perched lakes do not often have rimmed edges.

North Dakota's stagnant-ice environment between 11,000 and 9,000 years ago was similar to stagnant, sediment-covered glaciers in southern Alaska today. Much of the area consisted of aspen-spruce woodlands, with ponds and lakes in which fish, frogs, beetles, clams and other animals and plants thrived. Wooly mammoths, caribou, elk and other game roamed the areas of forested, debris-covered ice. As early as 12,500 ^{14}C yr BP[28], Paleo-Indians inhabited the areas of stagnant glacial ice, on land overlying the insulated glaciers, as well as areas on nearby lowlands, where the glacial ice had all melted.

Ice-thrust Moraine

Ice-thrust moraine, sometimes referred to as ice-thrust topography, consists of landforms resulting from thrusting by the glaciers. Such landforms are among the more unusual and interesting glacial features found in North Dakota. They are found in places where the glacier extracted large chunks of rock and sediment, transported them short distances, and set them down nearby. The simplest example resulted in a hole, usually flooded by a lake or slough, and a nearby hill, where the material was dropped, near the hole. However, larger and more complex thrust features involve several square miles of disturbed materials, hundreds of feet thick, moved by the glacier, coupled with broad depressions which today mark the locations from which the materials were derived.

Areas of ice-thrust moraine tend to overlie buried aquifers, which are rock or sediment layers that contain water. Such aquifers may be confined, that is, buried beneath, and surrounded by layers of impermeable materials. Water cannot move through impermeable sediments. When a glacier advanced over a confined aquifer, the weight of the ice sometimes caused the pore-water pressures to build up in the beds of sediment in and overlying the aquifer. The resulting pressures pushed the overlying sediments upward and into the path of the advancing ice. As a result, large, intact blocks of material became incorporated in the advancing glacier. Confined ground water is necessary for thrusting to occur because, if the water is able to easily escape ahead of an advancing glacier, pressure cannot build. However, if the ground water is trapped, thrusting is likely.

In 1963, soon after I first became aware of ice-thrust features, I

;ned the task of examining the ICBM Missile Site locations in eastern North Dakota, as the initial excavations were made (see my comments in the Preface to this book). During two field seasons (and one winter), I inspected the geology of 165 excavations. In half of those locations sited west of the Glacial Lake Agassiz plain in the Red River Valley, I found some evidence of ice thrusting. It soon became apparent that, far from being a rarity, the thrusting process was common.

▲ This is an excavation for an ICBM site in 1963. Ice-thrust landforms are not hard to recognize if they 1) have significant surface expression or 2) are exposed by an excavation. However, many occurrences of materials that the glacier shoved into their current positions are buried and entirely concealed.

This photo of a Minuteman ICBM excavation (see discussion in the Preface to this book) shows an inclusion of bedded sand, turned on end, encased within a layer of till (the tan-colored sediments). Dark gray Cretaceous Pierre Formation shale is at the bottom of the excavation. The bedded-sand inclusion has no surface expression; it would be impossible to know it was there except the ICBM excavation happened to cut into it. The sand, originally flat lying and deposited by water in a stream, was probably shoved into its current location as a frozen mass, pushed into place by thrusting glacial ice so that it now stands on end.

JPB Photo. Scan of 1963 slide

Widespread glacial thrusting occurred in central North Dakota, too. One reason thrusting was extensive there is that two glacier lobes collided south of Turtle Mountain. The collision of the glacier lobes eliminated the possibility of ground water trapped beneath the ice escaping, ensuring that high ground water pressures—and ice thrusting—would occur beneath one or both of the ice lobes.

As the glacier in north-central North Dakota melted and became thinner, it could no longer flow over the Turtle Mountain upland. The part of the glacier over the upland stagnated, resulting in dead-ice moraine, but the part of the glacier on the lower land on either side of the upland continued to flow. It split into two distinct parts—separate lobes—one flowing around the west side, and the other around the east side of Turtle Mountain. The lobes expanded toward each other in the area south of Turtle Mountain and, when they met, the result was intense thrusting through most of Pierce, Sheridan, western Benson, and parts of McLean, McHenry and Wells counties.

Colliding glacial lobes, or even colliding continental-scale glacial ice sheets, are not uncommon. Perhaps the best large-scale example is the "collision" of the glaciers of the Keewatin and Cordilleran ice caps in Alberta and Montana. As the two ice sheets expanded, they eventually coalesced. While these two major North American glaciers were in contact, they formed a continuous expanse of ice—the Laurentide ice cap, which stretched from the Atlantic to the Pacific oceans. Although glacial collisions are not rare, they are seldom combined with ground water conditions comparable to those that developed south of Turtle Mountain.

A good example of a simple ice-thrust hill can be seen at the town of Anamoose, where Steele Lake floods a depression formed when the glacier lifted a chunk of material, carried it forward as a single piece, and set it down a tenth of a mile to the southeast.[29] The lake is due south of Anamoose; the materials that once were where the lake is today form a hill, with ridges, at the southeastern edge of the lake. Examples of thrust features essentially identical to the one at Anamoose include Egg Lake and an adjacent hill five miles southeast of Harvey in Wells County, and Medicine Lake and the adjacent Grasshopper Hills seven miles south-southwest of Kensal in Stutsman County.

I have identified several kinds of ice-thrust features in most parts

TURTLE MTN
MISSOURI ESCARPMENT
PRAIRIE COTEAU ESCARPMENT
Missouri R.
Williston
Minot
Grafton
Devils Lake
Grand Forks
Dickinson
Bismarck
Jamestown
Valley City
Fargo
Wahpeton
DIVIDE
BURKE
RENVILLE
BOTTINEAU
ROLETTE
TOWNER
CAVALIER
PEMBINA
WILLIAMS
WARD
MOUNTRIAL
MCHENRY
PIERCE
BENSON
RAMSEY
WALSH
NELSON
GRAND FORKS
MCKENZIE
MCLEAN
SHERIDAN
WELLS
EDDY
FOSTER
GRIGGS
STEELE
TRAILL
DUNN
MERCER
OLIVER
BILLINGS
GOLDEN VALLEY
STARK
BURLEIGH
KIDDER
STUTSMAN
BARNES
CASS
MORTON
SLOPE
HETTINGER
GRANT
LOGAN
LA MOURE
RANSOM
RICHLAND
BOWMAN
ADAMS
SIOUX
EMMONS
MCINTOSH
DICKEY
SARGENT
0
50 mi

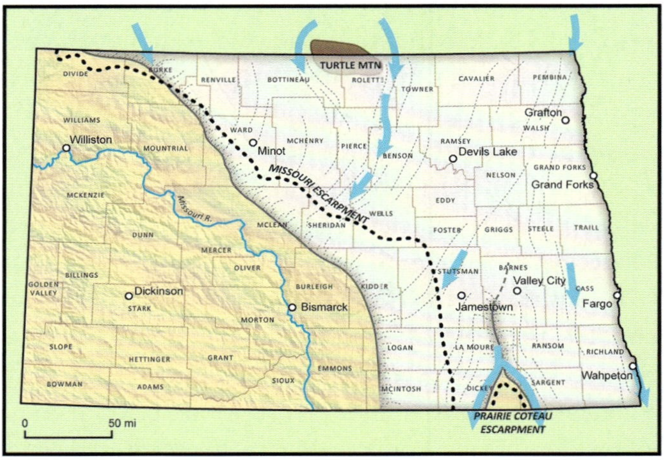
TURTLE MTN
MISSOURI ESCARPMENT
PRAIRIE COTEAU ESCARPMENT
Missouri R.
Williston
Minot
Grafton
Devils Lake
Grand Forks
Dickinson
Bismarck
Jamestown
Valley City
Fargo
Wahpeton
0
50 mi

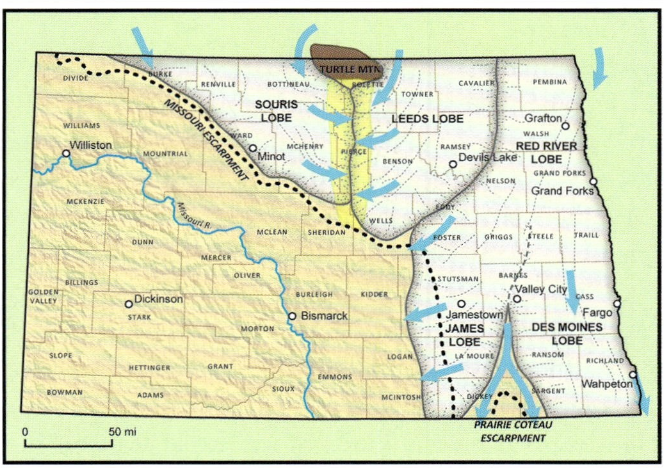
TURTLE MTN
SOURIS LOBE
LEEDS LOBE
RED RIVER LOBE
JAMES LOBE
DES MOINES LOBE
MISSOURI ESCARPMENT
PRAIRIE COTEAU ESCARPMENT
Missouri R.
Williston
Minot
Grafton
Devils Lake
Grand Forks
Dickinson
Bismarck
Jamestown
Valley City
Fargo
Wahpeton
0
50 mi

◀ Upper map—Stage 1. When the Late Wisconsinan glaciers first advanced into North Dakota, they were thick and flowed over the land surface essentially unaffected by the underlying local topography.

Center map—Stage 2. As the ice thinned, beginning 18,000 years ago, the flow directions became increasingly directed by the underlying topography, such as the Turtle Mountain upland, Missouri Coteau and Prairie Coteau.

Bottom map—Stage 3. By 12,800 years ago, the glacier had become lobate, with several discrete, active glacial lobes, each flowing in its own direction. One lobe (the Souris Lobe) flowed around the west side of the Turtle Mountain upland and another lobe (the Leeds Lobe) flowed around the east side. The two lobes collided in an area south of Turtle Mountain (yellow area), resulting in massive ice thrusting.

A somewhat similar situation developed (but without the thrusting) in southeastern North Dakota/northeastern South Dakota). The Red River Lobe (which likely coalesced with the Leeds Lobe) split into two parts as it was directed around the Prairie Coteau. The James Lobe flowed around the west side of the Prairie Coteau, while the Des Moines Lobe flowed around the east side.

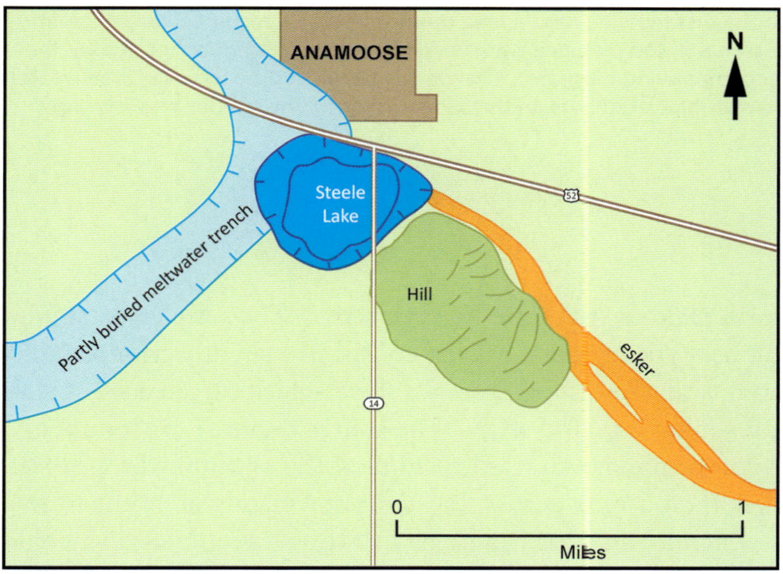

▲ This map shows the geologic conditions that allowed the ice-thrust feature at Anamoose to form. The glacier moved southeastward over the partly buried meltwater trench, which contained an aquifer. As it did so, the overlying pressure of the glacial ice caused the water pressure in the aquifer to rise, which in turn forced overlying materials to be pushed up into the advancing glacier. The glacier pushed the materials southeastward, resulting in a hole (now flooded by a lake), coupled with the hill where the materials were set down.

▲ This ice-thrust feature is at Anamoose in McHenry County. The photo shows hills in the distance, beyond a lake that formed as a result of glacial ice thrusting. The material in the distance was derived from the area now flooded by the lake (see accompanying diagram). The view is to the southeast, over Steele (Stelow) Lake, down-glacier, that is, in the same direction the glacier was moving.
JPB Photo: 8/21/2009

of North Dakota that have been glaciated. The simplest examples (the Anamoose-type features) usually consist of a slightly crumpled block of material, situated a short distance down-glacier from a source depression of similar size and shape. The source depressions commonly contain lakes or ponds. An ice-thrust hill of this type may be 100 feet high and less than a mile across, located immediately adjacent to the hole from which it came, or it may be as far as three miles away from the hole. If the glacier continued to advance over an area after it had thrust the material, the feature tended to become smeared out. For this reason, many ice-thrust features are buried beneath younger glacial sediment and may be difficult to recognize. The most prominent ice-thrust hills, large or small, are found in places where the glacier stopped advancing as soon as it completed the job of thrusting the materials.

▲ **This ice-thrust topography is in Eddy County six miles east of the town of Sheyenne. A wetland is in the middle distance. Casual viewing of the land, while walking through such an area of ice-thrusting, usually reveals little or no apparent differences between ice-thrust topography and dead-ice moraine. Both types of landforms commonly have moderate relief with scattered ponds and sloughs. However, closer inspection of the ice-thrust topography may indicate long, aligned ridges, with elongate ponds between the ridges—the ponds and ridges positioned normal to the direction of glacier flow. In contrast, ponds and sloughs in areas of dead-ice moraine are haphazardly positioned.** JPB Photo: 7/10/2009

Ice-thrust moraine features are abundant in central North Dakota and they are common in the Prairie Provinces of Canada; the largest ones I know of are in Alberta. Most areas of ice-thrust topography are much more extensive and complex than is the one at Anamoose. They range up to several hundred feet thick and may cover several square miles. An outstanding example is the Prophets Mountains, an eight-square-mile area west of McClusky in Sheridan County, where several dozen slabs have been stacked up, *en-echelon*, like shingles on a roof. Another, much larger example in northern Sheridan County includes an area of 120 square miles that the glacier quarried, lifting the underlying preglacial (Cretaceous Fox Hills and Hell Creek formations) bedrock materials at least 300 feet, emplacing them in several ice-thrust masses around the quarried area.

I drilled a test hole in Dogden Butte, an ice-thrust mass in northeastern McLean County, and documented several layers of glacial and non-glacial sediment. Dogden Butte stands 400 feet above the Glaciated Plains lowland to the north and 150 feet above the dead-ice moraine to the south on the Missouri Coteau. It consists of multiple blocks of ice-thrust material configured as a series of east-west ridges. The test hole, drilled near the top of the butte, penetrated, first, 100 feet of glacial till, then 212 feet of Bullion Creek bedrock, then 44 feet of glacial till and

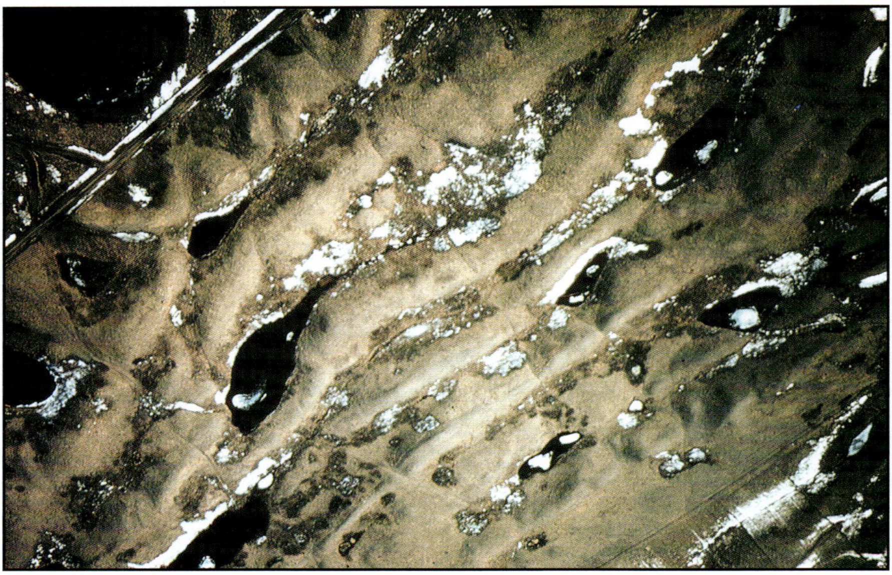

▲ **This is an air view of the Prophets Mountains in western Sheridan County. This ice-thrust topography shows several slabs of bedrock (sandstone) and glacial deposits stacked up, somewhat like shingles or fallen dominoes. The difference in elevation between ridge tops and intervening troughs is about 80 feet, but local relief exceeds 300 feet in places.**

The glacier was moving from the lower right, from the east. The top of the photo is northwest; the road in the upper left-hand corner goes north-south, with north at the top. The snow accentuates the troughs between individual slabs that were thrust by the glacier. Ponds flood the depressions between thrust ridges.

This is a scanned slide. I am unsure who took the photo, but I think it was taken in the early 1970s. JPB Photo: Scan of early 1970s slide glaciofluvial gravels, and finally more bedrock. The 212-foot section of Bullion Creek bedrock consisted of several slabs of the same section of material, repeated several times. Most of the ice-thrust bedrock was pushed up and over the 44 feet of glacial deposits beneath it. The main source area for the Dogden thrust appears to be a lowland flooded by Cottonwood Lake, three miles to the northeast of Dogden Butte.[30]

Other outstanding examples of ice-thrust topography can be seen near Lincoln Valley and Denhoff (both in central Sheridan County), in western Divide County near Alkabo and in the area of the Antelope Hills in Pierce County.

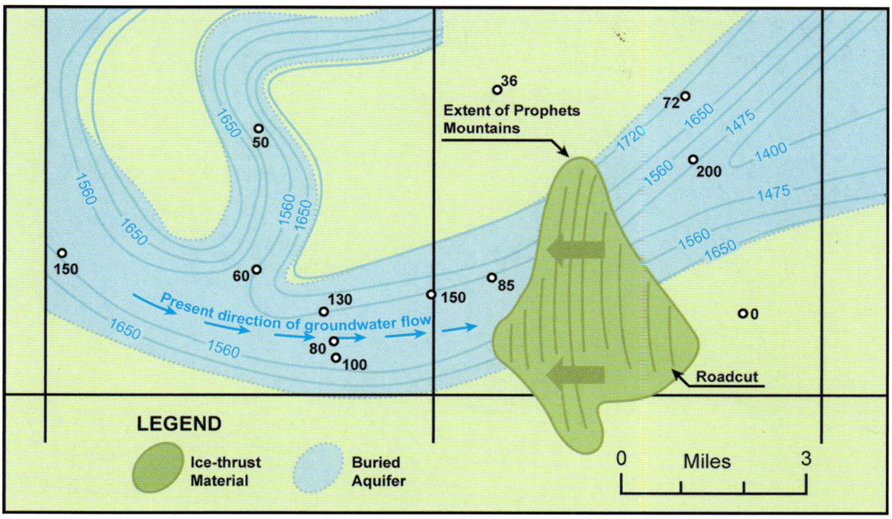

▲ This diagrammatic map shows how the ice-thrust Prophets Mountains formed over a large aquifer contained in a buried river channel, the old McLean River Valley (see chapter three). Ice-thrusting occurs most commonly over buried aquifers because high pore-water pressures are more likely to form within and above layers of permeable materials (sand and gravel) that alternate with impermeable materials (till, lake silt or bedrock such as shale). Probably 90 percent of the ice-thrust features in North Dakota are closely associated with specific glacial or preglacial aquifers.

The blue-shaded area represents the extent of the Lake Nettie aquifer; the darker green area is the Prophets Mountains. The numbers in blue on the contours within the aquifer (thousands of feet) are the elevations on the buried, preglacial bedrock surface. Individual numbers (tens or hundreds of feet, in black) represent aggregate gravel and sand thicknesses of the aquifer.

The most spectacular of all ice thrusts in North Dakota is the one responsible for forming Devils Lake. The materials taken from the ice-thrust holes that are today flooded by Devils Lake are piled up along the south side of the lake where Sully's Hill dominates the landscape. I will discuss the geology of the Devils Lake area in detail in chapter six.

Hundreds of areas of ice-thrust moraine are found in North Dakota, but the landform is not common in all glaciated areas. Why? A partial answer is that regional, preglacial drainage here was northward and the glaciers advanced southward, against the direction of regional ground water flow. In places where glaciers were moving in

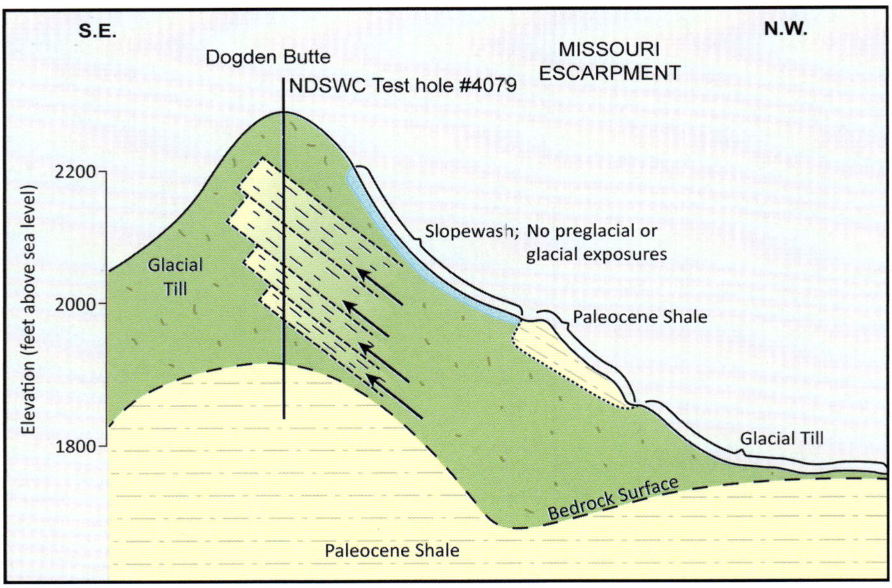

▲ This is a cross section through Dogden Butte four miles east of the town of Butte in northeastern McLean County. Dogden Butte consists of several slabs of Paleocene shale bedrock, along with slabs of glacial materials, which have been pushed southward. Most of the materials in the butte were derived from a lowland three miles to the northeast. The lowland is partly flooded by Cottonwood Lake. The vertical exaggeration in this diagram is approximately 50 times. NDSWC test hole #4079 is an exploratory test hole I had the North Dakota State Water Commission drill to determine the internal structure of Dogden Butte. *(Please see text for a detailed description of the layers of material that were involved in the thrusting.)*

▲ **This ice-thrust topography occurs three miles west of Denhoff in Sheridan County. In this location, the glacier was flowing eastward, in the same direction as the camera view. It deposited (thrust) a series of parallel ridges, aligned normal to the direction of ice flow. The lake in the foreground floods the depression that formed during the thrusting event, the material in the hills in the distance beyond the lake having been derived from the area now the lake. The area shown here is only a small part of a two- to three-mile-wide band of ice-thrust topography that extends 20 miles from north to south, from six miles north of Denhoff to two miles into Burleigh County on the south end. Two ice lobes collided along the 20-mile-long band, which coincides with the buried route of the Cannonball/Heart river valley. Some of the ice thrusts in the band were deposited by an east-flowing glacier lobe (the landscape shown here), while others were shaped by a west-flowing lobe. The thrusting was facilitated by high pressures that built up in the water-saturated sediments contained in the buried river valley beneath the zone where the two glacier lobes collided.** JPB Photo: 7/24/2011

the same direction as the regional drainage, the ground water tended to move ahead of (along with) the advancing ice, and escape. When this happened, thrusting was unlikely. Examples of such areas, where ice thrusting was rare, include much of the eastern U.S. and Canada, Scandinavia and the British Isles. However, in places where glaciers advanced against the direction of ground water movement, the ground water became trapped, making thrusting more likely. Large-scale glacial thrusting occurred in Alberta and Saskatchewan and in northern Europe (Germany and Poland), where glaciers advanced southward against the regional ground water flow systems.[31]

Landforms Shaped by Running Water

Eskers and Kames

Glacial ridges called eskers are among the best known of the various features formed by glaciers and by the running water associated with melting glaciers. Eskers come in all sizes: ridges snaking across the countryside ranging from a few hundred feet to several miles long, and up to 50 or 100 feet high. Kames may be cone or pyramidal-shaped hills as high as 100 feet, or they may be simply small mounds of material. Kames and eskers are found in most parts of North Dakota that were covered by the Late Wisconsinan glacier.

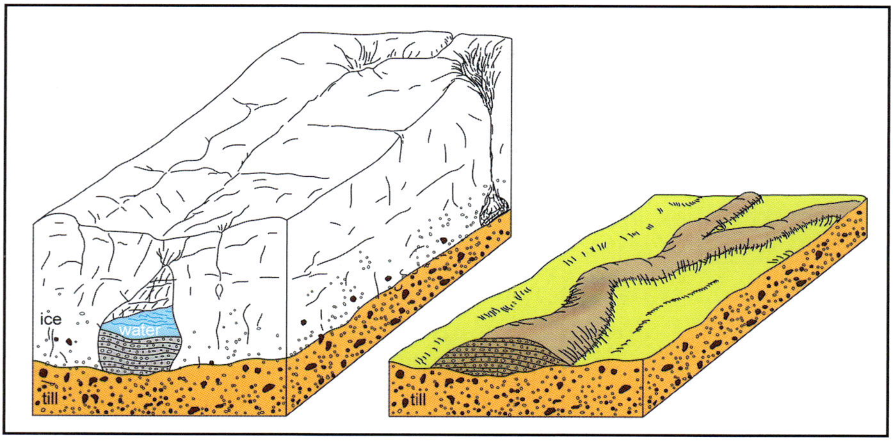

▲ This diagram shows how an esker forms when a stream of meltwater carrying sediment develops beneath a glacier or, in some instances, in an open crack in the glacier. Such a stream may meander a little as it winds its way underneath the glacier, making its way to the front of the ice. The sand and gravel are eventually deposited when the ice melts, the meltwater flow gradually slows and its load is dropped. After the ice has melted away, the fluvial sediment remains as a long, narrow, winding ridge, which marks the course of the former subglacial, ice-walled stream.

An esker stream may form in a crack rather than in a tunnel, and it may flow on solid ground in places beneath the glacier and over ice in other places. In places where the stream flowed over ice, the esker ridge may have gaps or dips in its crest because, when the ice beneath the stream bed melted, the overlying stream deposits collapsed.

Eskers were deposited by streams and rivers flowing on the surface of a glacier, in cracks in the glacial ice or, sometimes, in tunnels beneath the ice. The word "esker "is derived from an Irish Gaelic word, *eiscer*, meaning "ridge." Imagine a river flowing in a valley or crack in the glacier. The banks of the river were formed of ice and, in some places, the river floor might have been on ice, too. These Ice Age rivers and streams deposited gravel and sand in their ice-walled valleys, just as a modern stream deposits sediment in its valley. However, the ice banks of esker rivers eventually melted away, leaving the gravel that had been deposited on the bottoms of ice-walled valleys standing as ridges above the surrounding countryside.

North Dakota has thousands of esker ridges. Most are small and nondescript, apparent only from a height (air view). Some, however, are impressive. One of the best examples of an esker in the United States

▲ **The Dahlen esker is at the junction of Grand Forks, Walsh and Nelson counties. This view is to the south, in the direction the stream that deposited the esker was flowing. The esker is four miles long and 400 feet wide, and rises between 50 and 80 feet above the land on either side. It has a generally accordant crest level, but the crest appears irregular because of numerous minor gaps. The esker consists of a mixture of sand, gravel, till and boulders: a combination of materials deposited by running water and debris in the glacier sliding from the ice walls into the esker stream.** JPB Photo: 7/26/2009 is the Dahlen Esker, located midway between Fordville and Dahlen in the northeastern part of the state. It can be seen as a prominent ridge off to the west where State Highway 32 crosses the Grand Forks-Walsh County line. If the weather is dry, you can drive a mile on an unmaintained section line trail to the crest of the Dahlen Esker (if the fields on either side of the trail are being farmed, be careful not to drive on the crops).

The Dahlen Esker was deposited by a meltwater stream flowing in an ice-walled channel, possibly a tunnel, near the edge of the glacier. The stream flowed toward the margin of the glacier. The esker is four miles long, 400 feet wide, and as high as 50 to 80 feet. In some places, native prairie covers the surface.[32]

Other large eskers include one that is six miles long near Benedict in northeastern McLean County, another 10 miles southwest of Carrington in Foster County, two near Hannaford and Dazey in Griggs and Barnes counties, and an unusual one in Frankhauser Lake in northern Sheridan County. The Frankhauser Esker is particularly interesting because it winds its way through a lake, which floods a depression that was formed by ice thrusting. The esker apparently formed during the ice-thrusting process.

Kames are glacial landforms similar in many ways to eskers. Like eskers, they consist largely of gravel and sand, but they are conical or irregularly shaped hills, rather than long ridges. *Kame* is an Old English-Scottish term, *comb* or *camb*, meaning "crest." Water flowing on the surface of the glacier, or in esker valleys, plunged into holes in the ice, filling them with a mixture of materials. When the ice eventually melted, the materials slumped down, forming mounds and conical hills. Kames occur in several places on the Glaciated Plains, associated with ground moraine. Examples may be seen in southwestern Richland County, south of Lidgerwood, and to the west in the Cayuga area of Sargent County. A prominent kame, visible from several miles away, is located four miles south of State Highway 200 in eastern Foster County along the west side of the James River. Half of the 60-foot-high feature has been mined for sand and gravel.

Most eskers and kames are composed of coarse, poorly sorted materials, a mixture of sedimentary textures ranging from silt and sand, up to large cobbles or boulders. As they were forming, the flowing water deposited flat-lying beds of sand and gravel, but later, when the ice walls melted, the bedding collapsed, often resulting in contorted bedding. While a stream was flowing in an ice-walled tunnel or valley, or water was flowing into a hole in the glacier (a "moulin"), cobbles and boulders fell from the melting ice into the water-lain deposits, into the stream or down the moulin. Some eskers have a continuous covering of boulders on their surfaces. Most kame or esker deposits contain too many cobbles to be useful for construction purposes, but some consist of materials suitable for rough fill work.

▲ **This kame (high point in center of photo) is 10 miles southeast of Balta in eastern Pierce County. Like eskers, kames were deposited by running water. Some kames mark places where water plunged into large holes *(moulins)* in the glacier. Other kames consist simply of materials that were washed into low places in the ice. Most kames consist of coarse gravel and sand, similar to the materials found in eskers. Kames sometimes occur at the ends of eskers, where meltwater plunged into a hole in the ice, and from there flowed away in a stream through a tunnel or crack in the glacier.** JPB Photo: 8/20/2009

Veblens

Small, low, kame-like hills, found in parts of southeastern North Dakota, have a superficial resemblance to volcanoes, and they formed in a way that is somewhat analogous to how volcanos form. These hills, known as "veblens,"[33] formed as a result of geologic processes known as hydrodynamic blowouts. Unlike volcanoes, veblens are not

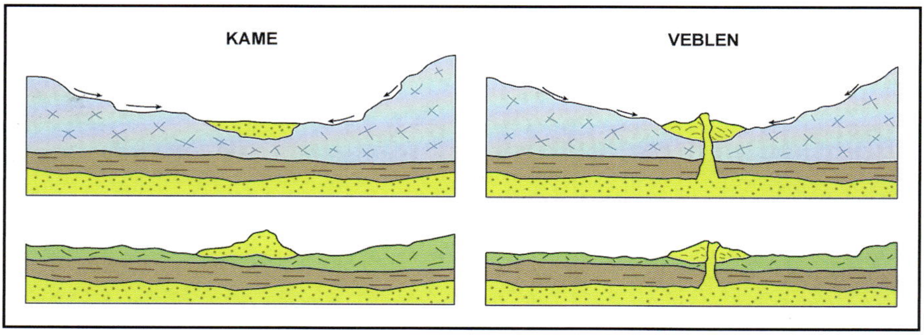

▲ Kames and veblens may be indistinguishable when viewed in the field. Both are hills of sand and gravel. However, kames and veblens are fundamentally different in the ways they form. These cross-section diagrams show the difference in the way the two features form.

In these diagrams, blue represents glacial ice; yellow is sand and gravel; tan is buried bedrock shale; and green is till—glacial sediment left on the land surface after the glacier melted.

The two diagrams on the left show a kame, formed when sand and gravel are carried by water on top of or within the ice and delivered to a low area on the glacier. When the ice melts, the sand and gravel slump down to their angle of repose, often forming a cone-shaped hill. The small arrows show the direction of gravel-laden water flowing toward and into a low area or hole in the ice.

The diagrams on the right show a veblen, formed when sand and gravel are delivered upward via a blowout from beneath, in a manner similar to a volcanic eruption. Beyond this point in their formation, the two features are essentially identical. When the ice melts, the sand and gravel of a veblen slumps down, forming a hill that may be indistinguishable in appearance from a kame.

Kames and veblens may be outwardly indistinguishable. The difference between the two features is in the source of the gravel and sand, that is, how it was delivered. The fluvial materials in kames were delivered from above, whereas the materials in veblens were delivered from below, via hydrodynamic blowouts. Veblens have cores—clastic plugs—that extend downward to the underground source of gravel and sand ("clastic" means material composed of fragments such as particles of sand).

the result of molten rock erupting from the Earth's interior. Rather, they formed when ground water flowed upward, with force, from pressurized aquifers, carrying large amounts of sand and gravel with it as it rose. In places where the sand-and-gravel laden water exited through cracks in the glacial ice, gravel and sand piled up on the ice and, when the ice melted, the sand slumped into cone-shaped hills. Most of the veblens in southeastern North Dakota are from 40 to 70 feet high and 0.1 to 0.2 mile across. Forty or 50 of them are found in the area near Lidgerwood and Geneseo, North Dakota, and Veblen, South Dakota.

▲ Devils Heart (or Devils Heart Butte) is in Benson County two miles west of Tokio. Devils Heart has been considered to be a large kame, but it is more likely that it is a veblen, a sand volcano, formed during the ice-thrusting event that formed the Devils Lake Basin and the hilly, ice-thrust topography to the south of the basin. At the time of thrusting, ground water flowed rapidly upward, from the underlying aquifer and through the glacier, bringing up large amounts of sand and gravel. When the glacier melted, the sand and gravel slumped down to form the hill that is Devils Heart.

The name of Devils Heart can be traced back to the name the French fur traders gave it, *Butte de Coeur,* which is a French translation of the Lakota *Miniwakan Cante Paha,* which can itself be translated directly into English as Heart Hill at Spirit Lake. The name "heart" likely means that it is at the physical center of the area, but the Dakota considered it the center of spiritual knowledge as well. Before the Dakota, the Hidatsa lived in the Spirit Lake area, and their name for Heart Hill was *Minixopa Nata Sh* (Heart of the Holy Water). The Hidatsa left about 1550 and moved to the Missouri River area. JPB Photo: 7/10/2009

Veblens occur in places where pressurized ground water escaped from artesian aquifers and flowed to the ground surface. While glaciers covered North Dakota, the ice sealed the ground surface beneath, preventing the ground water from escaping. Before glaciers existed, ground water flowed from southwest to northeast, from the area of the Black Hills toward the eastern Dakotas and southern Manitoba. During the thousands of years that glaciers covered the area, regional ground water flows were halted or reversed and hydraulic pressures in the aquifers built up greatly. Devils Heart Butte, near Tokio in Benson County, is a possible example of an exceptionally large veblen.

Tunnel Valleys

Tunnel valleys form when water flows beneath a glacier, carving a trench in the materials under the glacier. The water, which is pressurized, has high energy and is very erosive. Tunnel valleys form in a way that is analogous to how eskers are deposited, but the features are eroded downward into the material beneath the glacier ice, whereas esker valleys are eroded into the ice itself. Tunnel valleys represent only the basal parts of long (flat-lying) cylinder conduits. The upper part of the cylinder, which was cut in the glacier ice itself, no longer exists.

Tunnel valleys can result from rapid discharges of meltwater that had been trapped under the ice and was suddenly released. The water might have been ponded in lakes beneath the ice, held in place in a variety of ways, such as behind a dam of frozen soil and rock near the margin of the glacier. If a channel eroded headward from the margin to the reservoir of trapped water, it could quickly drain an under-ice lake. When the subglacial water escaped, its rapid flow eroded a tunnel beneath the glacier as it rushed toward the glacier margin.

Tunnel valleys tend to be aligned parallel to the direction of glacier flow, from areas of higher pressure to lower pressure (thicker ice to thinner ice). The valleys may contain lakes, sloughs, mounds of glacial sediment and erratics. Through-flowing streams seldom flow in tunnel valleys. In the final stages of formation of a tunnel valley, it is likely that the overlying glacier ice flowed into and partially or completely filled the valley.

Some tunnel valleys contain esker ridges. This suggests that the tunnels remained active for a time after erosion of the tunnel valley.

The valleys were gradually filled with sediment by streams flowing in the ice-filled valley.

Tunnel valleys may be common, but only a few have been identified in North Dakota. I have seen many valleys that might have formed as tunnel valleys, but many of them could as easily have formed in some other way. I mapped my first one in 1962, but at that time, I did not realize how it had formed, so I did not characterize it correctly (and I recognized it as a tunnel valley only years later).

A good example of a tunnel valley may be seen in southwestern Foster County, five miles south of Carrington. It trends southeastward, parallel to the face of the Missouri Escarpment, which is three miles southwest of it. The valley is straight, 12 miles long, and contains an esker ridge for nine miles of its length. Even though the esker ridge is 30 to 40 feet high, and 150 feet wide, it is contained entirely within the valley so its crest is apparent only when viewed from the edge of the valley, or from within the valley itself. The esker's crest is at the same elevation as the upland adjacent to the tunnel valley. A mile-long lake, Russell Lake, floods part of the tunnel valley to the northwest of the esker.

Another example of tunnel valleys can be seen in northeastern McLean County, north of the town of Turtle Lake The valleys, which run east of and parallel to North Dakota Highway 41 for 10 miles, are flooded by a series of lakes, which include Crooked Lake, Long Lake and Strawberry Lake. The tunnel valley (or valleys) that contain the lakes are a quarter-mile wide, and range from 60 to 100 feet deep.

Outwash Plains

Outwash plains are areas underlain by fluvial sediment (gravel, sand, etc.) that was deposited on flood plains by streams and rivers flowing away from a melting glacier. They may be broad, flat expanses or terraces confined within the sides of river valleys. If the rivers flowed on solid ground ahead of a melting glacier, they deposited flat-lying beds of sand and gravel. If they flowed on top of or within the melting glacier, or if they flowed over an area of stagnant glacial ice, the resulting plain may have hummocky topography similar to that of collapsed lake sediment or collapsed glacial sediment. The gravel and sand beds were deposited as flat layers, but when the underlying ice melted, the layers collapsed and became contorted. Collapsed river sediment is essentially the same in composition as uncollapsed river sediment, but its bedding tends to be faulted and tilted. Some outwash plains may be underlain by gravel several tens or even hundreds of feet thick. These thick gravel beds are often saturated with water and serve as aquifers, providing important ground water resources.

Notable examples of extensive outwash plains extending over tens of square miles include an area known as Tiffany Flats in western Eddy County north of New Rockford; in the Englevale area of western Ransom County, extending into northwestern Sargent County; in western Griggs County, south of Binford; and along the south slope of Turtle Mountain in Rolette County. An example of an extensive collapsed outwash plain is the broad, hilly area of gravel covering much of north-central Kidder County.

River Terraces

Some river valleys have flat, step-like terraces within the valley walls. Terraces are remnants of former river flood plains, either carved from the material in which the valleys are cut (cut terraces), or built of gravel and sand (fill terraces). Individual terraces can be traced for miles in meltwater trenches like the Sheyenne and James River valleys. A terrace level may have formed when a river maintained a stable flow for a sufficient period of time to build an extensive flood plain. Later, when the volume of flow increased or decreased, or when the load of sediment being carried by the river changed, the river responded either by cutting down to a new level or by building up its bed to a higher level, depending upon how much water was flowing and how heavy a load of sediment the river was carrying. Some fill terrace deposits, especially in the Sheyenne River Valley, are important commercial sources of gravel and sand.

Abandoned River Valleys

Hundreds of valleys, most of which do not carry modern rivers, are found throughout the glaciated part of North Dakota. Some of them may be tunnel valleys (see earlier discussion), but most such valleys were carved by rivers and streams flowing away from melting glaciers. In the few of these valleys that do carry streams today, the

▲ This image shows river terraces along the north side of the Sheyenne River Valley in Benson County five miles south of Esmond. I have drawn in dashed lines, each denoted by a letter, as follows: A—top level (with the house), marking the upland plain adjacent to the valley; B—upper terrace level; C—lower terrace level; D—the modern flood plain level of the Sheyenne River. Through much of its length, the Sheyenne River Valley contains two terrace levels (B and C on this illustration). Both are fill terraces, underlain by thick gravel deposits that have been developed for their gravel resource.

Following the initial formation of the Sheyenne meltwater channel, subsequent flows of gravel-laden water filled the deep channel to the level of the upper terrace level. A second flow (or possibly multiple flows) eroded into the gravel, removing much of it, forming the lower terrace level. Later river flows removed additional gravel and cut down to the level of the modern flood plain.

JPB Photo: 8/25/2009

relative sizes of the valleys and streams are often greatly mismatched; the streams seem much too small for such large valleys. Some larger valleys, like those of the Sheyenne, James, Des Lacs, Souris and Pembina rivers, are true meltwater channels, which carried great floods for short periods of time.[34] A meltwater channel is one that was filled, once or several times, bank rim to bank rim, by massive amounts of floodwaters released from lakes dammed by glacier ice, such as Lakes Regina and Souris. The largest meltwater channels in North Dakota are the size of the modern lower Mississippi River channel, and they carried comparable amounts of water, repeatedly, but not for long periods of time.

Most meltwater channels, large and small, have consistent bank heights, uniform widths, regular meander shapes and, in some instances, very narrow flood plains. The modern flood plains are limited in extent because the valley was formed quickly, not allowing time for a broad, flat valley floor to develop. One result of this characteristic is that, in times of modern flooding, the water cannot spread out. Such valleys may be quite deep and, during flood events, the water flows quickly through the narrow valley, thereby adding to the severity of the flood.

Meltwater channels differ greatly from valleys like that of the Missouri River, which varies greatly in width and depth. Although parts of the Missouri River trench were eroded by large rivers during the Ice Age, the entirety of the valley of the Missouri River is not a meltwater channel because it was never full of water; no river ever filled the valley rim to rim.

Landforms Shaped by Glacial Lakes

Introduction

Exceptionally flat areas, like the Red River Valley, are found in several places in North Dakota where lakes of glacial meltwater flooded the land. Sediment deposited in the lakes formed broad expanses referred to by geologists as glacial lake plains. Examples of glacial lake

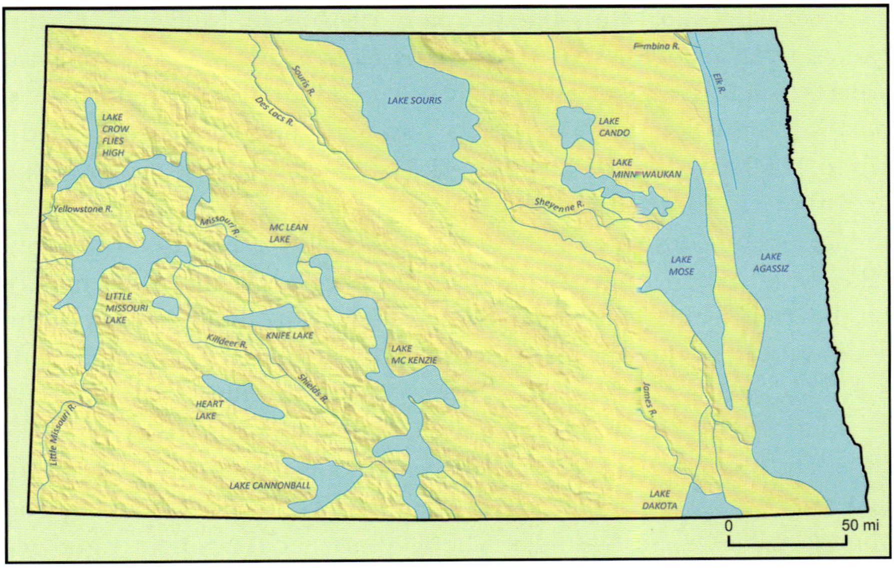

▲ This map shows some of the glacial lakes that formed in North Dakota during the Ice Age. Some of them may have formed more than once, each time glaciers blocked rivers such as the Heart, Cannonball, Little Missouri and Red. A large lake (Glacial Lake Agassiz) formed in the Red River of the North lowland each time glaciers advanced over the area and again when the ice melted.

The glacial lakes shown here are some of the largest that formed. Hundreds, even thousands, of small lake plains not shown on this map exist in the glaciated part of the state; all of them formed at the end of the most-recent glaciation, the Late Wisconsinan.

Evidence for glacial lakes includes the flat plains of lakes such as Agassiz, Dakota, Souris and Minnewaukan (Devils Lake), but not all glacial lakes left flat plains. Some of them can be identified by scattered deposits of sediment, mainly bedded silt and clay, which may be buried beneath younger glacial sediments. Examples of such lakes include Mose, McKenzie and Crow-Flies-High.

plains include the glacial Lake Souris plain in McHenry and Bottineau counties; the glacial Lake Dakota plain in Dickey and Sargent counties; glacial Lake Minnewaukan (the earliest version of Devils Lake) and glacial Lake Cando in Ramsey, Benson and Towner counties; the glacial Lake McKenzie plain in Burleigh County; and dozens of smaller lake plains. But the largest lake plain, by far, is that of Lake Agassiz, once the largest lake in North America. In view of its singular importance, I will discuss Lake Agassiz separately, in my discussion of the Red River Valley in chapter seven.

Glacial Lake Souris

Glacial Lake Souris flooded parts of Rolette, McHenry, Bottineau and Pierce counties, as well as an area in southwestern Manitoba, where a slightly smaller lake, Glacial Lake Hind, formed immediately north of Lake Souris (Lakes Souris and Hind may have been joined). Up to 125 feet of flat-bedded clay and silt beds cover the center of the area that was flooded by Lake Souris in North Dakota, but most of the lake plain consists of till that was washed by waves in the shallow lake. Small beach ridges mark the old Lake Souris shoreline, mainly on the eastern side of the lake plain in Bottineau County. Much of the southern half of the lake plain is covered by sand dunes associated with a delta located where the Souris River entered the lake.

Lake Souris existed at least twice, only a few hundred years each time. The first Lake Souris formed ahead of a glacier lobe 14,000 years ago as the ice withdrew from the area west of Turtle Mountain. The climate turned sharply colder 12,800 years ago, causing a rapid re-advance of the glacial ice. The climate shifted from warmth to full glacial again, probably within a period as short as 10 years, and the cold period lasted from 12,800 until 11,500 years ago. This cold period is known as the Younger Dryas. During this 1,300-year period, some of North Dakota's most important glacial features formed. The glacier re-advance during the Younger Dryas was rapid due to the lubricated surface over which the ice moved. This kind of glacial advance is known as a surge, and it can result in an ice advance of several miles in a single year.

The glacier then withdrew again, this time for good, and the second Lake Souris formed in essentially the same location that the first

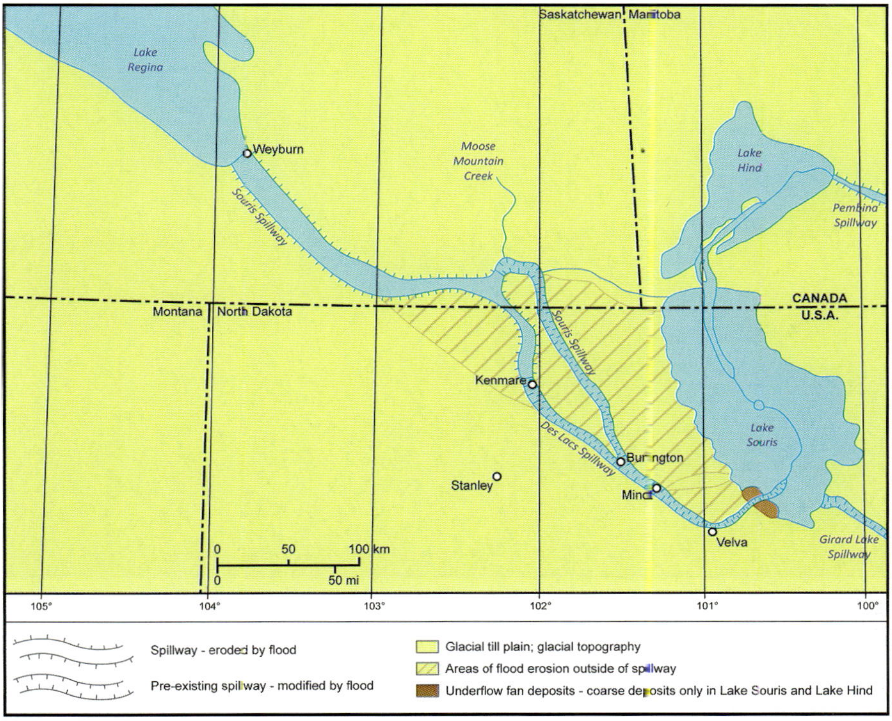

▲ This is a map of part of north-central North Dakota, southwestern Manitoba and southeastern Saskatchewan. Lake Regina, in Saskatchewan, was held back by glacial ice, which periodically receded or was overtopped by the lake water, resulting in massive floods of water, flowing first overland (hatch-lined area) in the area between the Canadian border and Lake Souris, and then down the Souris spillway in Saskatchewan and the Souris and Des Lacs spillways in North Dakota.

Lakes Regina (in Saskatchewan), Hind (in Manitoba) and Souris (in North Dakota) periodically overflowed, carving the Pembina spillway in southern Manitoba and northeastern North Dakota, and the Girard Lake spillway, which was at the northwestern end of the Sheyenne River Valley.

Lake Souris had been. The second Lake Souris was short-lived, but its short life was eventful. A huge flood of water flowed through the lake, the result of a catastrophic release of water from Glacial Lake Regina in Saskatchewan. Lake Regina had been held in place by a dam of glacial ice, but when the glacier in that area receded, removing the ice dam, a massive amount of water flowed southeastward. The water

▲ **Buffalo Lake is five miles west of Esmond in Pierce County. Buffalo Lake is contained in a 20-mile-long channel known as the Girard Lake Spillway. The spillway carried a large volume of water during a flood (or multiple floods) at the end of the Ice Age. Flood water from Glacial Lake Souris flowed to the southeast, through this spillway, and then eroded the Sheyenne River channel. The view is to the southeast, in the direction of water flow.** JPB Photo: 8/20/2009

quickly filled the Lake Souris basin because Lake Souris held only a fraction of the water from Lake Regina. When Lake Souris then overflowed, its floodwater carved the Girard Spillway in Pierce and Benson counties. As the water continued to flow southeastward, it carved the 200-mile-long Sheyenne River meltwater channel through eastern North Dakota.

Glacial Lake Dakota

Glacial Lake Dakota flooded an extensive area south of the glacier in eastern North and South Dakota. The northernmost end of the lake extended into North Dakota in the Oakes area of Dickey and Sargent counties, but the lake was mostly in South Dakota; approximately a 10th of the total area of the lake plain, 200 square miles, is in North

Dakota. The Lake Dakota plain in North Dakota is covered by sand, carried there by the James River, which flowed into the northern end of the lake. Only small amounts of layered lake sediment can be seen on the North Dakota portion of the Lake Dakota plain because most of the lake sediments are covered by sand. I was not able to identify any beaches on the old shoreline of Lake Dakota.

Wind-blown sand dunes with as much as 40 feet of relief cover the northeastern part of the Lake Dakota plain. Immediately east of the dunes, in southwestern Sargent County, is a hilly area known as the Lake Oakes Hills. The Lake Oakes Hills appear to be wind-blown dunes that were later overridden by a glacier, which deposited a thin layer of till on top of the dune sand.

Glacial Lakes Minnewaukan and Cando

Glacial lakes Minnewaukan and Cando are closely related, both existing for only a short time while the glacier margin was receding from the Devils Lake area. Both lakes were dammed, on the north, by the glacier. Lake Minnewaukan probably flooded a couple of hundred years earlier than Lake Cando, but it is unlikely that either lake lasted more than a few hundred years. I do not know how long the lakes lasted because they existed only while the glacier was receding from the area and I am unsure how long it took for the ice to melt.

Glacial Lake Minnewaukan is usually considered to be the ancestral Devils Lake, and most geologists have assumed that it was larger than Devils Lake. This is probably incorrect. Lake Minnewaukan may never have covered more than a few tens of square miles in Benson and Ramsey counties.[35]

An understanding of Lake Minnewaukan depends upon realizing the significance of its ice-marginal boundary. The lake formed immediately after a great ice-thrusting event formed the Devils Lake Basin. The area ahead and south of the receding glacier was immediately flooded as soon as the glacier margin began to withdraw. Lake Minnewaukan flooded the land between the glacier and the ice-thrust topography to elevations at least as high as 1,550 feet, but it was likely only a narrow body of water, which immediately began overflowing southward through the fresh, ice-thrust hills, carving several spillways to the Sheyenne River. It is unlikely that the southern boundary of Lake Minnewaukan had much integrity. Rather, it had an irregular configuration and the hills probably leaked in several places through a variety of spillways. The northern edge of the lake was bounded by the glacier and, as the ice margin receded, the lake level dropped quickly. When it got as low as 1,460 feet, the lake began to spread out over the area flooded by Devils Lake today. I have not found any beaches or bedded lake sediments above 1,460 feet elevation, and I believe that should be considered to be the maximum elevation level of Devils Lake.

Like Lake Minnewaukan, Glacial Lake Cando was bounded on the north by the receding glacier margin, but it covered a broader area than did Lake Minnewaukan, 150 square miles in Benson, Ramsey and Towner counties. Lake Cando was a shallow lake that had a maximum elevation of 1,500 feet and overflowed into the northwestern end of Devils Lake. The area flooded by Lake Cando is covered by hard, dark-colored lake clay in places, but the accumulations of lake sediment that I have seen are thin. When I mapped Towner County, I identified a few poorly developed, low beaches along what was probably the eastern shore of the lake, but the boundary of the lake is poorly defined. For most of the time it existed, Lake Cando's northern and eastern boundaries consisted of the ice at the edge of the glacier.

Glacial Lake McKenzie

Glacial Lake McKenzie formed when a glacier, advancing westward, dammed the Heart, Little Heart and Cannonball rivers in Burleigh, Emmons and Kidder counties. The lake, however, extended to the west into Morton and Sioux counties. Glacial Lake McKenzie flooded 150 square miles at elevations ranging from 1,700 to 1,730 feet. While it existed, the lake had a highly irregular shoreline. Layered sediments deposited in Lake McKenzie range from 60 to 230 feet thick, and average between 150 and 200 feet thick, but they are interbedded with thinner layers of gravel and till. Several sloughs, the largest of which is McKenzie Slough, occur on the poorly drained lake plain. Sand dunes, up to 20 feet high, cover 15 square miles at the southern and southwestern edges of the lake plain in southwestern Burleigh and northern Emmons counties.

The area that was flooded by Glacial Lake McKenzie differs from most of the other large glacial lake deposits in North Dakota in that it lacks the broad, flat plain we usually associate with lake plains. Rather, Lake McKenzie flooded an area of considerable relief, located where the Heart, Little Heart and Cannonball rivers had carved deep valleys in the preglacial bedrock. Thick layers of lake sediment were deposited in low areas, but they were not extensive enough and did not accumulate to thicknesses great enough to cover all of the existing landforms. As a result, islands of preglacial bedrock and older glacial sediment protrude above the Lake McKenzie plain.

The thick layers of lake sediment beneath the Lake McKenzie plain, along with the interbedded layers of glacial and fluvial sediments, suggest that the area was flooded several times, each time glaciers blocked the Heart and Cannonball rivers. These rivers, along with all other northeasterly flowing rivers, were dammed each time central North Dakota was glaciated. It is possible that only the uppermost few feet of lake sediment were deposited in lakes dammed by Late Wisconsinan glaciers. Much of the underlying cumulative thickness may have been deposited during numerous earlier damming events.

CHAPTER FIVE

OTHER GEOLOGIC FEATURES

Wind-shaped Landforms

Sand Dunes

North Dakota's sand dunes are pleasant places to spend a summer afternoon. Most of our dunes are in the glaciated parts of the state, especially where sand was deposited by ancient rivers flowing into lakes. When rivers, laden with sediment, flowed into glacier-dammed lakes, they deposited large amounts of well-sorted sand ("well-sorted" means that most of the sand grains are all nearly the same size). As the rivers built up fans at their mouths, the heavy, sediment-laden river water stayed near the bottoms of the lakes and spread out over broad delta-like areas. Coarser materials (gravel, coarse sand) remained farther upstream, on the stream beds themselves, while the finer sand and silt were carried into the lakes. The well-sorted sand that stayed in between—neither too coarse nor too fine—was later blown into dunes.[36]

Why did the wind blow the sand into dunes rather than spreading it out evenly across the landscape? The very finest materials picked up by the wind, the dust-size particles, were, in fact, carried long distances and are spread over much of the surface of North Dakota, or in areas to the east of North Dakota. This material is known as loess, and I will discuss it shortly. But the answer to why dunes formed in some places has to do with the nature of sand. One definition of sand is that it consists of particles small enough to be moved by the wind, but too heavy to be carried very far. When a sand grain is lifted by the wind, it bounces along on the ground surface. If the surface is hard, the sand grains just keep bouncing along as they are moved by the wind—like a basketball bouncing on a gym floor. However, if the surface is not hard, because it is already sandy, then the bouncing sand grains lose momentum—like bouncing a basketball on a sandy beach. Sand grains bouncing on sand slow down. If a small pile of sand has accumulated in the wind shadow of a bush or a rock, forming a barrier, it will tend to slow down the bouncing sand grains, which will stop on the pile. Eventually, the sand pile may accumulate enough material to form a dune.

Although the wind configures individual sand grains into dunes, it does not move an entire dune. As sand grains are blown over the crest of a dune, they fall into the dune's wind shadow and settle on the downwind side of the crest. So, why does a dune move forward? As sand continues to accumulate at the crest of a dune, it forms a steeper and steeper slope. When that steep slope, or slip face, becomes unstable, gravity pulls the sand down in a small avalanche. You can create slip faces of your own if you climb up the downwind side of a dune. Each step causes an avalanche. One slipping avalanche at a time causes the dunes to march steadily downwind, sometimes ending up miles from where they began. Some North Dakota dunes are as high as 80 to 100 feet and many have strong northwest-southeast linearity, parallel to the main direction the wind was blowing when it built the dunes.

Wooded areas of dunes occur south of Walhalla in Pembina County, where the Pembina River flowed into Lake Agassiz. Dunes near Wyndmere and McLeod in northeastern Richland and eastern Ransom counties coincide with the area where the Sheyenne River flowed into glacial Lake Agassiz. The Ransom County dunes are beautiful places to enjoy undisturbed native prairie in the Sheyenne National Grasslands, the only National Grasslands in the tallgrass prairie region of the United States. The Sheyenne National Grasslands account for most of the remaining native tallgrass prairie in North Dakota and the area is home to more than 40 sensitive plant and animal species, including the largest population of greater prairie chickens in the state.

The dune fields along the western edge of the Lake Agassiz plain began to form soon after the lake drained, 9,000 years ago. Most of these dunes have been intermittently active since then, stable during wet periods when vegetation covers them, and actively blowing and moving during periods of drought, when plants are scarce. They are constantly changing in shape and size.

The explanation for the origin of an area of dunes in northeastern-most Eddy and southeastern-most Benson counties, just east of Warwick, is not as straightforward as that for our other dune fields. I refer

▲ **This map of North Dakota shows areas of sand dunes. Most of the dunes occur in areas coinciding with glacial lake sands, which were deposited in places where rivers emptied into proglacial lakes. A few small areas of dunes are found in association with bedrock sand deposits (Hettinger and Grant counties, for example) and some dunes are the result of wind blowing over the Missouri River flood plain (Emmons and McLean counties). Numerous local areas of dune topography, too small to show, are found throughout the state.**

This map is modified from one in an article by Fred J. Anderson in Geo News, the newsletter of the North Dakota Department of Mineral Resources, 2011, vol. 38, No. 2.

to it as the Hamar Dune Field, named for the village site of Hamar. It consists of small dunes, which cover an area of a township in size. It is a pleasant place, covered by aspen woodland. Like several of the other areas of dunes I have mentioned, the Hamar dunes were built where a river once flowed into a lake, building a delta of sand.

The lake, which existed perhaps 1,000 years earlier than Lake Agassiz, is referred to as Lake Mose. It was trapped on its east between the edge of the glacier that filled the Red River Valley and on its west by higher land. Evidence for Lake Mose is much less obvious than for Lake Agassiz (no flat lake plain, no obvious shoreline) because the area Lake Mose flooded was overridden when the glacier in the Red River Valley re-advanced westward, covering most of the area with glacial sediment. However, part of the sandy delta was not reached by the re-advancing glacier, and that's where the Hamar dunes formed.

North Dakota's largest area of dunes, and the place where the largest individual dunes in the state are found, is along U.S. Highway 2 between Towner and Denbigh, where the Souris River flowed into glacial Lake Souris. These dunes cover almost 1,000 square miles and have as much as 100 feet of local relief in places, but most of the dunes are between 10 and 20 feet high.[37]

An area of dunes near Oakes in southwestern Sargent County (the Brampton Dunes), where the James River flowed into the north end of glacial Lake Dakota, is unusual in that, soon after the dunes formed, they were overridden by glacial ice, which advanced from the east, depositing a covering of sandy till on the dunes. In most places, the till cover has protected the Brampton dunes from additional wind erosion, but in some places, the till cover has eroded or slid off the dune sand, exposing it to the wind and, in such places, some of the dunes have been reactivated.

A small area of dunes known as Devils Nest is easily accessible off Interstate Highway 29 in Richland County. Turn west onto State Highway 11 at exit 8 and drive a mile to where the road curves to the northwest toward Hankinson. Then drive a mile south on 169th Avenue Southeast. The hilly, wooded area to the west is Devils Nest. The dunes are developed in an area where the Sheyenne River once flowed into the early Lake Agassiz (through a valley known as the Milnor Channel).

▲ **These sand dunes are five miles southeast of Sheldon in Ransom County. Some of these dunes are as high as 80 feet, the largest on the Sheyenne Delta. Even higher dunes occur near Denbigh, in McHenry County, on the delta of the Souris River (where the river flowed into glacial Lake Souris).**

The area shown here is part of the Sheyenne National Grasslands, which comprise 70,180 acres in Ransom County. The area is the only remaining tall-grass prairie in public ownership in the United States. It is associated with the 64,769 acres in private ownership located in both Ransom and Richland counties of North Dakota. Because tall-grass prairie in eastern North Dakota was so productive and easily converted to agricultural use, it is now one of the most endangered habitat types in the world. The area has marked trails that provide excellent hiking opportunities.

The *Geologic Explorer's Guide to the Sheyenne National Grassland* by Ann M.K. Fritz is an excellent publication on the area. The reference is:

- Fritz, Ann M.K., 2001, *Geologic Explorer's Guide to the Sheyenne National Grassland,* North Dakota Geological Survey Educational Series 27, 73 pages.

JPB Photo: 7/17/2009

A few small areas of dune topography may be seen southwest of the Missouri River, where glacial deposits are mostly absent. Some of these dunes occur near modern rivers, downwind from sandy flood plains. Dunes 15 to 30 feet high cover the slopes of the Knife River Valley west of Stanton in Mercer County, and prominent dunes occur immediately east of the Missouri River flood plain in Emmons County.

Dunes are present in several places on tracts of sandy, preglacial bedrock, too. Quite small, such dunes may be seen near Fort Yates, Carson and in several other locations in southwestern North Dakota.

Most of North Dakota's dune-building activity took place between 8,000 and 5,000 years ago, during a time geologists refer to as the Altithermal Period, one of many episodes of global warming that have occurred throughout the millennia. The Altithermal (the word means "high temperature") was the warmest and driest time since the end of the Ice Age. Conditions during the Altithermal were similar to those during the 1930s on the Great Plains, when the cover of vegetation became too sparse in places to effectively control wind erosion.

Loess

At the same time the wind was shaping sand into dunes, it was blowing dust and silt-size particles greater distances, spreading them over broad areas. Much of North Dakota has a veneer of such windblown silt, known as loess (loess is a German word that rhymes with "bus" and translates, roughly, as "loose"). In places along the Missouri River Valley south of Garrison Dam in McLean County, in northern Mercer County and in parts of Emmons County along the Oahe Reservoir, the loess is up to 25 feet thick. Most loess in North Dakota is buff to yellowish brown in color. It may contain shells, bones and teeth of mammals, as well as a variety of invertebrate fossils. Exposures of loess may be marked by small, narrow, vertical tubes, left by successive generations of grass roots.

In 1967, I noticed deposits of loess exposed in the wave-cut bluffs adjacent to Lake Sakakawea, near Riverdale in McLean County. The loess exposure contained two black bands: fossil soil horizons (paleosols). Since then, geologists have done at least a dozen scientific studies on the Riverdale loess deposits and paleosols, using what they have learned to help understand how the climate has changed during Holocene (post-glacial) time.

The lowermost loess at Riverdale lies directly on Late Wisconsinan till. This part of the loess exposure is overlain by the lower of the two paleosols. The lower paleosol is a fossil forest soil that formed while glaciers were still melting from the area. More loess overlies the forest soil, which is capped by yet another paleosol. This higher paleo-

▲ This exposure of wind-blown loess is two miles north of Riverdale along the shore of Lake Sakakawea in McLean County. Wave action along the lake shore is eroding the materials into near-vertical cliffs. Various layers in the loess have been assigned names by geologists studying the deposits, and I will use those names in this caption. At the base of the cliff is Late Wisconsinan till, which extends upward to the light-colored silt, the Mallard Island Member (MI). The Mallard Island is probably a subsoil zone that developed on the last till deposited by the glacier.

The black zone immediately above the Mallard Island is the Aggie Brown Paleosol (AB). The Aggie Brown is the darkest-colored zone in the exposure. The lower part of the paleosol is a fossil forest soil, and the upper part is a fossil prairie soil. The paleosol formed between 11,000 and 8,500 years ago.

The Aggie Brown Paleosol is overlain by the Pick City loess (PC), wind-blown silt deposited between 8,500 and 5,000 years ago. The loess is overlain by another black layer, the Lower Riverdale Paleosol (LR), which formed after 5,000 years ago. Two more paleosols, the Middle Riverdale Paleosol (MR) and Upper Riverdale Paleosol (UR), are at the top of the exposure. The Upper Riverdale grades into modern soil in places. The sequence of materials exposed here provides an excellent, detailed record of climate change during post-glacial time.

This photo was taken in 1975 by Lee Clayton.

sol was formed in a sparse grassland environment 4,500 years ago.

The climatic history is considerably more complex and detailed than I have just described. The Riverdale loess and interbedded paleosols are important because they provide a detailed and nearly complete history of the post-glacial environment in central North Dakota. Further studies of pollen and other microfossils in the loess in places like Riverdale where the loess is thick may help us better understand how the post-glacial climate has changed during the past 11,000 years.

Loess accumulations, most of them much thinner than those at Riverdale, tend to be located downwind from river flood plain deposits or downwind from any place the wind could sweep over a broad area of fine-grained sediment. Some of the loess deposits in North Dakota probably accumulated during Altithermal dust storms. Elsewhere, deposits of loess up to 200 feet thick are found in the Loess Hills of western Iowa, just east of the Missouri River Valley. Parts of the Gobi Desert in Mongolia are mantled by loess deposits more than 1,000 feet thick.

Wind-sculpted Features

The best-known wind-sculpted feature in North Dakota is Wind Canyon, in Theodore Roosevelt National Park. Wind Canyon was initially carved from sandstone by a small tributary stream of the Little Missouri River. The stream valley happened to be aligned northwest-southeast, parallel to the prevailing northwest winds. Northwest winds whipping across the broad expanse of Little Missouri River bottomland picked up sand from the river bars, blowing it into the canyon, channeling it and creating a sand-blasting effect. The sand blasting has carved the rock into smooth, wind-eroded surfaces and shapes. Another prominent wind-sculpted feature is Castle Rock near Mott, in Hettinger County. Wind-blown sand polishes and facets glacial erratics, giving them a burnished appearance.

▲ This wind-sculpted slope in Sentinel Butte Formation sandstone is in northern Billings County along Blacktail Creek (five miles southeast of the Elkhorn Ranch site). As it buffets the sandstone, the sand-laden wind "sand-blasts" the surface and tends to remove the less-well cemented materials, accentuating the bedding. Black to reddish-brown bodies of ironstone are left behind after the finer sand matrix is removed by the wind. JPB Photo: 9/14/2009

Field Stones

Field stones are common throughout North Dakota. Throughout much of the southwestern part of the state, most field stones had local sources unrelated to glaciation, but in parts of the state that were glaciated, the field stones are mainly erratics that were deposited by the glaciers. Glacial erratics lie scattered in pastures, occasionally surrounded by a crop of wheat, gathered along fence rows or collected into piles. The main exception is the Red River Valley, where they are rare.

Glacial Erratics

Geologists use the term "erratic" to refer to boulders and cobbles left behind by glacial ice. Erratics are rocks that differ in lithology from the surface on which they are found. The term "erratic," as it refers to rocks, dates to 1779, when Horace de Saussure, a Swiss naturalist, described granite boulders lying on top of limestone in the Jura Mountains in Switzerland. His term, *terrain erratique*, is derived from the Latin *erratus*, "to wander," and means, literally, "ground that has wandered." De Saussure recognized that the boulders had been moved and "did not belong" where he found them, but he did not realize the role glaciers had played in their movement. The first scientific paper noting the glacial origin of erratic boulders was published in 1833 by Ignace Venetz, a Swiss engineer.

The glaciers delivered various kinds of erratics to North Dakota: rocks and boulders composed of materials that would not otherwise be found here. Most of the erratics came from Canada, but only a few rock types are sufficiently unique to enable us to accurately determine the exact location of their source. If you pick up a piece of granite in the central part of the state, you can be sure it had a Canadian source, but usually you cannot precisely determine from what part of Canada it came.

In some instances, the source area of an erratic can be pinpointed. North of Winnipeg, for example, distinctive Paleozoic carbonate (limestone) formations are quarried and, by comparing an erratic with the Manitoba carbonate, we can determine from which formation it was derived. We may even be able to say with some precision where it originated. Several years ago, Bob Biek, a former North Dakota Geological

Survey geologist, found a number of unusual erratics along Lake Sakakawea – dark-colored stones, which he identified as "omars." The rocks are found in-place (where they originally formed) today only in the Belcher Islands in southeastern Hudson Bay, so it is possible that the Lake Sakakawea omars originated in the Belcher Islands. The Belcher Islands are located 1,000 miles northeast of Lake Sakakawea.

Erratics have been used as exploration tools in the search for ore deposits. Copper mines were opened in Finland after copper-bearing erratics were traced back to their source. Analysis of gold-bearing erratics in Maine resulted in the discovery of gold ore deposits in Quebec. Such tracking of erratics is one way that geologists can help determine the history and source of glacial deposits. I have occasionally found erratic boulders in North Dakota containing traces of gold. Such erratics were probably transported 700 miles by glaciers to North Dakota from the metal mining districts of Manitoba and Saskatchewan.

In contrast to the long-distance travelers, glaciers moved some erratic sandstone boulders only short distances, tens or hundreds of feet, from nearby outcrop areas. Sandstone is less well consolidated than granite or limestone, and any long-range transport of pieces of sandstone would break them down into smaller fragments, or simply reduce them to sand. I have seen a few shale boulders, too, but because they are so fragile, the glacier probably moved them only a few tens of feet from their original source. Some shale boulders may have been frozen at the time they were transported; otherwise they might have fallen apart quickly.

Large erratics, 3 or more feet in diameter, tend to be hard rock such as granite or gneiss. Such rocks are more resistant to abrasion and fracturing than are sedimentary rocks such as limestone or sandstone. Large granitic erratics (5 to 10 feet in diameter) are numerous in places: along the slopes of the Sheyenne River Valley near Fort Ransom; on the hills south of Tappen in Kidder County; on the valley slopes along the Souris River near Velva and in the valley at Minot; and along the White Earth River Valley in Mountrail County. Both large and small erratics are particularly abundant near Venturia and Zeeland in McIntosh County. A few miles west of Mandan, on "Old Highway 10," a large erratic lies a few feet north of the highway. Several large boulders, all of the same pink quartzite, and all once part of a single boulder, are scattered close together in a field in northeastern McKenzie County. I estimate that the largest piece weighs approximately 100 tons; the original, intact boulder (all the pieces combined) was at least three times the size of the largest piece. The largest erratic I have seen is located 11 miles south of Calgary, Alberta. Composed of quartzite, and known as the Okotoks Erratic (aka Big Rock), it weighs 16,500 tons, stands 30 feet above the surrounding area and is billed as the world's largest glacial erratic.

Erratics tend to be abundant in places where the ground surface was washed by flowing water or winnowed by the action of waves along the shores of glacial lakes and modern reservoirs. Wave action or running water removed the finer materials, leaving behind a concentration of cobbles and boulders. Boulders are abundant along the

▲ This large glacial erratic sits along Old Highway 10, 10 miles west of Mandan. Several pieces of quartzite, some of them rising 10 feet above the ground surface, are all part of what was once a single erratic, which weighed at least 350 tons (but I do not know how much additional rock may be buried beneath the ground surface). Numerous large granite and quartzite erratics occur in this part of eastern Morton County. They are all that remains of an old (pre-Wisconsinan) glacial advance. Other evidence of the glaciation (landforms, till deposits, etc.) has been eroded away. JPB Photo: 6/11/2009

Lake Agassiz shoreline near Hankinson in Richland County and on the wave-worn shore of glacial Lake Agassiz near Pisek in Walsh County.[38]

Some erratics are grooved or polished, a result of abrasion by the moving ice. Others have beveled or faceted surfaces. During the course of their journey, the rocks were jostled against one another within the glacier, or against the rock over which the glacier was flowing. Coarse sand and gravel within the ice scraped against the boulders, scratching or striating them as they moved along with the advancing glacial ice or when the glacier flowed over a stationary rock without moving it. Striated boulders with straight grooves (striae) may be found in buried "boulder pavements," which are lag deposits of boulders that were later covered by younger glacial deposits. If the boulders have not moved since they were deposited, the trend of the striae can sometimes help

▲ This "buffalo boulder" is nine miles north of Glen Ullin in Morton County. This quartzite boulder, which glaciers carried here from the Canadian Shield in northwestern Ontario perhaps 80,000 years ago, is the only remaining evidence of glaciation in this vicinity; everything else was eroded away long ago.

Today, cattle use the boulder as a rubbing stone. Before the cattle arrived, bison used it for the same purpose. Foot (hoof) traffic has dug into and loosened the ground surrounding the boulder, and the wind has blown away the loose soil, leaving the rock standing in a depression. JPB Photo: 9/29/2009 determine the direction the glacial ice was flowing.

Large, isolated erratics are sometimes surrounded by depressions created by animals such as bison using them as rubbing stones. The depressions formed when bison rubbed against the rock, trampling and loosening the soil, which then washed or blew away. Sometimes, given enough time, the depressions grew in size, becoming small wetlands areas, even buffalo wallows, still seen today. Many "buffalo boulders" were polished by the animals' dirty, gritty hides. Buffalo boulders have been recognized on the northern Plains for a long time, though today they might better be termed "cattle rubbing stones."

Some erratics are famous. Everyone has heard of Plymouth Rock in Massachusetts where the Pilgrims set foot in the New World on December 21, 1620. In North Dakota, we have the Standing Rock, located in Ransom County near where State Highway 46 crosses the Sheyenne River Valley. The explorers Nicollet and Fremont noted Standing Rock Hill on their maps in 1839. Another large erratic, Writing Rock, near Alkabo in northwestern North Dakota, was known by the Sioux as *Hoi-waukon* or Spirit Rock.

Seasonal freezing and thawing causes buried rocks to work their way upward from below the plow zone to the land surface. Each spring, a new crop of stones must be removed from the fields. Smaller rocks can be picked up by hand or with rock-picking equipment. Larger erratics are sometimes blasted with dynamite and the pieces hauled away.

In his book, *Blue Highways: A Backroads Tour of Rural America*, William Least Heat-Moon captured the resignation of farmers to a continual crop of boulders:

> East of Fortuna, North Dakota, just eight miles south of Saskatchewan, the high moraine wheat fields took up the whole landscape. There was nothing else, except piles of stones like Viking burial mounds at the verges of tracts and big rock pickers running steely fingers through the glacial soil to glean stone that freezes had heaved to the surface; behind the machines, the fields looked vacuumed. At a filling station, a man who long had farmed the moraine said the great ice sheets had gone away only to get more rock. "They'll be back. They always come back. What's to stop them?"[39]

▲ Standing Rock is located in Ransom County eight miles north of Fort Ransom along State Highway 46. The rock is metamorphic gneiss, a combination of quartz, feldspar and mica brought to North Dakota from Ontario by glaciers. The hill on which Standing Rock is located is an ice-thrust block consisting of Niobrara Formation shale that was pushed three miles from the northeast to this location. It stands 110 feet above the surrounding area.

A State Historical Society sign at the site of the Standing Rock reads as follows: "The Inyan Bosendata of the Dakota Indians used this site as a place for ceremonials and offerings. In 1839, the explorers Nicollet and Fremont noted this landmark on the maps of that date. The state geographic board has designated this place as one of the most sightly and historic in North Dakota."

JPB Photo: 7/17/2009

▲ **Glacial erratic boulders are abundant in this field two miles north of Arena in Burleigh County.** JPB Photo: 9/4/2010

Siliceous Rocks

Silcrete

Silcrete is found in parts of North Dakota that were not glaciated. It goes by a variety of names: gannister, pseudoquartzite, orthoquartzite, duracrust and silcrete (the name I prefer). It can occur on the surface as angular (sharp-cornered) blocks of hard, weather-resistant quartzite in parts of southwestern North Dakota (see chapter two for a discussion of silcrete-capped buttes). Boulders of silcrete mantle the tops and sides of small buttes and ridges, forming an armor of resistant material that remains, even after all or much of the surrounding softer materials have been removed by erosion.

Several varieties of silcrete are found in western North Dakota. They include the early Paleocene Rhame Bed, the late Paleocene Rainy Butte Chert, the early Eocene Taylor Bed, Hard Siliceous (HS) Bed and Knife River Flint (the ages I have listed are the times the materials are thought to have formed). Ed Murphy, writing in the North Dakota Geological Survey's newsletter, has provided a comprehensive description of several of the kinds of siliceous rocks found in the state (Murphy, 2014).

The different varieties of siliceous rocks vary in color and other characteristics. Most are unbedded, and have irregular fracture patterns (when broken, they form irregular shapes). Wind has sand-blasted some silcrete boulders so they have smooth, burnished surfaces. Silcrete paves many hilltops near the Cannonball River in Hettinger and Grant counties and throughout Adams County where, in places, silcrete boul-

▲ This silcrete boulder is near Haynes in Adams County. The holes in the rock, which range from a half inch to 2 inches in diameter, are the molds of plant stems that grew in the sediment, which eventually hardened into silcrete. The interior of some of the holes have ridges, suggesting they once contained plants such as horsetails.

Silcrete is very hard rock that likely formed when slough sediment, in which the plants grew, was transformed to silcrete after the sloughs dried up. Because of its resistance to erosion, silcrete beds commonly form caprock on buttes. Boulders of silcrete remain, strewn over the surface, after the surrounding materials have eroded away. JPB Photo: 6/24/2010 ders are so thickly strewn, they make walking difficult.

The weathering and soil-development processes that formed silcrete are not well understood. The various occurrences do not represent any single event. Rather, they result from recurring geologic processes, or a group of similar processes. Most silcrete probably formed as a product of soil formation in a hot, dry climate. Silica was dissolved and transported downward from overlying materials and carried by ground water, in solution or colloidal form, to be re-deposited as a greatly concentrated layer lower in the soil profile. Some silcrete may have formed when buried beds of volcanic ash were weathered and altered to clay. This reaction produces silica as a byproduct. The silica was carried through the soil to swamps and sloughs, where it provided a cementing matrix for decomposing plant matter.

Knife River Flint

William Clark was surely referring to Knife River Flint when he mentioned the occurrence of black flint at the mouth of the Little Missouri River in his journal entry for April 12, 1805. He wrote: "I walked out on the lower Side of this river and found the country hilley the soil composed of black mole & a small perportion of sand containing great quantity of Small peable some limestone, black flint, & sand Stone."

The Knife River (the term is the translation of an American Indian name) was so-named because flint suitable for stone tools was quarried near the river. The exact age of the quarries is not known, but Knife River Flint is common in projectile points made during the Woodland, Plains Village and Paleo-Indian Periods. It was certainly utilized for several thousand years, perhaps since the end of the Ice Age.

Artifacts made from Knife River Flint were widely traded by the American Indians, and artifacts have been reported as far away as Alberta, Ohio and Missouri. The Knife River Flint has been used and traded for thousands of years. A spear point dated at 10,500 years old, beautifully crafted of Knife River Flint, was found in west-central Manitoba, far from its source in western North Dakota. I was once on a bus field trip in Massachusetts and sat next to an archaeologist who said she had seen artifacts made from Knife River Flint found near Boston.

▲ All three of these photos show Knife River Flint. The piece on the left, which I collected in the 1970s, measures about 5 inches by 6 inches (quarter coin for scale). Knife River Flint was used as raw material for tools by early humans in the northern Plains and Midwest.

Knife River Flint originated when a lignite bed, or beds, silicified. It ranges from cream-colored to dark brown—essentially the color range shown on these photos. The flint has a conchoidal fracture, making it an excellent material for stone tools. Darker-hued specimens of flint were preferred for tool manufacture as they are less likely to splinter along bedding planes.

Knife River Flint apparently occurred as part of the HS (hard siliceous) bed of the Eocene Golden Valley Formation. The HS bed is still widespread, but erosion has removed it over much of the area where the flint is found—and flint has not been found in places where the HS bed occurs today. The flint is always found as a lag material; it was eroded out of its original position in the HS bed and left lying on the ground when less-resistant surrounding materials eroded away. The original extent of the flint-bearing part of the HS bed is not precisely known.

Notice the light-colored weathering rind or patina on top of the left-most photo. Such patina layers (known also as corticated layers) are commonly found on fracture surfaces of the flint.

The two artifact photos are courtesy of the State Historical Society of North Dakota. The center photo, of a point collected in Golden Valley County, was provided by Wendi Field Murray. The right-most photo, of a biface cutting tool was found in Ward County. It was collected and photographed by Paul R. Picha. Both Murray and Picha are with the Archaeology and Historic Preservation Division of the State Historical Society of North Dakota. Murray is Archaeology Collections Manager, and Picha is Chief Archaeologist with the Division. I took the photo on the left.

For more information on lithic materials found in North Dakota, please see:

- Murphy, E.C., 2014, *Knife River Flint and Other Siliceous Rocks in Western North Dakota,* North Dakota Department of Mineral Resources Geo News, vol. 41, no. 1, p. 1-7.

The flint is found as pebbles, cobbles and boulders as large as 2 feet in diameter. All known occurrences of Knife River Flint are lag deposits; the flint has not yet been found in the position in which it originally formed. A lag is a zone of hard, erosion-resistant materials remaining after all of the surrounding, softer sediments have been removed by erosion.

The flint litters hill slope surfaces in Dunn and Mercer counties especially at and near the quarry sites, where Indians and Paleo-Indians dug shallow pits into it. Knife River Flint is a nonporous, layered, dark brown rock with a conchoidal fracture, excellent for making tools. The brown color is due to fine-grained, dispersed organic material. Some of the layers contain a cellular microstructure, indicating that they consist of silicified plant fragments. The flint likely formed when a lignite bed was petrified. However, all of the softer sediments that once surrounded the lignite bed were removed by erosion long ago. It has not been determined for sure just when the flint formed. A likely candidate for its source is the upper member of the 50 million-year-old Golden Valley Formation (the Camels Butte Member).

While I was with the North Dakota Geological Survey, we routinely got inquiries from amateur flint knappers concerning the availability of Knife River Flint. Much of it has been removed from the area of flint quarries in Dunn County, some areas entirely denuded of large pieces of flint. Where there had been large rock piles of flint, the flint is gone, and only a few granite boulder erratics are left. Local ranchers told Survey geologist Ed Murphy in 2000 that the flint was hauled away by the semi-truck load.

Because it is unique to North Dakota and an important cultural material, I once had a bill introduced in the state Legislature that would have designated Knife River Flint as North Dakota's State Rock.[40] A legislator objected, noting that the rock is a nuisance, hard on tires when it is mixed in gravel on roads. The bill was defeated. However, on June 9, 2012, the Lynch Knife River Flint Quarry National Historic Landmark was dedicated near Dunn Center, North Dakota. Perhaps the landmark designation will help to protect the remaining resource.

Chert

Varieties of other kinds of chert nodules, a few inches in diameter, are scattered on the surfaces of buttes such as Rainy Butte and Sentinel Butte. Chert is lighter in color, more homogeneous than Knife River Flint, and lacks visible internal structure. It is mostly light brownish gray with a uniformly dull luster on fracture surfaces and it is more translucent than Knife River Flint. Native Americans used chert for arrows and spear points in much the same way they used flint.

Stones Used in Construction

Field stones of all kinds have been used for virtually every conceivable construction purpose: for homes, churches, retaining walls, decorative purposes, landscaping and as riprap on the faces of dams, shorelines and roads susceptible to wave action. Early settlers used field

▲ **The Schulstad House, a stone home, was built in Forbes, Dickey County, in 1907. It is constructed of glacial erratic field stones.** JPB Photo: 7/8/2010

stones for the foundations of their homes and farm buildings, and they sometimes built whole structures of them. Skilled stone masons shaped the rocks and precisely fit them together. Where they were available, cobble-size river stones were often used for construction. Large numbers of field stones have been used to guard against erosion along the

▲ A sandstone bed caps many buttes over an extensive area of Emmons County. The sandstone is part of the Linton Member sandstone of the Cretaceous Fox Hills Formation. It was used in the construction of several buildings in the Linton area and for home construction in various parts of the county. This photo shows some of the stone work on an old bank building.

By late 1902, a stone quarry was operating a mile southeast of Linton, but other quarries were located in several places in all directions from Linton. At the time it was used for construction in the area, the dark gray building stone was available for 25 cents a wagon load. It was easily worked with stonemason tools.

In recent years, several of the sandstone buildings in Linton have been refurbished, the sandstone cleaned and re-grouted. Some buildings have been refaced with a sandstone "veneer" 3 to 4 inches thick. However, the original blocks, still in place in most of the buildings, were 1 to 2 feet thick.

I thank Kristy Llerenas and John Beck of Linton for providing information about the stone buildings. Both of these people operate businesses in buildings constructed of stone. In addition, some of the sandstone was quarried on land belonging to John Beck. JPB Photo: 11/7/2012 shores of Devils Lake, the reservoir lakes Sakakawea and Oahe, and many road and railroad embankments.

In addition to field stones, dimension stone (rock quarried and cut to specifications) has been used for construction. Preglacial bedrock formations containing beds suitable for construction purposes include the Golden Valley, Fox Hills, Hell Creek, Bullion Creek and Sentinel Butte. Linton, in Emmons County, with several buildings built of a local sandstone bed in the Fox Hills Formation, probably has more dimension-stone construction than any other town in North Dakota. Many towns boast one or more stone buildings or monuments, built of either field stones or dimension stone during the 1930s as Civilian Conservation Corps (CCC) projects.

▲ Smith Butte, in the southwestern corner of Emmons County, is capped by the Linton Member of the Cretaceous Fox Hills Formation. This sandstone has been used as a dimension stone in the construction of buildings locally (see photo of stone building in Linton). The exposure below the top of the butte is part of the same sandstone bed as that on top of the butte, but it has slid part way down the slope. JPB Photo: 5/17/2010

▲ **Coghlan Castle is a field stone construction of shaped glacial erratic boulders. It is located in Rolette County three miles east of St. John. The exterior of the building is being restored, but the most recent information I have on the restoration is that, due to insufficient funding, the project is at a stand-still.**
JPB Photo: 7/29/2010

Coghlan Castle

Coghlan Castle, located five miles north of the town of Rolla in Rolette County, is an outstanding example of a building constructed of field stones. Maurice Coghlan, an Irish-born entrepreneur from Boston, had the house built between 1906 and 1909 for himself and his family on land he had homesteaded in 1889-1890. The house was built by a Manitoba mason in a style resembling Richardson Romanesque (but without rounded arches) with massive, rough-cut stone walls, deeply recessed windows, a turret and turret-like extensions. The 2-foot-thick walls are in-filled with cemented field-stone rubble and are supported by a full stone basement. Portions of the exterior stonework still show evidence of tuck pointing, a mortaring technique intended to create the illusion of regular, rectangularly jointed stonework.[41]

In addition to a diversity of crystalline (igneous and metamorphic) glacial erratic rocks, the walls of Coghlan Castle contain a few shaped sandstone blocks (dimension stone). These are well-indurated, slightly calcareous, most of them either buff or reddish colored. Pieces of the same sandstone show up in some stone buildings in Bottineau and Dunseith. Two good examples of the reddish yellow sandstone can be seen above one of the ground-floor windows on the west side of Coghlan Castle. Someone must have liked them because it is obvious they were deliberately placed there. The nearest local source of sandstone of this kind that I know of is near the west end of Turtle Mountain, between Carbury and the Canadian border, where Cretaceous Hell Creek Formation bedrock occurs at and very near the ground surface.

Gold in North Dakota

The earliest mention of gold that I know of in North Dakota is in a diary kept by Dr. William Denton Dibb, a physician and surgeon for an expedition of emigrants led by James Fisk through Dakota Territory in 1864. The expedition was on its way to the Montana gold fields. Dibb wrote that he had discovered a hill of gold-laced quartz along Deep Creek (probably in central Slope County). I am unsure what Dr. Dibb saw, but it certainly was not gold.[42]

During the years I mapped geology in North Dakota, the only gold I saw was contained in glacial erratics, brought here by glaciers from Canada. The best example of gold traces in a boulder that I ever came across in North Dakota was in a large erratic near the town of Niagara, in westernmost Grand Forks County. The erratic likely originated in the Flin Flon area of Manitoba.

In addition to the Dibb diary account I just mentioned, I have read several more stories of gold found in North Dakota. Some are (or may be) credible, some are hoaxes; most tend to be embellished somewhat by the newspapers that reported them. One thing all of the discoveries have in common is that little or no gold resulted from them. I will comment on only a few of the more interesting of them.

Ransom County

The earliest "gold rush" in North Dakota, apparently involving actual gold, occurred in 1883–1884 when gold was discovered in the Sheyenne River Valley nine miles northwest of Lisbon. The gold was discovered on land owned by H. W. Griswold, a Chicago businessman, who had bought 340 acres of land, for cattle pasturage, along the Sheyenne River. Mr. Griswold found gold in a mound of dirt brought to the surface by gophers. He took a sample to an assayer, who told him it was "rich in gold."

The Lisbon Gold Rush followed an article in a Lisbon newspaper. Articles about the discovery appeared in newspapers in Grand Forks, Bismarck, St. Paul, Chicago and New York, among others. Some sample headlines: "Rich Gold Discoveries" (*St. Paul Globe,* October 21,

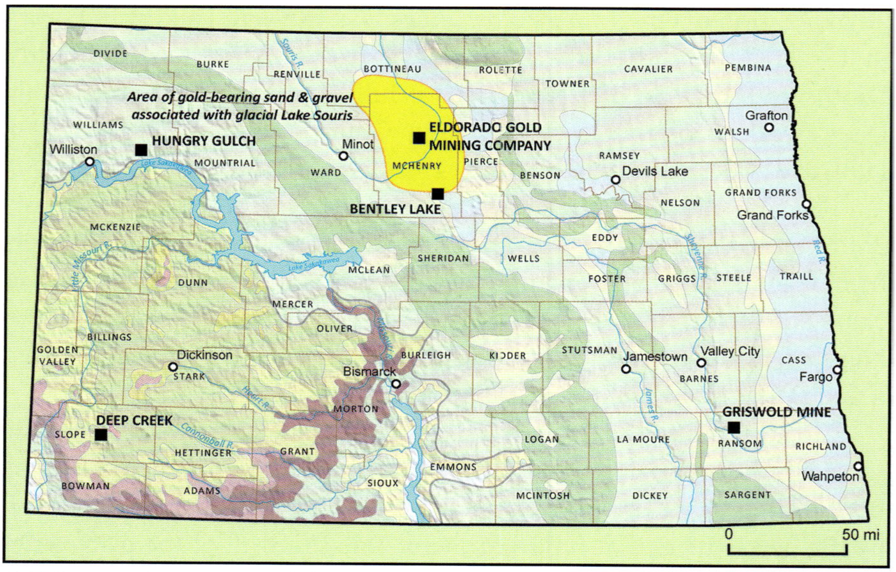

▲ This map shows some of the locations where gold has been reported in North Dakota. Each of these locations is discussed in the text. Of the locations shown on the map, gold was actually present at three of them (Griswold, Eldorado and Bentley Lake), one was a hoax (Hungry Gulch) and one was probably the result of an overactive imagination (Deep Creek). It is unlikely that North Dakota will ever be a major gold-producing state.

1883); "Gold Found in Dakota" (*New York Times*, October 21, 1883); "Reported Discoveries of Rich Deposits of Gold at Lisbon, Dakota" (*Chicago Tribune*, October 22, 1883); "'Lisbon's Wild,' Immense Gold Discovery Just Made" (*Grand Forks Herald*, October 27, 1883); "The Hegira for Lisbon" (*Grand Forks Herald*, October 26, 1883); "Gold Excitement Still Intense, and Lisbon Fast Becoming the Mecca of the Northwest" (*Grand Forks Herald*, November 2, 1883).

Griswold enlisted a group of Chicago investors, bought rock-crushing machinery, and their company immediately began processing ore in October 1883. Other speculators bought up land along the Sheyenne River, for at least 20 miles from Lisbon to Fort Ransom. Since the discovery was in October, only a small amount of time was available before winter shut down mining operations and, on January 8, 1884, the *Grand Forks Herald* reported that Griswold and his associates had "abandoned all further mining enterprises until spring." In April, Griswold resumed work with an ore-crushing machine to smash an accumulated 2,000 tons of raw ore. During the summer, though, the gold rush fizzled—the ore did not contain sufficient gold to mine.

When I mapped the geology of Ransom County in 1970, I visited the mine site and found a few pieces of rusted machinery that had been used in mining operations, but I am unsure if it had been used in the Griswold mine or by someone else attempting to mine at the time. I spent some time studying the apparent mine site and determined that any gold was contained in what is probably a cemented placer deposit, consisting of gravel that was deposited by a preglacial stream or river. The river probably had its origin in the Black Hills, the likely source area for the gold, and it must have flowed through the area sometime prior to 3 million years ago.

When glaciers later advanced over North Dakota, they completely buried the gold-bearing stream deposits with a thick covering of sediment. The modern Sheyenne River is currently cutting its way downward through the area, removing the covering of glacial sediment, so that, at the point where the Sheyenne River Valley crosses the route of the old preglacial river, the gold-bearing gravel deposit is being exhumed. The gravel is cemented by minerals, mainly calcium carbonate, deposited by ground water that long ago seeped through the gravel, resulting in a concrete-like conglomerate. Springs occur where the cemented gravel is exposed along the walls of the Sheyenne River Valley.

Stream gravels similar to those found near Fort Ransom are found along the Oahe Reservoir in Emmons County as well as on a few butte tops and ridges in northwestern South Dakota, in areas that were not glaciated. All of these gravel deposits were probably deposited by pre-glacial rivers, or complex systems of rivers, that flowed northeastward from the Black Hills, through the Fort Ransom area, in Miocene or Pliocene time.

According to an engineering report published in 1936, the gravels there assayed between $5 and $11 of gold per cubic yard, but I am unsure whether the assay was done at the time the mine was in operation or whether it was done as part of the 1936 study. The price of gold in 1884, when the Griswold Mine was operating, was $18.94 per troy ounce; it was $34.87 in 1936.

The idea of gold in North Dakota, and the hope that it might be found, is so persistent, that I have commented on the Griswold mine story several times through the years. The material here includes information included in an article, dated January 8, 2013, by Dr. Steve Hoffbeck of the Minnesota State University Moorhead History Department. Hoffbeck's article provided useful new information about the Griswold mine that I had not seen before.

Lake Souris–Denbigh Deposit

Since the early 1900s, gold has been known to occur in the sand and gravel deposits associated with glacial Lake Souris in McHenry, Bottineau and Pierce counties. The first discovery of gold that I know about in that area dates to May 1908, when mining claims were recorded near Denbigh in McHenry County. I do not know how much gold was involved in the Denbigh discovery, but L. E. Mills, D. B. Kaufman and others claimed mineral rights for placer gold on 240 acres three miles east of Denbigh. In August of 1908, they founded the Eldorado Gold Mining Company. Their articles of agreement stated that the participants would "cooperate and purchase a dredge or mining boat of sufficient capacity to do the work in the best possible manner." Several similar companies were formed in the area at the same time, but the search for gold was unsuccessful and all the companies soon went out of business.

The gold-bearing deposits in McHenry County were "rediscovered" in 1931 when reports from Chicago claimed that (presumably small) gold nuggets had been taken from the crops of turkeys being processed from the Denbigh area. True or not, as soon as this information became general knowledge, a new rush was on. Individuals and organizations filed placer claims and took mining leases until every available piece of ground was leased throughout a large area surrounding Denbigh and Towner.[43]

From what I can learn, it appears that no one conducted a systematic method of prospecting or sampling in the Denbigh area in the 1930s, but rumors affirming and denying the existence of gold in commercial quantities were widespread. Miners used a variety of methods of processing, such as jigging, tabling, sluicing, cyanidation and flotation. All operated unsuccessfully for short periods at various places throughout the area.

Where did the gold come from originally? The gravel in the glacial Lake Souris deposits was derived from rocks of the Canadian Shield to the north. The glaciers that transported the gold to north-central North Dakota flowed over what are today the Gods Lake, Flin Flon and Lac La Ronge mining districts of northern Manitoba and Saskatchewan. This part of the Canadian Shield is situated 600 miles north of the Denbigh area and, based on our understanding of the flow directions of the glacial ice, it is the likely location to have served as a source for the gold.

After the glacial sediments were dropped by the melting glacial ice, they were washed by streams of meltwater, which helped concentrate the gold by removing the lighter-weight materials. Finally, the gold-bearing sediments were deposited on the floor of glacial Lake Souris as undercurrent fans at the mouths of the Souris and other rivers.

In the summer of 1934, the School of Mines at the University of North Dakota obtained assistance from the Federal Emergency Relief Administration (FERA) and conducted a Gold Investigation project in the Denbigh area. Of approximately 1,500 samples that were assayed, 200 showed traces of gold. Forty samples had values ranging from \$0.17 a ton (a value corresponding to a gold content of 0.005 oz. per ton) to \$2.10 a ton (0.06 oz. per ton). These values are based on the then-current price of \$34.695 an ounce for gold; the samples with these values were scattered throughout 15 different plots of ground, and they failed to indicate anything of commercial importance in the area. As I write this, in May of 2015, gold is trading at $1,179 an ounce. At that price, the Denbigh gold-bearing gravel could range in value from $5.90 to as much as $70.75 a ton. Is that a high enough value to justify attempting to mine the gold commercially today, presumably using more efficient mining methods?

The particles of gold found in the Denbigh area were small, well-rounded, pitted and flattened grains. Their well-worn condition suggests that the gold had been subjected to stream erosion and weathering, similar to other placer deposits. The largest nugget found in the FERA survey was twice the size of a grain of wheat, and that, I believe, is still the single largest piece of naturally occurring gold ever found in North Dakota.

Bentley Lake Deposit

John Alden Bethke, of Drake, reported to Ms. Kay A. Buri that gold was once mined near Bentley Lake, three miles west of the town of Drake in McHenry County.[44] The lake at the time was known as Brush Lake, but the name was changed to Bentley Lake to avoid confusion with another Brush Lake near the town of Mercer, McLean County.

Mr. Bethke noted that the gold mined at Bentley Lake was known as flour gold, extremely fine-grained placer gold that is the consistency of flour. Flakes of gold of this consistency are known, too, as gold dust. An unsuccessful mine was attempted by T. E. Sleight at the same time that the Denbigh area gold was being mined, around 1908.

The gold that occurred in the Bentley Lake area likely had the same source as the gold found near Denbigh. The water that washed and sorted the glacial deposits and deposited sediments in glacial Lake Souris flowed southeastward out of Lake Souris, over the Bentley Lake area. As it flowed through the area, the water deposited gravel and sand, which included small amounts of finely divided gold. As in the Denbigh area, only small amounts of gold were recovered.

Hungry Gulch

I have mentioned three legitimate North Dakota gold "discoveries," none of them profitable. Other discoveries of gold besides those I have described have been reported through the years, as well as a number of scams designed to incite gold fever and separate people from their money in various ways. One of the more interesting scams happened in the fall of 1910 at Hungry Gulch, 20 miles east of Williston.

Hungry Gulch is a deep, wooded coulee leading to the Missouri River Valley. A spring-fed stream runs through the coulee. Hungry Gulch was owned by a couple of rancher brothers, Joseph and James Jackson. After they reported the discovery of gold-bearing rocks, the *Williston Graphic* carried a story headlined "Gold Discovered in Hungry Gulch." I will quote from a 1981 story in the *Grand Forks Herald* by Lloyd W. Sveen in his column, "Potpourri."[45]

> That's all it took to set off the Great North Dakota Gold Rush. Every able-bodied man who could possibly go . . . rushed to the scene, ready to make his fortune. They combed the thick, wooded maw of Hungry Gulch from one end to the other, and pounded innumerable stakes into the earth, setting forth the boundaries of their gold claims. The discovery of more gold-colored rocks added increased incentive and momentum.
>
> Most of the new miners did not bring shovels or picks. They stopped at the Jackson's homestead ranch near Hungry Gulch, and asked to borrow tools. Miners discovered the brothers had no tools available for the borrowing, but they did have tools for sale.
>
> A few days later, when miners began to run out of food, they discovered that the brothers at the ranch had fortunately stocked up on large quantities of side pork, flour and salt, all of which were for sale. Later when the miners thirsted for whiskey, they found that the Jackson brothers, luckily, also happened to have a sufficient stock of that for sale too.
>
> It went on like this for several days. Somehow the rancher brothers were able to supply all the needed tools, food and drink. They even had some old army tents and blankets . . . to sell.

> . . . A few days later, the assay office reported that the samples were not gold, but iron pyrite. Instantly, the gold rush collapsed, miners returned sheepishly home, and not a single further word was published about the episode in the *Williston Graphic.*

Summary

North Dakota has none of the rock types normally associated with economic gold-bearing deposits and it is unlikely that economic amounts of gold will be found here. Most of the gold that is known to occur naturally is distributed evenly through such a great amount of sediment that it has not been cost-effective to mine it. Even so, the fact that small amounts of gold have been found in the state has helped to maintain at least some interest through the years in prospecting for it. If the value of gold rises high enough, someone may figure out a way to extract it profitably. All of which is a way of saying, Never say never!

Aquifers and Artesian Flows

When precipitation falls on an upland area and infiltrates into the subsurface, it may accumulate in pores in buried sediments such as gravel, sand or sandstone. Such water-saturated permeable materials (permeable means water can flow through it) are referred to as aquifers. If the water-bearing material is at the land surface, it may be said that the water is contained in an unconfined aquifer. The top of the buried water surface is referred to as the water table. However, if the water-bearing sediments are overlain by a material that water cannot flow through easily, an impermeable material, the aquifer is confined. Aquifers, whether they are confined or unconfined, provide important sources of ground water for a variety of uses, ranging from irrigation to drinking water supplies.

Precipitation that falls over an area such as Turtle Mountain or the Missouri Coteau will infiltrate to some depth and tend to flow laterally through the subsurface, away from the main upland area. It seeps downward to lower elevations and comes out of the ground near the base of the Turtle Mountain or Missouri Escarpment. When the ground water reaches the surface, it evaporates, leaving a white, salty crust of minerals. Such salty accumulations are found along the base of the Missouri Escarpment in several places and at the bases of some of the larger buttes in western North Dakota. If sufficient water is present, the result may be swampy land, or perhaps an area of saturated sediment, quicksand. A long, narrow area of quicksand borders the southern base of Turtle Mountain, extending from Bottineau to Dunseith to Belcourt.

The kinds of situations I have described involve small areas, covering only a few square miles. Larger areas of natural springs, or seeps, are found in the Red River Valley of eastern North Dakota where slightly salty ground water is escaping to the surface. Kelly Slough, just east of Grand Forks, is an example. Lake Ardoch, between the towns of Ardoch and Warsaw, is another (see the discussion of Glacial Lake Agassiz in chapter seven).

Springs like Kelly Slough are the result of artesian flows. The water escapes from deeply buried aquifers, which are under hydrostatic pressure. If a well is drilled into the aquifer, the water will rise in the drill pipe, and if the pressure is sufficient, it can flow out the top of the well. Or, if the ground is porous enough, it may seep naturally to the surface over a broad area.

A simple example of an artesian system might involve precipitation entering the ground at a certain elevation, a recharge area, and flowing through an aquifer to a lower elevation, to a discharge area, where it escapes from the ground. The situation is analogous to putting water in a pipe on top of a hill and letting it run through the pipe to the bottom of the hill.

Several years ago, a person from Hope, North Dakota, sent me an article from the July 30, 1896, issue of the *Hope Pioneer* newspaper. It describes "that famous Mud Springs, in the Sheyenne bottom." The mud spring is "about 15 miles south of Hope, across on the other side of the river." As nearly as I can tell, that would place it beneath Lake Ashtabula today. According to the article, the spring "has been sounded fifty feet down, and no bottom found," and goes on to say: "During the summer months it is full of nice clean ice below the mud, which is about four feet deep and very sticky with pure but salty water below the mud."

The writer for the *Hope Pioneer* does not offer an explanation for the Mud Springs, but simply describes them. Through the years, I have heard of a number of deep or "bottomless" pits in various places, mainly in eastern North Dakota. Most are located where a well was drilled to the Cretaceous Dakota Formation, an artesian aquifer that underlies much of North Dakota. Large numbers of wells were drilled to the Dakota Formation in the late 1800s in search of water supplies. Many of these wells flowed salty water to the surface and, in some of them, the water was initially under considerable pressure. Sometimes it was impossible to control the flows from the wells after they were drilled and they continued to flow indefinitely.[46] Some of these old wells still flow, although at much slower rates than when they were drilled. Because water flowed continually upward from these wells, they gradually broadened into deep pools at the surface. They are deep, but not bottomless.

The *Sheldon Enterprise* newspaper told a story of how drillers lost control of a well (the Sheldon "Outlaw Well") in Sheldon, in eastern Ransom County, in April of 1906:

> [W]hen the well came in, an unbelievable amount of water came up from the outside of the casing. John Dahm was at the well site when the well came in. John stated that the water shot 150 to 200 feet into the air and mushroomed out. He said that you could work around the well and actually stay perfectly dry since the water formed an umbrella-type effect.
>
> Since they were unable to force most of the water up through the drill pipe, it was then decided to force the next larger size casing. It was imperative that the water be forced through the drill pipe, knowing that this huge excess amount of water coming up from around the outside of the casing had to be stopped or it would wash away the entire well site in a short period of time. The drilling company made two more attempts with larger pipe coming in via the railroad to force this water inside. It finally had to be abandoned. At this time, a ditch was dug measuring approximately 13 feet wide by 4 to 5 feet deep, from the well site to the Maple River, taking advantage of the existing low-lying areas on the way to the river. The ditch was necessary so the water would be drained from the area.
>
> After the well site was abandoned, there was great concern amongst the townspeople as to whether this hole would continue expanding underground causing concern for the safety of the town itself.

However, according to a book published for the Sheldon Centennial in 1981, after a fire in the town around 1948, a considerable amount of water was drawn from the hole; and according to Loval Good, a resident of Sheldon, it was discovered that the hole tapered inward rather than branching out, consequently eliminating any danger of the surrounding area collapsing. The well is still running today from an open hole."[47] A lawsuit ensued:

> The suit against the Village of Sheldon instituted by a number of the neighboring farmers north of town claiming damages owing to the overflowing of their land due to the artesian well, will be amicably settled at the present term of court. A conference was held last week between the plaintiffs and the Village

> Board at which an agreement was reached as to the amount of damages to be paid. So all that is now required is to have the court pass upon it in a formal way.
>
> – The *Sheldon Progress*, May 8, 1908

Large numbers of wells were drilled to the Dakota Formation aquifer in eastern North Dakota during the late 19th and early 20th centuries. The wells were allowed to flow freely for so long that the pressure in the aquifer was depleted and little danger remains of drilling a runaway well today. Even so, some water continues to seep upward naturally in areas such as Kellys Slough in Grand Forks County and Lake Ardoch in Walsh County. Areas of white, salty deposits can be seen in such places during periods of dry weather. Where the salty water reaches the surface, salt-loving plants (halophytes) are common. Some of these include salt grass, alkali grass, cordgrass and salt wort. The red plants that cover the surface in many areas where artesian water is discharging are salt worts (*Salicornia rubra*). The presence of halophytes is an indicator for salty soil.[48]

CHAPTER SIX

THE DEVILS LAKE BASIN

Introduction

Devils Lake covers approximately 300 square miles within a 3,810-square-mile basin. This is a closed subbasin that lies entirely within the Red River of the North drainage basin.[49] The area does not drain to the Red River; rather, all of the land in the subbasin drains to Devils Lake itself. Water from the Devils Lake Basin flows to the Red River only if Devils Lake gets high enough to overflow to the Sheyenne River.

The original land survey of the Devils Lake area was conducted in 1887. However, when Joseph Nicollet explored the Devils Lake area 50 years earlier, in 1839, he already discerned the character of the region. He noted, for example, that the area between Devils Lake and Stump Lake consisted of deep coulees that "are dry at the moment, but in time of high water seem to receive water from Devils Lake." When Nicollet first saw the area, Stump Lake stood between 1,420 and 1,425 feet above mean sea level. His comments imply his belief that water had flowed from Devils Lake to Stump Lake sometime between 1820 and 1839. Nicollet was correct, and when Devils Lake reached 1,447.1 feet above sea level in the summer of 1999, water from Devils Lake began to trickle into Stump Lake for the first time in 175 years. Stump Lake continued to rise so that it is now at the same level as Devils Lake and may be considered to be a bay of Devils Lake.

In recent years, weather patterns in North Dakota, including the Devils Lake area, have resulted in precipitation heavier than the long-term average (dating to 1867, when Devils Lake's surface elevation was first measured). Since 1990, the area has been substantially wetter, with the years 1990 through 2009 ranking as the wettest 20-year period in a century and a half of recordkeeping.

Since 1867, the level of Devils Lake has fluctuated between a minimum of 1,400.9 feet in 1940 and a maximum of 1,454.3 feet in 2011.

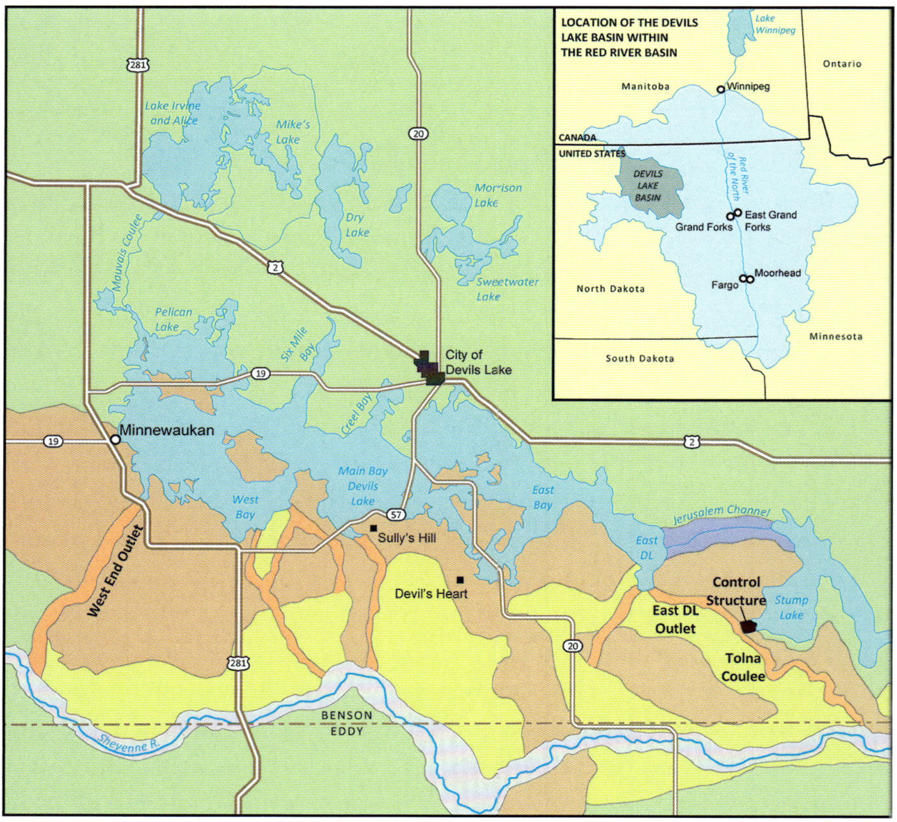

Were it to reach a level of 1,458 feet (and in the absence of a control structure), Devils Lake would spill through the Tolna Coulee into the Sheyenne River, a tributary of the Red River.[50]

Late Preglacial History

Devils Lake is largely a product of glaciation, but the geologic events that set the stage for the lake began long before the first glaciers advanced into North Dakota. During millions of years, a variety of geologic events culminated with the Ice Age and the formation of the

◀ This generalized geologic map of the Devils Lake area shows the several bays of the lake (blue), ice-thrust topography (tan), glacial outwash (yellow), natural outlets from the lake (narrow orange areas), the recently completed control structure at the southwest end of the Stump Lake Bay of Devils Lake (black), the Sheyenne River Valley and its flood plain (gray area on either side of the river), and areas of ground moraine (green), which surround the Devils Lake area.

Each of the several bays that comprise Devils Lake is a flooded depression that formed as an ice thrust. All of the bays directly overlie the Spiritwood Aquifer. The advancing glacier pushed materials south-southwestward from the bays to areas south of the lake, resulting in rugged, hilly topography. When the early lake (Glacial Lake Minnewaukan) flooded and overflowed, the overflowing water cut valleys (shown as the orange strips) in the ice-thrust masses. Several outlet valleys formed. South of the ice-thrust masses, broad plains of gravel and sand (yellow areas) built up until the overflow water from the lake coalesced into valleys that served as outlets for the water (West End Outlet and Tolna Coulee). The last outlet valley to form was Tolna Coulee.

When the earlier outlet channels formed, the rapid flow of water through them carried a heavy load of gravel and sand, which it deposited south of the ice thrusts as outwash plains (yellow areas). Later channels were incised in deeper valleys, and these carried their loads farther south, some of them discharging into the Sheyenne River.

North of U.S. Highway 2, lakes Irvine, Alice, Mike's, Dry, Morrison and Sweetwater are the result of minor, shallow ice-thrusting, which occurred following the Devils Lake thrusting. Numerous, similar shallow thrust depressions occur to the north of the area shown, mainly in Towner County.

The inset map shows the location of the Devils Lake Basin within the Red River of the North drainage basin. The Devils Lake Basin becomes an active part of the Red River Basin when Devils Lake overflows into the Sheyenne River.

lowland that is now the Devils Lake Basin.

Before it was glaciated, the landscape in the Devils Lake area was much different than it is today. Erosion during the long Tertiary Period, beginning 65 million years ago and intensifying 15 million years ago, had shaped the Pierre Formation shale surface of eastern North Dakota into a gently rolling plain, well drained toward the northeast.

Two rivers drained the Devils Lake area shortly before the Ice Age began. The Grand River flowed from South Dakota northward through McIntosh, LaMoure, Barnes, Griggs and Nelson counties to the Devils Lake area, where it joined another river near Churchs Ferry. This second river was probably the downstream extension of the combined preglacial Cannonball, Heart and Knife rivers, a river system that drained half of North Dakota. The Missouri River Valley, as we think of it today, did not yet exist. The combined Cannonball/Heart/Knife and Grand rivers flowed northeastward, all of them far beyond the points where they today enter the modern Missouri River (see the discussion of the Missouri River in chapter three).

The merged river flowed northward past Churchs Ferry into Canada just east of Turtle Mountain and probably joined the Assiniboine River in Manitoba. The ultimate destination of these combined rivers was the Labrador Sea off the northeastern coast of Canada. An awareness of all of this drainage history is important because the routes these rivers followed through the area prior to glaciation played an important role in determining the eventual development of the Devils Lake Basin.

Glacial History of the Devils Lake Basin

Glaciers have covered the Devils Lake area at least a half-dozen times during the Ice Age. Every glacier, as it has flowed over the area, has deposited its load of debris. The earliest glaciers laid their sediments on preglacial bedrock and the later ones deposited new layers on top of existing glacial sediment. Multiple glaciations have resulted in a cumulative thickness of glacial deposits averaging from 50 to 150 feet over much of the Devils Lake region. The maximum thickness, measured by test drilling, is more than 600 feet in a deep, buried valley in northern Ramsey County. Water flowing from the melting glaciers deposited gravel in the ancient Cannonball and Grand River valleys. These thick layers of gravel, now buried beneath the youngest glacial sediment, serve as aquifers providing important ground water resources.

The Spiritwood Aquifer, the largest buried aquifer of glacial origin in North Dakota, is a several-hundred-mile-long body of buried gravel and sand. It extends northward from South Dakota, past Valley City, beneath the town of Spiritwood (hence the name), to near Tolna in Nelson County, where it turns toward the northwest and passes directly beneath the chain of bays that comprise Devils Lake. At Churchs Ferry, the buried aquifer turns north again and continues past Turtle Mountain into Canada. The Spiritwood Aquifer exists in a complex network of buried, interconnected, preglacial and glacial valleys.

Events preceding the latest glaciation set the stage for the formation of Devils Lake. When the Late Wisconsinan glacier re-advanced, flowing over the Devils Lake area 12,000 years ago,the area was already covered by thick deposits of older glacial sediment.[51] The glacier flowed over a tundra landscape, an area of permafrost, land that was frozen to a depth of tens to hundreds of feet.

As the glacier advanced straight southward through Ramsey, Nelson and Benson counties, it overrode the Spiritwood Aquifer. The ground water in the aquifer became trapped, a result of being confined by a combination of 1) the underlying shale, 2) the surrounding glacial sediment and shale, 3) a covering of frozen ground (permafrost ahead of the advancing glacier), and 4) the covering of advancing glacial ice. Possibly, the water in the aquifer was itself partially frozen. The combination of these circumstances made it impossible for the ground water to escape to the surface. The great weight of the advancing glacier raised the ground water pressure in the materials beneath and ahead of the glacier (see additional discussion of glacial thrusting in chapter four).

Alternating layers of permeable and impermeable sediment (sand, silt, gravel, clay, etc.) beneath the glacier caused pore-water pressures at depth to build to high levels directly below the base of the glacier sediment and well down into the underlying shale. The only direction anything could "give" to relieve this pressure was upward. Beds of glacier sediment and preglacial shale were forced upward beneath the leading edge of the glacier and into the path of the moving ice.

The glacier shoved these beds a few miles to the south, in the process excavating a lowland where the ice-shoved materials had originated. As soon as it had shoved the materials southward, the glacier

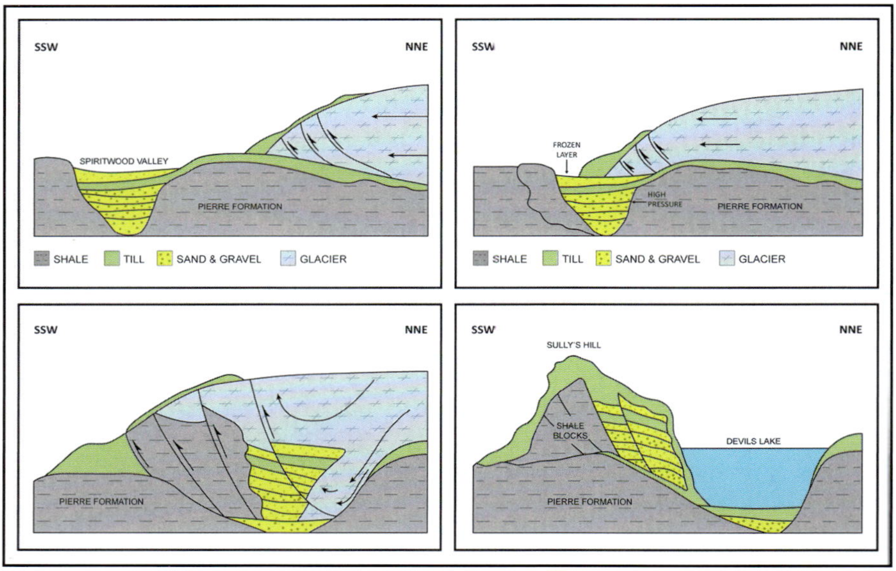

▲ This four-part cross-section diagram shows how the Devils Lake basin and the range of hills to the south of the basin formed. The upper-left diagram shows the glacier advancing from north-northeast to south-southwest, toward the gravel-filled valley that contains the Spiritwood Aquifer.

As the glacier continued to advance (upper-right diagram), pressure built up in the ground water in the aquifer because the glacier sealed off the top of the aquifer, preventing water from escaping. When this happened, thrusting began in the shale ahead of the glacier.

The continued advance of the glacier (lower-left diagram) resulted in massive thrusting in both the shale and aquifer sediments (gravel and sand). The thrusting resulted in the excavation of a broad area and movement of the excavated materials southwestward.

The final diagram (lower right) shows the excavated depression resulting from the thrusting flooded by the newly formed Devils Lake, along with prominent hills, including Sullys Hill and others, to the south of the Devils Lake depression. The hills south of the lake include large amounts of Pierre Formation shale that were moved by the glacier.

stopped advancing. The halt in the advance of the glacier was probably unrelated to the thrusting. Rather, the glacier stopped advancing simply because of changing climatic conditions. Had the climate not warmed at that time, halting the advance of the glacier, the ice might have continued to flow southward, overriding and filling in the lowland. Devils Lake would not have formed or, perhaps, a much shallower lowland would have remained in the area.

It was the ice thrusting in the Devils Lake area that formed the lowland depression now flooded by the lake, as well as the range of hills to the south of the lowland. Slabs of glacier sediment and pre-

▲ Ice-thrust material, mainly Pierre Formation shale, till and glacial gravel, was crushed and mashed during the process of thrusting that emplaced the hills south of Devils Lake and excavated the Devils Lake Basin. This exposure was made during construction work at the Spirit Lake Casino, Benson County, in 2004.

The chunks of shale are angular and were not transported by running water. It is likely they were scraped up by the ice when it gouged the underlying shale surface. The shale was ground between the glacier and underlying bedrock surface. JPB Photo: 10/21/2004 glacial shale beds, stacked like fallen dominoes, form the hills. Even though the slabs consisted mostly of soft, incompetent (loose) materials, some of them remained largely intact during the thrusting event. They were probably frozen when the ice moved them or else they would have fallen apart. The highest elevations and the most rugged topography in the hills immediately south of Devils Lake are at Sully's Hill, where slabs of shale, interbedded with glacier sediments, are exposed in several places. Bordering the range of ice-thrust hills along their north side is a 50-mile-long, east-west source depression, not a single feature, but a string of several poorly connected ice-thrust depressions now flooded by the several bays of Devils Lake. The several depressions may have formed as separate ice thrusts, but they all formed at nearly the same time.

When the glacier thrusting occurred, large volumes of sediment-laden water escaped rapidly to the surface through the source depressions where the ice-thrust materials had been. This escaping water carried large amounts of gravel and sand upward through cracks in the glacier in some places and then to the edge of the ice sheet. Thick layers of outwash gravel and sand accumulated south of the glacier in places like Warwick and New Rockford. Devils Heart, near Tokio, may have formed when the gravel and sand that had been carried to the surface of the glacier by escaping ground water slumped down, forming a large cone-shaped hill, a kind of "sand volcano" (see the discussion of veblens in chapter four).

Summary of Circumstances Resulting in the Formation of the Devils Lake Basin

1) The ground ahead of the advancing glacier was probably frozen, although this may not have been a critical factor. Other geologic circumstances might have interacted to provide the alternating permeable and impermeable beds required for thrusting.
2) The glacier advanced across (not parallel to) the large, pressurized Spiritwood Aquifer.
3) Pressurized ground water forced subglacial materials up into the path of the advancing glacier, which then shoved them a few miles.
4) The glacier stopped advancing immediately after it had moved the ice-thrust materials.

The Devils Lake Basin would not have formed had the glacier not advanced over the pressurized Spiritwood Aquifer. Without the aquifer, thrusting could not have occurred. Furthermore, the presence of the buried aquifer continues to be a factor in the modern behavior of the lake.

Fluctuations in the Level of Devils Lake

A lake formed ahead of the ice, in the newly formed basin, as soon as the glacier margin began to recede from the ice-thrust hills south of the source basin. The lake was dammed between the glacier on the north and the ice-thrust hills on the south. Geologists have named the lake Glacial Lake Minnewaukan (see discussion in chapter four). As soon as it formed, Lake Minnewaukan began overflowing through Seven-Mile Coulee southeast of Fort Totten. As the glacier receded, the level of Lake Minnewaukan dropped quickly as the lake spilled southward through several other outlets, including the Stony Lake-Long Lake channel, the Twin Lake channel and, most recently, the Tolna Coulee.[52]

Throughout its history, since the Devils Lake Basin formed, Devils Lake has risen and fallen between two extremes: 1) a dry lake bed and 2) a full lake, overflowing into the Sheyenne River drainage at times. The accompanying time/event diagram depicts a generalized history of the fluctuations in the level of Devils Lake since it formed. During the early centuries following its origin, frequent changes in the climate caused the level of Devils Lake to fluctuate within a considerable range every few hundred years. The lake likely dried up completely by 8,300 years ago. Since it formed, Devils Lake has overflowed to the Sheyenne River several times. It most recently spilled into the Sheyenne River between 1,800 and 2,500 years ago.

Devils Lake has risen high enough to overflow into Stump Lake numerous times, certainly more often than my diagram shows, because I cannot identify every intermediate high-water level. Succeeding high-water levels obliterate most of the earlier high-water marks. Prior to its current overflow, Devils Lake most recently overflowed into Stump

Lake in the mid-1820s. Sometimes when the lake rose high enough to overflow into Stump Lake, it probably continued to rise several feet higher, but not necessarily high enough to spill into the Sheyenne River. I do not know how often this happened.

Devils Lake dried up frequently between 8,500 and 4,000 years ago, a time known as the Altithermal Period, when the climate was

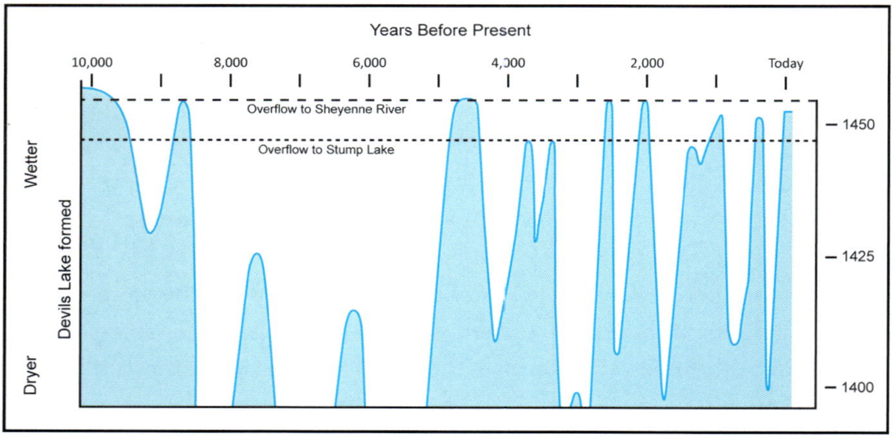

▲ This time/event diagram shows how the level of Devils Lake has fluctuated since the lake formed 10,000 years ago. The vertical axis shows the relative level of the lake and the horizontal axis represents time since the lake formed. The lake level is constantly changing due to climatic fluctuations that have affected all of North Dakota since the end of the Ice Age. Higher levels, including times when the lake overflowed, correspond to times of moist climatic conditions, and lower levels mark times when the climate was drier. Farthest to the right is the present condition, with the lake level rising toward overflow.

The diagram is necessarily generalized, reliant on the availability of accurate (or any) data. It becomes more accurate nearer the present (to the right), less so longer ago. During the past 3,000 years, Devils Lake overflowed to the Sheyenne River and out of its basin twice. It also overflowed at least once 5,000 years ago and probably several times between 9,000 and 10,000 years ago. During a prolonged period of warm, dry climate, between 8,500 years ago and 4,500 years ago, I have shown the lake bed as being dry. It is likely, though, that the lake existed sporadically during that time (data range from sparse to absent for the period), but the climate was generally dry.

I drew the first version of this diagram for Devils Lake in about 1980. I have since revised it several times, but the overall idea remains the same: Devils Lake has fluctuated between dry lake bed to overflowing several times during its history.

warmer and drier than it is today. The closest the lake has come to completely drying up in recent times was during the 1930s and 1940s, when it shrank to a small pool a few feet deep. It dropped to its historic low in 1940, when it stood at a level more than 50 feet lower than it does now, in 2014. At the 1940 level, the lake covered 14 square miles, compared with its current 295 square miles.

Since it formed, Devils Lake has *always* been either rising or falling, never remaining long at the same level.[53] Rising or falling is the lake's natural behavior, the result of constant climatic cycling and the closed-basin nature of the topography. Except for the past 130 years or so, since European immigrants settled the area, the lake, and drainage to the lake, were not subject to any kind of manipulation such as farming practices, wetland drainage, road construction, diking or any other human activities. The lake's overall behavior has always been driven, and continues to be driven, by changes in climate. Without any kind of engineered solution, such as an artificial inlet or outlet (depending upon the immediate concern), Devils Lake would continue to rise and

▲ A levee was built at Minnewaukan to help protect the western shore of Devils Lake from erosion. The shoreline is riprapped with glacial erratics. JPB Photo: 9/13/2011

fall. Without intervention, the lake could rise until it overflowed, already to Stump Lake, and eventually to the Sheyenne River. Or it might reverse its trend and begin to drop.

Puzzling Behavior of Devils Lake

Devils Lake does not always seem to react logically to short-term climatic conditions. The lake sometimes keeps rising, even after the area has entered a period of drought. Sometimes it falls, even during wet years.

Barring heavy withdrawals through wells, the amount of water in the Spiritwood Aquifer, or any other aquifer, is determined entirely by the same climatic cycles that affect Devils Lake. Aquifers, however, respond to precipitation and drought much less quickly than does water in rivers and lakes at the surface. Some deep aquifers seem to respond during periods of centuries to changes in climate, others after only a few years.

In 1982, when I was mapping geology in the Rock Lake area of Towner County, I noticed a large number of springs in places immediately overlying the Spiritwood Aquifer. Farmers there told me this is not always the case, but that year followed an extended period of heavy annual precipitation. The same thing was happening in other areas overlying the aquifer, indicating that the aquifer was full to capacity and leaking water to the surface, resulting in springs. This was true during much of the first decade of the 21st century as well.[54]

Periodic interaction of the water in Devils Lake with the ground water in the Spiritwood Aquifer may explain why the lake seems to misbehave at times. During times of heavy precipitation, the aquifer could become pressurized, forcing water to transfer from the aquifer to the lake. Conversely, when the aquifer is not full to capacity (not fully charged), and therefore not under pressure, water may pass downward from the lake directly into the aquifer.

After a prolonged wet spell has completely recharged the aquifer, water seems to escape upward to the ground surface and to the lake. The amount of escaping ground water is noticeable only when the lake is at low levels. Even though ground water may move from the aquifer to the lake when the lake is high, transfer of water under those conditions might not have a perceptible effect on the lake level because so much more water is needed to raise the lake level. Even so, at high levels, any amount of additional water can be an important consideration as it will keep the lake from falling during short-term drought conditions, compounding the flooding problem.

Similarly, the lake level sometimes tends to fall during a wet cycle, or at least at the beginning of a wet cycle. Following a prolonged drought, the aquifer is only partly full, and it is not pressurized, so the lake is draining into the aquifer. As a drought ends and precipitation increases, the lake should logically begin rising immediately, but instead it falls for a while, until the aquifer becomes completely charged. Following that, the lake will rise normally as long as the wet cycle continues, or it might rise slightly more quickly than "normal" as it receives additional water from the aquifer.

Discussion

As I write this in September of 2014, Devils Lake stands at 1,452.6 feet, down from its recent high of 1,454.7 feet. Discussion centers on how much the lake might still rise. Barring human intervention and completion of an engineered solution to the current high-water problem, the ultimate limit to the amount the lake can rise is set by the elevation of the threshold of the Tolna Coulee: 1,458 feet.[55] Apart from this limiting factor, it has been suggested that the potential for further rise should be somewhat self-limiting. Increasing amounts of water are required for each additional foot of rise because the lake is so much larger and contains a much greater volume of water at higher levels. For example, much less water is required to raise the lake level from 1,425 to 1,426 feet than from 1,450 to 1,451 feet because the lake is much larger at the higher level.[56]

Notwithstanding my comments in the preceding paragraph, Devils Lake has often risen to levels higher than today, and the current (2014) level of the lake is evidence enough that the amount of precipitation in recent years has been sufficient to cause the lake to rise to its current level (and higher). Given sufficient precipitation, the lake could rise essentially any amount in a given year (up to 1,458 feet). If the Devils Lake drainage basin continues to have above-average snow and rainfall for several years, the lake will overflow into the Sheyenne River by way of the recently completed outlet through Tolna Coulee.

Total annual inflows of runoff to Devils Lake ranged from near zero during the drought years of the 1930s to 400,000 acre-feet in 1993. Inflows averaged 65,500 acre-feet annually between 1950 and 1993, but rose to 317,000 acre-feet annually between 1993 and 2000, a five-fold increase. The years 1993 through 1995 contributed 24 percent of all inflow to Devils Lake between 1950 and 1995.

Although Devils Lake can rise substantially in any given year, it is unlikely to fall more than 2½ feet in a year because the main way the lake loses water is by evaporation. Evaporation amounted to 86 percent of total inflow in 1986 and 73 percent in 1987. In 1988, evaporation loss exceeded total inflow by more than 200 percent, resulting in a net lake-water loss of approximately 100,000 acre-feet.

The average annual evaporation from Devils Lake is 34 inches. The most water that Devils Lake has lost in a season to evaporation in recent years was in 1976 and 1988. During those years, between 40 and 45 inches of water evaporated off the lake surface, but the level of the lake did not fall by that amount. The lake is unlikely to fall more than 2½ feet in any given year because evaporation is partially offset by precipitation and inflow into the lake.

Summary

a. Fluctuations in the level of Devils Lake are caused entirely by climatic changes, long-term cycles of wet and dry weather. These cyclic fluctuations are inevitable and they can be extreme. Had an outlet structure not been built, the level of the lake could continue to fluctuate as it always has.
b. The actions of humans have not been the defining factor in determining the behavior of Devils Lake. The lake rose and fell, often and dramatically, long before European settlers arrived on the scene and began altering the landscape. Agricultural practices, drainage of wetlands, road construction and other modifications are not responsible for the *overall* behavior of the lake.[57] The impact of draining wetlands in the Devils Lake Basin has been debated for a long time. Some people contend that such draining has little or no effect on the level of Devils Lake, whereas others believe that wetlands drainage is

almost totally responsible for the high lake levels today. Even those in the middle are unsure whether restoring drained wetlands would result in much lowering of the lake level.

c. The natural condition for Devils Lake is to be either rising toward overflow or falling toward a dry lake bed. The lake should not be expected to maintain a stable level or to remain long at any given level. This is purely a consequence of climatic cycling. Such cycling is long term insofar as humans experience it, but short in terms of geologic time. Devils Lake cycles between dry and overflowing in terms measurable in human lifetimes or shorter periods of time.

CHAPTER SEVEN

THE RED RIVER VALLEY

Glacial Lake Agassiz

William Keating, a geologist attached to the U.S. government's first expedition to the Red River Valley in 1823, provided the earliest geological report on eastern North Dakota. Among his observations, Keating wrote: "The flatness of the surface that almost uniformly prevails throughout the Valley may be regarded as a defect in its character that cannot be easily remedied."[58]

It could be argued whether the flat surface in the Red River Valley is, in fact, a "defect," but you have to admit it is a characteristic that would be hard to remedy! Keating observed, or at least suggested, that the entire region had been covered by water: "The whole of the country may be considered to be an immense lake, interspersed with innumerable barren and rocky island."

Keating was apparently the first person to anticipate the discovery of an ancient glacial lake. Twenty five years later, in 1848, another early geologist, David Dale Owen, realized that the Red River Valley had been formed by a large glacial lake. However, the existence of Lake Agassiz was not proven until Warren Upham surveyed its shorelines in

1881.[59] It is unlikely that either Keating or Owen had any idea of the origin of the body of water to which they alluded. The very idea of an Ice Age, to which the lake owes its origin, was not proposed until 1840 by Louis Agassiz.[60]

Glacial Lake Agassiz was the latest in a series of proglacial lakes that formed in the Red River Valley several times during the Ice Age. Lakes formed ahead of the ice sheets each time they spread southward out of Canada, blocking north-draining rivers, and they formed again each time the ice retreated. Buried, alternating layers of glacial and lake sediments in the Red River Valley attest to the fact that the valley was repeatedly affected by glaciers, lakes and rivers. Sediments deposited in earlier versions of Lake Agassiz are today buried beneath younger ones. In some places, earlier lake deposits were partially removed by later glaciers.

Like its namesake, Louis Agassiz, the father of glacial geology, Lake Agassiz's influence was felt far and wide. Evidence of Lake Agassiz occurs over roughly 365,000 square miles, an area five times the size of North Dakota, although the lake never covered such a large area at any one time. Ice-margin positions and lowering of outlets by erosion combined to limit the size of the lake at any given time.

Based on an assumed average depth for Lake Agassiz of 300 feet, I think Lake Agassiz might have contained as much as 20,000 cubic miles of water at its maximum. In comparison, Lake Superior, the largest and deepest of the modern Great Lakes, with an average depth of 500 feet, contains 2,900 cubic miles of water.

Evidence for glacial Lake Agassiz is found principally in:

1) its stratigraphy, the layers of flat-lying lake sediment near the surface
2) its border of shoreline features (beaches, wave-cut scarps, etc.)
3) its flat topography

History of Glacial Lake Agassiz

The history of glacial Lake Agassiz involves fluctuating water levels, recorded by beaches and wave-cut scarps that correspond to short periods of lake-level stability. Changes in the lake level were caused by

▲ This map shows the approximate total extent of glacial Lake Agassiz (light blue area, although the lake never flooded the entire colored area at any one time). During the time the southern part of the area in North Dakota and Minnesota was flooded, beginning 11,700 years ago, the northern regions in Canada were still covered by glacial ice. By the time the glacier in northern Manitoba and Ontario had melted, and the land was flooded, parts of the southern portion of the lake plain had already drained.

Most of the area that Lake Agassiz flooded in Canada (north of Winnipeg) is a wave-washed plain with little lake sediment. The main area of thick lake deposits is along the North Dakota-Minnesota border.

The map also shows the locations of several smaller glacial lakes that existed at various times. These lakes were not all contemporaneous with one another, or with Lake Agassiz. Many of them existed prior to the formation of Lake Agassiz.

a variety of circumstances. Alternating advance and retreat of glaciers caused changes in the water level, as did opening and closing of different overflow outlets, rates of glacial melting, periodic catastrophic inflows from meltwater floods, and postglacial rebound of the Earth's crust when the glaciers melted.

Lake Agassiz began to form when the glacier margin at the southern end of the Red River Valley retreated to the north of the drainage divide between the Hudson Bay and Mississippi River drainage basins between Ortonville and Browns Valley, Minnesota. The lake that formed ahead (south) of the retreating glacier was small at first and overflowed southward through the Browns Valley outlet, which separates Big Stone Lake and Lake Traverse, both on the Minnesota-South Dakota state line. The outflow created a giant river, referred to as the Glacial River Warren. Today, Big Stone Lake and Lake Traverse flood part of the valley that was carved south of the outlet by the Glacial River Warren. The Minnesota River flows in the valley, but its flow is only a small fraction of the volumes of water that once flowed in the Glacial River Warren.

LAKE AGASSIZ I

Glacial Lake Agassiz existed twice in most of the North Dakota Red River Valley. The first phase of flooding has been referred to as Lake Agassiz I and the second as Lake Agassiz II. The earliest flooding along the edge of the receding glacier in the Red River Valley resulted in land being inundated at the southeastern-most corner of North Dakota as early as 13,800 years ago. By 13,500 years ago, the edge of the glacier had probably melted back so that no active glacier remained in the Red River Valley south of the Canadian border. However, the glacier re-advanced prior to 13,000 years ago, likely to a point midway between Grand Forks and Fargo. When that happened, Lake Agassiz flooded the Red River Valley to the Herman level, the highest continuous shoreline (the shoreline gets its name from the town of Herman, Minnesota). While the lake was at the Herman level, the glacier still occupied the central part of the northern Red River Valley in North Dakota, and the lake flooded the area between the ice and the higher land to the west.

As the edge of the glacier receded northward, the lake south of it expanded, flooding the Red River Valley. During that time, erosion at the south end of the lake lowered the threshold of the Lake Traverse outlet so that the lake level gradually fell. By 11,700 years ago, the erosion of the southern outlet had lowered the threshold to a granite floor that resisted further erosion. From then on, the lake never rose above that level, known as the Campbell level.[61] The Campbell is the longest, continuous shoreline of Lake Agassiz and it, too, is named for a Minnesota town.

By 11,500 years ago, the North Dakota and Minnesota shorelines were several hundred miles long and lined by forests of spruce and deciduous trees interspersed with sagebrush-dominated openings. I have found caribou and bison bones and mammoth teeth in the beach sands along the uppermost (oldest) shorelines of Lake Agassiz. As early as 11,500 years ago, nomadic bands of humans may have hunted these and other game. I have found fish (perch) fossils in Lake Agassiz sediments. Other species of fish probably inhabited the lake, too, and perhaps people living along the lake utilized the fish for food.

By 10,000 years ago, grassland had replaced much of the forest area and, when that happened, bison moved in and became an important source of game for early hunters. With the change to grasslands, the Paleo-Indian population living along the shores of Lake Agassiz increased.

Lake Agassiz I persisted until 10,900 years ago, when the ice sheet in northwestern Ontario deteriorated to the extent that lower routes, to the east, were cleared for Lake Agassiz to drain into Lake Superior. The lake then drained from the North Dakota part of the Red River Valley. From 10,900 until 9,900 years ago, Lake Agassiz was restricted to areas north of the Canadian border, and perhaps the northeasternmost corner of North Dakota.

LAKE AGASSIZ II

The glacier re-advanced in Ontario 9,900 years ago, blocking the eastern outlets, and the lake again flooded the Red River Valley to the Campbell level. This resulted in the formation of the second Lake Agassiz, which lasted until 9,300 years ago, when the glacier melted from the eastern outlet, allowing the lake to drain permanently from the North Dakota Red River Valley.

Beginning 9,300 years ago, the rate of deglaciation accelerated. The glacier margin damming Glacial Lake Agassiz on its southern margin retreated northward at a rate of 1,000 feet a year (a mile every five years). By sometime between 9,000 and 8,500 years ago, Glacial Lake Agassiz had completely drained from North Dakota.

▲ This exposure of Lake Agassiz sediment is at Mayville in Traill County. The lake sediments are evenly banded, consisting of 6- to 8-inch-thick layers of light brown silt separated by layers of dark clay ¼ inch to ½ inch thick. The light-colored silt layers are summer deposits and the dark layers are winter deposits. At this location, approximately 20 years of lake history are recorded.

Bands of this kind, found in glacial lake sediment, are known as *varves* (a Swedish word for "layers"). A glacial varve normally includes two distinct layers: (1) a lower "summer" layer of relatively coarse-grained, light-colored sediment (usually silt or sand) produced by rapid melting of glacial ice in the warmer months and delivered by streams to the lake; and (2) an upper, thinner winter layer consisting of fine-grained, often organic, dark clay that was deposited slowly in quiet water while streams that flowed into the lake were ice-bound and not delivering sediment to the iced-over lake.

The 2-foot-diameter boulder must have been rafted (by an iceberg) from the edge of the glacier to the north. It dropped into the lake at this point and sank into the soft lake-floor sediment, contorting the light and dark bands at its edge as it pushed them downward.

I took this photo in September 1966. JPB Photo: Scan of 1966 slide

As the glacier receded, Lake Agassiz expanded in Canada, even as it was draining from North Dakota. When the glacier margin finally receded far enough so that it no longer formed a dam holding Lake Agassiz in place, the lake drained quickly northward into the Tyrrell Sea, which by that time (8,300 years ago) inundated the Hudson Bay Lowland in Ontario.

Surface Features

Over most of its area, the Lake Agassiz plain is essentially level, the result of flat-bedded sediment accumulating on the floor of the lake. Exceptions include areas of sand that were deposited where rivers, such as the Sheyenne and Pembina, flowed into the lake. At least 30 such sandy river deltas formed in Lake Agassiz in North Dakota, Minnesota and Canada. The largest is the Assiniboine Delta, formed where the Assiniboine River flowed into Lake Agassiz in southern Manitoba. Sand dunes have developed on the sandy areas in some places (discussion of dunes in chapter five).

Another deviation from the table-top-smooth lake-floor surface occurs in eastern Ransom County, from Enderlin south to Lisbon and east to Sheldon. In this area, the land surface is pitted, much like low-relief dead-ice moraine. The early Lake Agassiz flooded the area before all of the glacier ice had completely melted, and the lake sediments were deposited on both bare ground and ice-covered ground. When the ice melted, some of the lake sediments collapsed, resulting in a pitted landscape.

GROOVES

Viewed from an airplane, the central part of the Red River Valley appears terribly scratched. The scratches are grooves made by icebergs as they were blown over the surface of the shallow lake during spring breakup, the bottoms of the icebergs dragging in and gouging the soft lake-floor sediment. Viewed from the ground, the grooves are almost imperceptible, but they are responsible for the beaded shelterbelts, which cross the grooves. The shelterbelts consist of alternating tall and short trees whose growth characteristics are controlled by soil fertility and soil moisture differences between the land in the grooves themselves and the higher areas between the grooves. Taller trees grow in the grooves, which tend to be more moist; shorter ones grow on the slightly higher, less fertile (more salty) and drier land between the grooves.

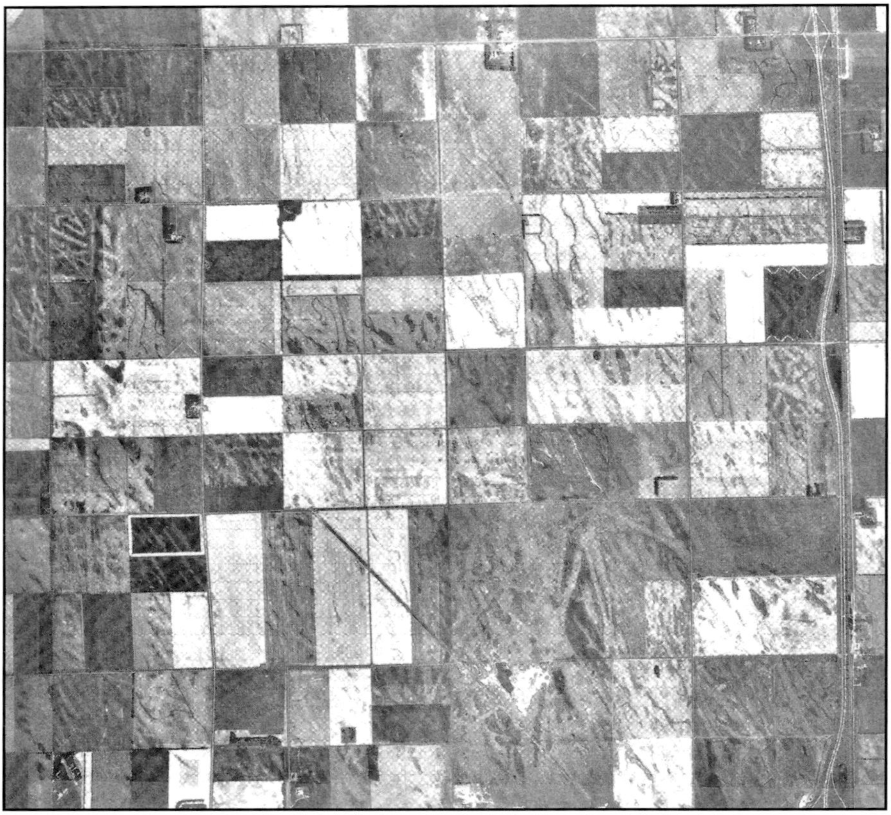

▲ This air photo of part of the glacial Lake Agassiz plain in northeastern Pembina County shows an array of ice-drag markings. North is at the top of the photo. The photo shows an area approximately five miles square extending from six to 11 miles south of the town of Pembina. Although obvious on the photo, the grooves are almost imperceptible when viewed on the ground. The grooves were probably formed during a spring season when icebergs or floe ice were propelled by strong winds. The grooves trend mainly from north-northwest to south-southeast; it is likely icebergs were propelled toward the northwest by winds blowing in that direction. Interstate Highway 29 is shown along the east edge of the area. The I-29 interchanges are at Joliet on the south and McArthur in the northeast corner. Scanned air photo

▲ A beaded shelterbelt can be seen three miles north of the Grand Forks International Airport in Grand Forks County. Trees growing in shelterbelts in parts of the Red River Valley, where salty water is coming to the surface, vary in height, depending upon whether they are growing in one of the shallow grooves or on a high area between grooves. Taller trees grow in the grooves, shorter ones on the slightly higher land between the grooves. The grooves, which are imperceptible on the ground, were formed when icebergs or floe ice floating in the shallow water of glacial Lake Agassiz dragged on the lake floor, gouging the soft lake sediment (see text for more complete explanation). JPB Photo: 9/7/2012

COMPACTION RIDGES

During the time that Lake Agassiz was partially drained, from 10,900 to 9,900 years ago, several rivers established their routes across the former lake plain. Rivers like the Pembina, Park, Turtle, Elm, Maple and Sheyenne, all of which today flow into the Red River, followed routes different than their modern routes across the then-drained lake bed. They cut their valleys into the lake sediments that had been recently deposited. During the thousand-year period that rivers flowed over the area, the time between Lake Agassiz I and II, vegetation on the drained lake plain likely began as tundra and taiga (a pine-hardwood boreal forest), with spruce, and various hardwood trees, gradually changing to prairie with scattered pine.

The routes of some of the rivers that crossed the lake plain during the time it was drained are today marked by low ridges, most of them too subtle to notice unless you are looking for them. This may seem counter-intuitive. It might seem that, after a river had eroded a valley in the lake plain, depositing gravel and sand in its channel, and later abandoned its route, the valley it cut should be lower—a valley partly filled with gravel and sand.

However, after the rivers had cut their valleys, depositing river sediment in them, Lake Agassiz again flooded the area. A new layer of lake sediment was deposited on the floor of the new lake. This layer of

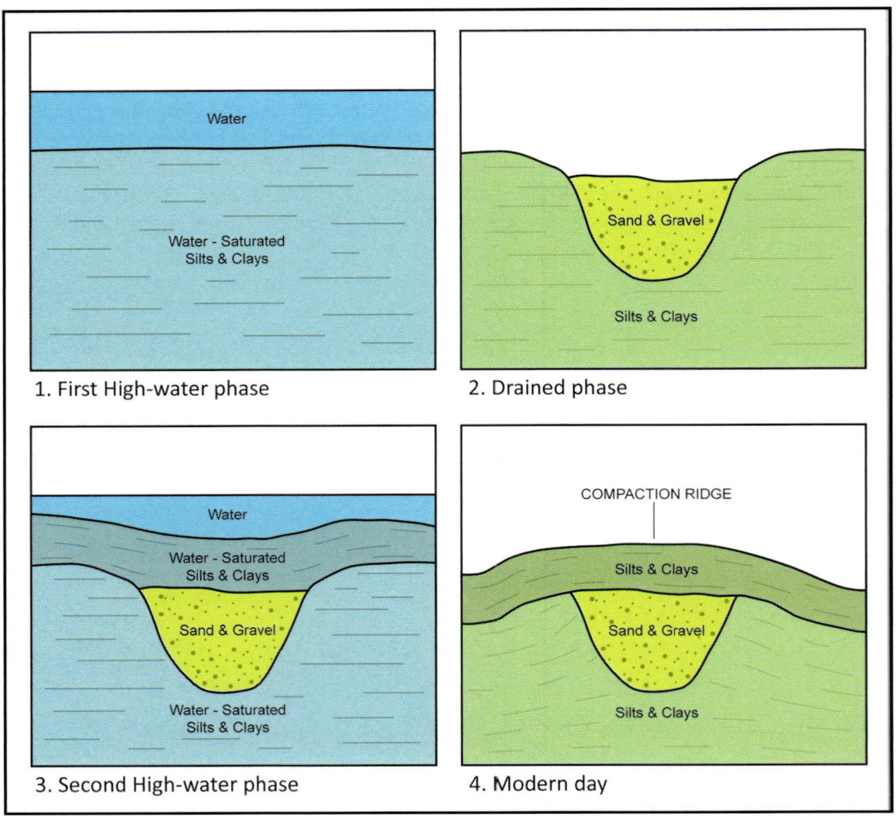

silt and clay covered both the older lake sediment and the river gravels that had been deposited when the lake had been drained. Lake sediments were deposited by the new Lake Agassiz for approximately 600 years, but then the lake drained from the area for the last time.

When everything eventually dried out, the water drained from both the lake sediments and the buried river deposits. The lake sediments, because they consisted of fine-grained silt and clay, settled substantially (when the water came out of the silt and clay, the fine-grained particles packed much more tightly). The water contained in the stream gravel and sand deposits drained out, too, but the more-solid gravel framework did not settle. Where once there were valleys, today we find ridges of sand and gravel mantled by a layer of lake silt. The former river valleys became low ridges (compaction ridges) on the glacial Lake Agassiz plain.

◀ This four-part diagram shows the steps in the formation of compaction ridges, such as those that formed on the glacial Lake Agassiz plain during the thousand years it was drained and rivers flowed across it.

The first cross-section diagram shows lake clay and silt being deposited in "Lake Agassiz I" (first high-water phase).

The second cross-section diagram (drained phase) shows a cross section of a river valley cut into the lake clay. Gravel and sand, deposited by the river, partially fill the valley.

The third cross-section diagram (second high-water phase) shows conditions when "Lake Agassiz II" flooded the Red River Valley. All the features that had formed during the drained phase were flooded and buried beneath a layer of new lake sediment. The new lake deposits covered the valley, the gravel it contained and the nearby older lake clays. During the approximately 600 years that Lake Agassiz II flooded the area, all of the sediments beneath the lake were saturated with water.

The fourth cross-section diagram (modern day) shows conditions after the saturated sediments had drained and dried out. As the lake deposits drained, they settled because the clays become more compact and settle when they dry. The gravel in the valley, though, did not settle. When water drains from gravel, its framework remains intact.

Today, former routes of the buried river valleys remain as subtle ridges. Most of the larger streams and rivers in North Dakota that flow to the Red River today (Sheyenne, Park, Elm, Maple, Pembina and others) have one or more corresponding compaction ridges nearby that mark former routes of the rivers across the lake plain when it was drained.

A good example of compaction ridges can be seen just southwest of Fargo, where several ridges trend north-south. The ridges are situated just east of and parallel to the modern route of the Sheyenne River and they were probably formed by early versions of the Sheyenne River. Another river formed a compaction ridge across the southern part of Traill County. This ridge, which is most apparent near and east of Kelso, was probably the route of the eastward-flowing Elm River.

SPRING PITS

In eastern Walsh and Grand Forks counties, salty water is seeping to the surface today, escaping from buried sandstone formations. As it comes to the surface, the water slowly loosens and removes the surface sediments as it flows away, a process referred to as "sapping," resulting in slightly lower areas. In some places where the salty water is surfacing, a white crust may be seen after a summer dry spell. When

▲ **Red samphire *(Salicornia rubra)* is concentrated in a drainage ditch in an area of saltwater discharge near the Grand Forks International Airport in Grand Forks County. Red samphire grows in salty soil in places where salty water is escaping to the surface. This drainage ditch is in Kellys Slough, which is a salt-water discharge spring.** JPB Photo: 9/7/2012 the rate of evaporation far exceeds the rate of water escaping at the ground surface, salt-encrusted soil forms when the salty water evaporates, leaving behind the salt deposit. In such places, the surface may be vegetated by salt-loving plants, which thrive on the sodium chloride in the soil water they absorb. The most common plant that grows in the salty soil in the Red River Valley, as well as in other salty areas in North Dakota, is known as red samphire (see discussion of artesian flows in chapter five).

Most of the large spring pits on the glacial Lake Agassiz plain, such as Kellys Slough in Grand Forks County, and Salt Lake and Lake Ardoch in Walsh County, formed as the glacier was melting from the area. Ground water had seeped eastward through deeply buried preglacial sandstone beds and escaped to the surface in eastern North Dakota for millions of years prior to the Ice Age. However, during the time the area was glaciated, the land surface was sealed by ice, keeping the

▲ Red samphire grows on salty soil. The salt-loving plants are growing in areas where salt is seeping to the ground surface. The white background is salt that accumulates when the salty water seeping out of the ground evaporates.

JPB Photo: 8/4/2010

ground water from escaping. This caused pressure to build up in the water-saturated sediments beneath the ice. When the glacier melted, removing the seal, large volumes of ground water flowed to the surface at a rate much greater than it had escaped prior to glaciation (and faster than it is escaping today). The rapid upwelling of ground water carried sand with it as it rose (see the discussion of veblens in chapter four). Most of the sand-laden water escaped into Lake Agassiz while it still flooded the area, and the sand spread out over the lake floor. Some such areas of sand, now saturated where the salty water still rises to the surface, are quicksand, where a person might sink. In most places the sand is less than 2 or 3 feet thick.

Lake Shorelines

The old shorelines of the larger glacial lake plains, such as Agassiz and Souris, are bordered by beach deposits of sand and gravel or by wave-cut scarps. The force of the waves sorted the near-shore deposits into sand, silt and clay. Currents carried the silt and clay offshore to deeper water. The coarser sand and gravel were deposited along the shore as beach ridges, some of which can be followed for hundreds of miles. Although beach characteristics vary, most are a few hundred feet across and from 5 to 25 feet high. The beaches and scarps along the shore of Lake Agassiz in Minnesota tend to be larger because wave action was driven by prevailing westerly winds. As a result, wave activity along the eastern shore of the lake was more intense than in North Dakota.

SCARPS

In places where only small amounts of sand and gravel accumulated, the old shoreline may consist simply of a wave-cut scarp, a notch with little or no accumulation of beach deposits. Bouldery, wave-worn deposits of glacial sediment are common along such scarps. A good example of a wave-cut scarp can be seen between Mayville and Portland where State Highway 200 climbs several feet as it crosses the Campbell shoreline scarp, which rises abruptly to the west.

In some places, waves washing over the shoreline of glacier sediment (till), removed most of the fine-grained materials, leaving behind bouldery surfaces. Such wave-worn surfaces may be seen in eastern

Walsh County between Voss and Pisek, and in northern Traill County near Buxton. Wave-worn surfaces may be covered by large numbers of stones left behind when the water removed the surrounding finer materials.

BEACHES

The levels of glacial lakes tend to rise and fall frequently: seasonally, as the supply of water from the melting glacier varies, as well as sporadically and catastrophically as new outlets and inlets open and close, depending on the position of the damming glacier. Well-defined beaches and scarps probably developed along the shore of Lake Agassiz during times that the lake level remained stationary for a while. However, intense storms lasting only a matter of days sometimes built large beaches quickly, so the size of a beach at any given location is not a reliable indicator of the length of time a certain water level persisted. Some of the millions of tons of sand delivered to Lake Agassiz and other glacial lakes became beach ridges extending great distances across the countryside. Where sand was sparse or lacking, erosion cut steep shoreline scarps.

The best-developed of the Lake Agassiz shorelines is the Campbell Beach, named for the village of Campbell, Minnesota. In some places, the Campbell Beach is a massive ridge of gravel and sand, such as near Arvilla in Grand Forks County. In other places, as at Mayville, only a wave-cut scarp marks its location. It can be traced for much of the distance along the old Lake Agassiz shoreline on both sides of the lake. The Campbell shoreline existed twice, once for several hundred years around 11,000 years ago, and again perhaps 9,500 years ago.

The oldest and highest shoreline that extends for an appreciable distance is the Herman Beach, named for Herman, Minnesota, where it is well-developed. The Herman Beach marks the limit of Lake Agassiz when it had first become a large and continuous lake, flooding the Red River Valley in North Dakota and Minnesota. West of the Herman Beach in North Dakota, glacial deposits predominate. In North Dakota east of the beach, all the landforms are related to Lake Agassiz.

Some small beach segments occur above the Herman Beach, but they are short and discontinuous, only two to three miles long. Short beach segments can be seen near Cuba, Fingal and Nome in Barnes

County, and at Alice in Cass County. These beaches probably formed early, along a lake or lakes ponded between the edge of the glacier ice on the east and higher land on the west while a glacier still filled most of the Red River Valley. They are as much as 380 feet higher than the Herman Beach and may have formed along shorelines of an early lake that formed between the glacier in the valley and uplands to the west. The lake is referred to as Lake Mose. In addition to the beach segments, widespread lake silts were deposited in Lake Mose over an area extending from Nelson to Ransom counties. The silts can be seen near the surface in many places, but they are covered by till in most places, deposited when the glacier re-advanced over them.

The beaches and scarps along the edge of the Glacial Lake Agassiz plain are easily observed from the air, but for the ground-based viewer, they are subtle features. The sand and gravel in the beaches have been a source of excellent-quality construction material. The gravel and sand in many of the beaches have been so completely mined that they no longer exist as beaches. All that remains are scars where once there were beach ridges.

BARS

Offshore bars and spits formed along the Lake Agassiz shoreline in places. Like beaches, they consist of sand and gravel, and they can be mistaken for beach ridges. One example is the so-called McCauleyville shoreline, which occurs a few hundred yards offshore of the Campbell Beach in several places. It may have formed as a bar, rather than a beach, in the Arvilla area near Turtle River State Park, continuing to the southeast as far as State Highway 15, to a point 10 miles west of Thompson. Another example is a hook-shaped gravel ridge just west of Fordville in Walsh County.

Deltas

Numerous rivers periodically carried great volumes of silt-laden water, which flowed into Lake Agassiz, creating thick layers of mud, known locally as gumbo. Several rivers flowed into glacial Lake Agassiz in North Dakota, Minnesota and Manitoba. The largest rivers flowing into the lake from the west were the Pembina and Sheyenne rivers in North Dakota and the Assiniboine in Manitoba. The rivers carried water that was released from large lakes to the west—Lake Souris in north-central North Dakota, and Lakes Regina and Hind in Saskatchewan and Manitoba. All of these lakes were periodically dammed by glacier ice. The ice dams repeatedly failed and, each time they did so, floods flowed down the Pembina and Sheyenne rivers, forming the valleys, eroding large amounts of sediment from the valley walls, and carrying it to Lake Agassiz. The sediment formed deltas at the mouths of the rivers, along the western shore of Lake Agassiz.

Another Ice Age river, the Elk River, deposited a thick layer of gravel and sand in Walsh, Grand Forks and Traill counties. The Elk River was probably a combination of the early Park, Forest and Turtle rivers, and it flowed southward between the edge of the glacier to the east and the upland to the west, from near Fordville on the north to beyond Galesburg on the south. The gravel deposits in the Elk Valley Delta are coarsest on the north end, and they become finer to the south. The gravel serves as an important aquifer near Larimore.

Along the west side of Lake Agassiz, the largest delta is the Assiniboine Delta in southern Manitoba. In North Dakota, three major deltas formed along the western shore of glacial Lake Agassiz. I list them here from south to north, which is the order in which they formed:

1) Sheyenne Delta in Ransom and Richland counties
2) Elk Valley Delta, in Walsh, Grand Forks and Traill counties
3) Pembina Delta in Pembina County
4) Assiniboine Delta in Manitoba

The deltas consist of thick accumulations of sand. After Lake Agassiz drained, sand that had been deposited on the Pembina and Sheyenne deltas was blown into dunes that are constantly changing shape and position, even today (see discussion of dunes in chapter five).

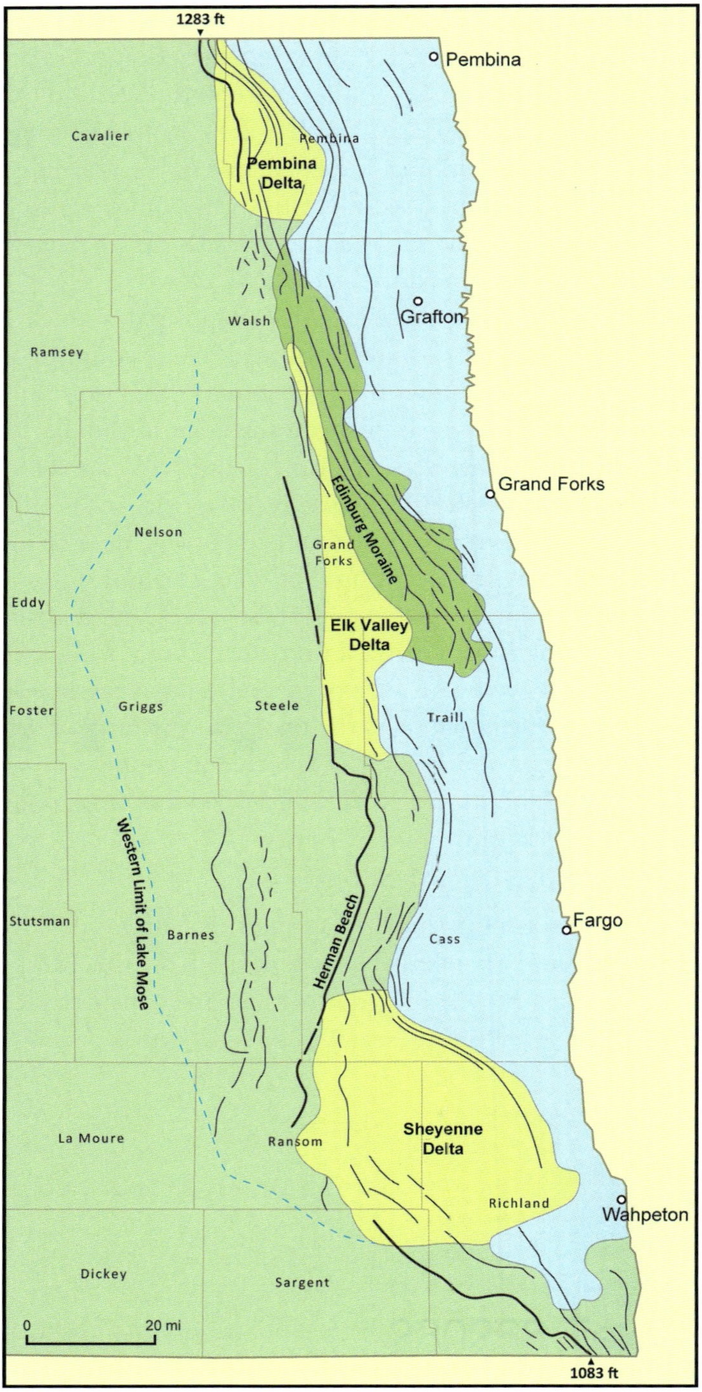
1283 ft
Pembina
Cavalier
Pembina
Pembina Delta
Walsh
Grafton
Ramsey
Edinburg Moraine
Grand Forks
Nelson
Grand Forks
Eddy
Elk Valley Delta
Foster
Griggs
Steele
Traill
Western Limit of Lake Mose
Herman Beach
Stutsman
Barnes
Fargo
Cass
La Moure
Ransom
Sheyenne Delta
Richland
Wahpeton
Dickey
Sargent
0
20 mi
1083 ft

◀ This map of eastern North Dakota shows features associated with the Red River Valley. In addition to showing the extent of Glacial Lake Agassiz in North Dakota, the map shows locations of the beach ridges along the western edge of the lake plain (black lines) and the areas covered by lake sediment (blue areas).

The uppermost continuous beach along the Lake Agassiz shoreline, shown by a heavier line, represents the Herman Beach. The elevation of the Herman Beach at the Canadian border is 1,283 feet; at the South Dakota border the elevation of the beach is 1,083 feet. The 200-foot difference is a result of crustal rebound (see explanation in text).

Three major deltas, the Pembina, Elk Valley and Sheyenne, are shown (yellow areas). The area to the west (light green) is glacial topography, mainly ground moraine.

The dashed line through Nelson, Griggs, Barnes and Sargent counties represents the approximate western limit of the mostly buried Lake Mose beds of silt and clay. The silt was deposited in an early lake that was dammed between glacial ice on the east (the ice still filled most of the Red River Valley) and higher land to the west. A subsequent westward re-advance of the glacier from the Red River Valley lowland overrode the Lake Mose beds.

The north-south lines shown in eastern Barnes County are beach ridges. They may have formed along a shoreline of Lake Mose or possibly along the shore of a very early Lake Agassiz.

Note the Edinburg End Moraine in Walsh and Grand Forks counties (darker green). The end moraine formed at the edge of the glacier in the Red River Valley as its margin was receding, but paused and stabilized for a time as the Ice Age was ending. Only a few actual end moraines can be confirmed in North Dakota; most features previously thought to be end moraines are really areas of ice-thrust topography.

Crustal Rebound

The oldest beaches bordering the Lake Agassiz plain are at the highest elevations, the younger beaches occurring lakeward, at lower elevations. It might seem logical to assume that any given beach, especially if it is a continuous feature, should be at the same level wherever it is found, but this is not the case. For example, the uppermost continuous beach, the Herman Beach, stands at an elevation 200 feet higher at the Canadian border than at the South Dakota border. The lower, younger beaches rise to the north, too, but the difference in elevation, from north to south, is less for the lower beaches than for the upper ones.

This discrepancy may seem odd. After all, wasn't the lake at the same level everywhere when any given beach formed? Yes, but as the Ice Age was ending, and the glacier was melting; the gradual disappearance of the ice was relieving a great weight from the land surface. The weight of the glacier ice had been so great that it had depressed the Earth's crust, which is an elastic structure.

Glaciologists have learned that once a glacier reaches a certain thickness, it depresses the crust a foot for every 3 feet of ice because ice weighs roughly one-third as much as crustal rock. The glacier was thicker in the north than in the south, so the amount it depressed the land surface was greater there. When the glacier melted, the land rose more in the north than in the south as it returned to its preglacial level. Studies of rebound in the Hudson Bay area show that the most recent glacier ice there was more than 6,500 feet thick. The thickest glacier ice in North Dakota was probably at the northeast corner of the state, where it may have been as thick as 3,000 feet.

Crustal deformation usually results in gentle stretching, subsidence and uplift (rebound), but sometimes it can cause faulting and reactivation of regional tectonic structures. Tectonic forces are those that alter the Earth's crust. When the glacier melted, the crust returned to the position it was at before the ice advanced over it. The rebound may have been smooth and gradual, or it may have happened in quick jumps, resulting in earthquakes along fault lines. I have seen 60-foot-high fault scarps, fresh in appearance, in Lapland in Sweden where deglaciation occurred recently and rapid rebound is still ongoing, spasmodically, resulting in small earthquakes. At Churchill, Manitoba, the crust is still rebounding at a rate of 4 feet per century. Stones used for mooring 17th century ships in the harbor at Churchill are now 20 or more feet above sea level.

A recent possible example of the ongoing rebound of the Earth's crust was recorded from April 29 to May 4, 2011, when dozens of tiny earthquakes were noted along the eastern coast of Maine. The quakes were all instrumentally recorded; none were felt. The microearthquakes may have been caused as the Maine shoreline, once covered by a thick glacier, continues to return to its preglacial level.[62] In North Dakota, I think that rebound is complete and no further adjustment is likely.

Much of the glacial rebound in North Dakota had already taken place by the time Lake Agassiz stood at its lower, more recent levels and, for this reason, less rebound occurred along the length of the lower beaches, from north to south, than it did for the higher ones. Another important effect of glacial rebound has been to make it increasingly difficult for the Red River of the North to flow north; the land has risen more in the north than it has in the south.

The comment I just made about the Red River may seem odd. After all, doesn't the crust simply rebound at the end of a glaciation in an amount comparable to the amount it sank earlier? Perhaps, but in the meantime, while it was depressed, a large lake existed and a layer of sediment accumulated in that lake. The rebound at the end of the Ice Age, when the glacier melted, was less than an amount equal to the amount of new lake sediment. Consequently, the post-glacial river has to flow over an even shallower gradient than before. In addition to contributing to flooding problems, the very flat gradient causes the river to meander in a tortuous fashion, flowing over a distance much greater than the straight-line distance to make its way to Lake Winnipeg in Manitoba.[63]

Pembina Escarpment

The Pembina Escarpment marks the western edge of the Red River Valley in northeastern-most North Dakota and southern Manitoba, where it rises 300 to 400 feet above the floor of the Red River Valley. Along the escarpment, hilly slopes with wooded ravines of aspen and birch contrast sharply with the flat lowland of the Lake Agassiz plain to the east. The Pembina Escarpment may have been one of the "mountains" referred to by the French fur trader La Verendrye in 1738.[64] By the time geologists first explored the region, the hills were rumored to be the source of silver ore. However, William Keating, geologist with the U. S. Government's first expedition to the Red River Valley, led by Major Stephen H. Long, had it right when, in 1823, he wrote: "We saw no ore of any kind. The prairies do not present any character that would leave us to anticipate the discovery of mines in this neighborhood."

Rumors persisted, long after Keating's visit, that coal and perhaps even gold, existed on Pembina Mountain. David Dale Owen, geologist for the fledgling United States Geological Survey, visited the area in 1848. Owen had traveled northward, up the Minnesota River, and then down the Red River to Lake Winnipeg. As Keating before him, Owen, in 1852, found no evidence for any deposits of economic significance.[65] He wrote:

> Pembina Mountain is . . . in fact no mountain at all, nor yet a hill. It is the terrace of table land - the ancient shore of a great body of water that once filled the Red River Valley. On its summit it is quite level, and extends so, for about five miles westward, to another terrace, the summit of which I was told is level with the great buffalo plains that stretch away towards the Missouri, the hunting-grounds of the Sioux and the half-breed population of the Red River.

The geologic materials exposed in the area of the Pembina Escarpment have their origins millions of years ago during the Cretaceous Period, long before glaciers flowed southward through the Red River lowlands (see chapter one). The Cretaceous shale formations, exposed today along the streams in the deep gullies and valleys of the Pembina Escarpment, were deposited in a seaway that connected the Arctic

Ocean to the Gulf of Mexico 90 million years ago. Fossil bones and teeth of fish, fossil mosasaurs (large marine lizards), giant squid and pterosaurs (flying reptiles) have been found in the Walhalla area in the shale exposures of the Pembina River gorge, and in the nearby Morden area of southern Manitoba. Beyond the western shoreline of the Cretaceous seaway, in western North Dakota and Montana, duck-billed dinosaurs, triceratops and tyrannosaurs roamed the land.

Exposures in the Pembina River Gorge reveal shale that is interbedded in places with layers of yellowish-colored bentonitic clay. These layers are weathered volcanic ash beds, transformed to clay by millions of years of weathering. In late Cretaceous time, clouds of ash, spewed from volcanoes erupting in the developing Rocky Mountains to the west in Wyoming and Montana, blew eastward over North Dakota, washing into and settling in the shallow sea that covered the area at the time.

The formation of the Red River Valley, and ultimately the Pembina Escarpment (or Pembina Mountain or Pembina Hills or Manitoba Escarpment—various terms have been used), began long before the Ice Age. Springs emanating from the Cretaceous-age Dakota Group sandstone and overlying shale formations resulted in gully development and eventually a small escarpment. The early Red River cut into the bedrock in what is now northwestern Minnesota, eroding the escarpment and causing it to gradually retreat westward. Glaciers later steepened and shaped the escarpment to its modern configuration. As glacier ice repeatedly flowed southward, up the Red River Valley during the Ice Age, it scoured and steepened the face of the escarpment.

When glacial Lake Agassiz flooded the Red River Valley, storm waves at the shore further accentuated the steep face of the escarpment. Periodic massive outbursts of floodwater from ice-dammed lakes to the west flowed along the Manitoba-North Dakota boundary, eroding and shaping the deep Pembina River Valley. Erosion since the end of the Ice Age has deepened the many small valleys in the face of the Pembina Escarpment and carved fresh exposures in the shale.

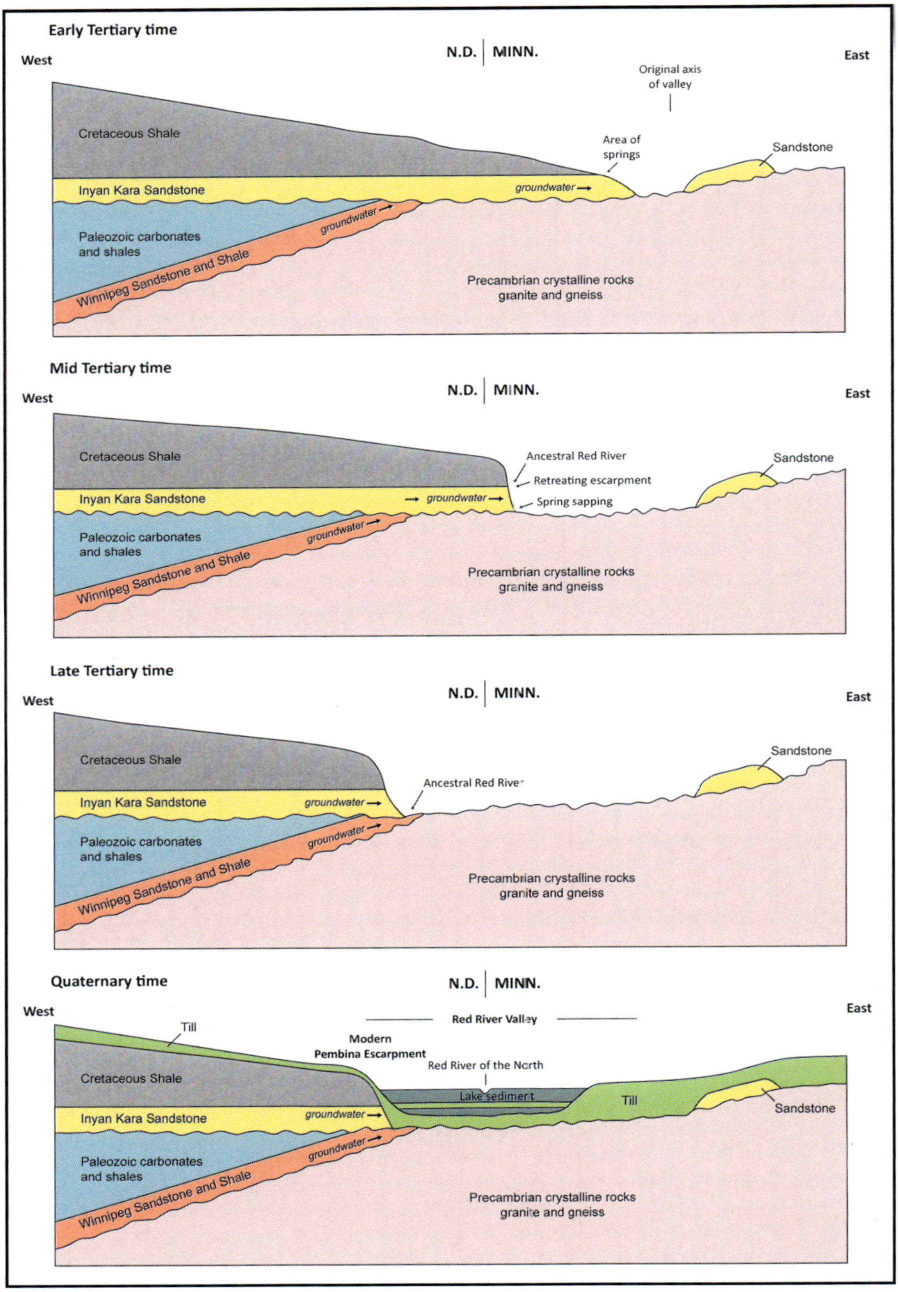
Early Tertiary time
West
N.D. | MINN.
East
Original axis of valley
Cretaceous Shale
Area of springs
Sandstone
Inyan Kara Sandstone
groundwater
Paleozoic carbonates and shales
groundwater
Winnipeg Sandstone and Shale
Precambrian crystalline rocks granite and gneiss
Mid Tertiary time
West
N.D. | MINN.
East
Cretaceous Shale
Ancestral Red River
Retreating escarpment
Sandstone
Inyan Kara Sandstone
groundwater
Spring sapping
Paleozoic carbonates and shales
groundwater
Winnipeg Sandstone and Shale
Precambrian crystalline rocks granite and gneiss
Late Tertiary time
West
N.D. | MINN.
East
Cretaceous Shale
Sandstone
Ancestral Red River
Inyan Kara Sandstone
groundwater
Paleozoic carbonates and shales
groundwater
Winnipeg Sandstone and Shale
Precambrian crystalline rocks granite and gneiss
Quaternary time
N.D. | MINN.
West
East
Red River Valley
Till
Modern Pembina Escarpment
Red River of the North
Cretaceous Shale
Lake sediment
Till
Sandstone
Inyan Kara Sandstone
groundwater
Paleozoic carbonates and shales
groundwater
Winnipeg Sandstone and Shale
Precambrian crystalline rocks granite and gneiss

◀ These four diagrams are sequential, the oldest on top, progressing downward through younger phases. They show a west to east cross section—a vertical slice—near the International Boundary, illustrating the formation of the Red River Valley lowland during Tertiary time. The vertical dimensions of the cross section are greatly exaggerated, compared with the horizontal. The dates (Early Tertiary time, etc.) are speculative.

The top diagram shows a small valley (labeled Original axis of valley), through which a stream flowed northward to Hudson Bay (on all four diagrams the view is northward, down the valley). The early stream may be considered to be the early Red River, which had already cut through the Inyan Kara Sandstone (known also as the "Dakota Formation"), leaving an area of the sandstone on its east side.

Ground water seeping eastward through buried sandstone formations (especially the Winnipeg and Inyan Kara formations) came to the surface (as it does today) along the west side of the evolving valley.

The two middle diagrams, Mid- through Late-Tertiary time, show the Red River gradually widening its valley by cutting away the upland to the west as it "slid" down-slope along the top of the Precambrian surface. As it gradually migrated westward, the river steepened the west valley wall and carried material that the escaping ground water was sapping, removing it as it exited the ground. Note that the river gradually progressed from its original route in Minnesota into North Dakota.

The lower diagram, Quaternary time (the Ice Age) represents conditions after glaciers had flowed southward through the Red River Valley. The glaciers: 1) further steepened the Modern Escarpment (Pembina Escarpment) and 2) deposited a layer of till over the entire area. After the glaciers receded, Glacial Lake Agassiz deposited layers of lake sediment on top of the till, leaving the flat Lake Agassiz plain.

Following glaciation, the Red River assumed its current route, which is situated slightly farther to the east than immediately prior to glaciation. The Red River's northward gradient to Hudson Bay is very low and the river struggles to flow north. The early Red River may have meandered widely over the lowland as it shaped the valley.

CHAPTER EIGHT

THE EFFECT OF GEOLOGY ON THE HUMAN ENVIRONMENT

Perhaps the first part of this chapter ought to be titled "How Geology Should have Influenced Where People Settled in North Dakota, but Didn't." The geology of the landscape distinctly influenced the history of human settlement in both objective ways ("This land is level and tillable, and free of rocks; it would make excellent farmland.") and subjective ways ("This looks just like home back in Ukraine—I think I'll settle here."). The occurrence of economic resources like lignite (determined by geology) might have helped miners decide where to settle. Other things that settlers might, or should have, asked include such things as: I wonder if this valley gets flooded often? How high would the water rise? or Might those steep hills up there slide down here where I'd like to build my house?

Early North Dakotans seldom chose a place to live based on conscious appraisal of the geology, but such things did indirectly influence their choices. They considered the terrain, the availability of water, natural resources and undoubtedly nostalgia for lands that somehow resembled the homeland.

Few, if any, settlers asked any of the questions I just mentioned and, unfortunately, most people today don't, either. Geologists have written hundreds of books on the geology of flooding, unstable ground, earthquakes, volcanoes, wave erosion and many other topics often classified as geologic hazards. Of course, we don't want to be obsessed with all the bad things that might happen, but sometimes a little common sense can be useful. In a book like this one, I cannot go into a lot of detail about how the land contributes to geologic hazards, but I will touch on a few ways.

Geologic Hazards

Introduction

Geologic hazards (geohazards) represent an important, but often unrecognized, threat to people and their homes or other structures. This is demonstrated by the additional costs incurred due to unforeseen ground conditions and by post-construction losses sustained by the insurance industry due to building damage. Because homeowners are becoming aware of the dangers, assessment of geohazards are being done more frequently during the house-buying process.

Where a potential geohazard exists, the use of the land can sometimes be modified to ensure that the hazard is not triggered, or its effects may be mitigated. I have listed several geology-related hazards found in various places in North America. Of these, the two most-common in North Dakota are landslides (or other kinds of unstable ground) and flooding.

Shrink-swell clays. Clays can dry out and shrink, or they may absorb a lot of moisture and swell, causing cracks to open in the ground or in structures built on clay-rich land. Local factors, such as tree roots or leaking pipes can cause problems where the shrink-swell potential is high. I have seen places in North Dakota where clays have shrunk as certain deep-rooted plants (alfalfa, etc.) have dewatered the upper several feet of an area during times of drought, causing deep cracks to form in fields. Similarly, streets in cities in the Red River Valley, unless they have an adequate amount of gravel beneath the paving material, heave when the underlying clay beds alternately shrink and swell.

Landslides. Rocks or soil can slide downhill in places with certain combinations of geology, steepness of slope and a raised water table. Increased rainfall can lead to elevated water saturation levels, thus triggering landsides. I will discuss landslides in more detail later in this chapter.

Compressible deposits. Deposits like peat or slough sediment can be compressed (squashed) through time under the load of buildings, road-fill, etc.

Flood damage. Stream, river or lake water can rise, causing damage to structures on flood plains. I will discuss the flooding problem in North Dakota in more detail.

Running sand. Sand can flow to fill an excavation, resulting in danger to construction workers.

Soluble rocks. Limestone and rocks composed of salt can dissolve, leaving underground cavities that can collapse. This may result in karst topography. Karst landforms are not a problem in North Dakota because soluble rocks rarely occur close to the land surface.

Collapsible deposits. Certain fine-grained sediments can collapse or flow when loaded for the first time. This is a problem in parts of the Red River Valley, and in some other places where glacial lake sediments or certain kinds of till occur. Deep pilings may be necessary to support large structures in such areas.

A similar collapse hazard occurs in some places where underground lignite mines have fallen in as the wooden timbers supporting the mine shafts have rotted away. The result may be isolated pits or lines of pits on the ground surface. It can be argued that such collapse is less a geohazard than a result of human activity: mining of lignite. However, it could be argued, too, that floods or landslides would not be hazards if people did not put themselves in harm's way by building on a flood plain or in a place where a landslide is probable.

Flooding

In the last several years, the reality of flooding in North Dakota has been demonstrated in many different ways. These include the slow-motion, on-going flood in the Devils Lake area; the floods in Minot and other communities along the Souris and Des Lacs rivers; the like-clockwork floods in the Red River Valley; the 2011 flood along the Missouri River in Bismarck and Mandan; and the flash floods along many of North Dakota's smaller rivers and streams.

Floods are natural, geologic occurrences. Severe floods occur every 15 to 20 years. After the 1997 flood in Grand Forks, I co-authored a book on Red River flooding (LeFever et al., 1999). The book included an exhaustive listing and analysis of geologic factors that contributed to that disaster, but it could have applied as well to Fargo and other communities along the Red River.

The flooding at Devils Lake is a result of the unique geologic circumstances that formed the lake, coupled with natural climatic cycling. The area has been flooded in the geologic past and, sometime after the current water recedes, it will flood again, although next time mitigated by engineered modifications.

The Souris River Valley was formed so quickly by floods of glacial meltwater at the end the Ice Age that the river did not have time to build a respectable flood plain. That circumstance contributed to the recent flood there, but the unprecedented precipitation at exactly the wrong time played a major role in the 2011 Minot flood. In addition, the valley slopes have not yet stabilized and landslides are common.

In Bismarck and Mandan, the Missouri River flooded regularly prior to construction of Garrison Dam. With the construction of the dam in the 1950s, people assumed a flood was impossible. In the 1990s, I gave a talk on flooding along the Missouri River. I mentioned that even though Garrison Dam had been built, we could still have a flood in Bismarck. A former governor of North Dakota was at my talk and, afterwards, he told me I was wrong; we could never have a flood in Bismarck. Nature (sometimes coupled with human error) can conjure up a flood regardless of any or all engineering precautions.

The American Indians, during the thousands of years they lived in North Dakota, developed lifestyles adapted to the geology and other aspects of the natural environment. Their ways of life, whether it was farming or hunting, were very much in tune with geology. Those who farmed did so in logical sheltered places along the river, as they were concerned about the suitability of the land and water for agriculture, as well as the availability of building materials, food and fuel. Floods were not a concern for them because they maintained no permanent structures in areas that flooded.

We can mitigate flooding problems in a variety of ways, but we cannot solve the problem completely. We can build diversions around populated areas; we can build dams to hold back the water. Most of the time, diversions and dams do the job for which they are intended. We can devote more flood-prone lands to parks and greenways.

Landslides

If you are surrounded by it, the topography in an area of landslides lacks apparent rhyme or reason, but instead seems to be a jumble of haphazard hills, troughs and assorted shapes. Understanding such areas is important, though, if only so you can avoid them. I have seen homes built on areas of landslide ridges in North Dakota, usually in a river valley, situated to take advantage of the view. This is sometimes a mistake. If the houses have not yet begun to disintegrate, they may do

▲ **This landslide is on the north side of Sentinel Butte in Golden Valley County. The landslide, which occurred in early June 2011 following heavy rains, blocked the road to the top of Sentinel Butte. Note the blocked road in the foreground and the road continuing on the far side of the slide. The materials in the landslide consist of chunks of bedrock. The slide moved as a unit so that much of the soil cover on the slide surface is essentially undisturbed.** JPE Photo: 7/7/2011 so soon enough, or perhaps they may move down closer to the river for an even better view!

In spite of the overall gentle topography, landslides are common in North Dakota, even more common than in some states with steeper, mountainous landscapes. Part of the reason for this is that the soft, poorly consolidated silt, sand and clay deposits in glaciated areas are still fresh. They were deposited during the Ice Age, geologically very recently, so they have not had sufficient time to stabilize. In unglaciated parts of the state, layers of swelling clays and other soft sediments are unstable. Add to this that human activities such as road and levee construction often disturb and reactivate already unstable land, and the potential for failure can be considerable.

Landslides dot the flanks of buttes in western North Dakota where the bedrock has been deeply and intricately eroded. Sentinel Butte, the Chalky Buttes, Black Butte, Slide Butte and the Killdeer Mountains are bordered by continuous landslides. The valley walls of the Little Missouri River near the North Unit of Theodore Roosevelt National Park have slid extensively, and are continuing to do so. Trees may help stabilize such areas, but they can grow in all directions, resulting in "drunken" forests as the land beneath them continually shifts. During the summer of 2011, record rainfall and floods helped cause large numbers of landslides in many places. The North Dakota Geological Survey has published an extensive series of maps showing areas of landslides. These valuable maps are available in traditional paper or digital format.

Landslides are common in the Des Lacs and Souris river valleys, and they border valleys like the James, Pembina and Sheyenne in eastern North Dakota. Landslide topography may be seen along the Sheyenne River at Valley City, adjacent to and along the south side of Interstate Highway 94. All of the valleys I mentioned, as well as many smaller ones, were eroded quickly by large rivers of glacial meltwater late in the Ice Age so recently that their steep valley walls have not yet reached a state of equilibrium.

If the upper part of the soil zone becomes saturated with water, it can separate from the underlying subsoil and glide downslope. Fence posts on steep slopes may be shifted and tilted as the soil in which they are anchored slides downslope. Soil slides (known as "soil creep") are common in hilly parts of the state, but they can occur in level areas of glacial lake sediment, especially along streams in the Red River Valley where beds of lake sediment are interbedded with layers of glacial sediment. When the saturated Lake Agassiz sediments and interbedded till deposits are cut by stream meanders, or where drainage ditches have been dug, they slump or slide easily.

Treppe

Treppe (German for "stair" or "stair steps"), are step-like ridges that develop along the sides of hills throughout the state. They are another kind of unstable ground that forms as a result of a combination of soil creep and the action of animals walking along the hillsides. Treppe may form in any hilly area where livestock are pastured.

▲ In this scene north of Lehr in central Logan County, the step-like ridges on the hills left of the lake are treppe, which form due to a combination of soil creep and animals walking along the hillsides, making horizontal paths. Treppe can form in any hilly terrain, on either bedrock or glacial slopes. JPB Photo: 5/30/2010

Geology and Settlement Patterns

Among the first European explorers in North Dakota, Pierre la Verendrye explored the region that is now North Dakota in the period 1738 through 1743 while looking for a trade route from Canada to the Mandan and Hidatsa settlements. La Verendrye and his sons traveled from Fort de la Reine (near present-day Portage la Prairie, Manitoba) to the Indian villages on the Missouri River and back. Geology played an important role in La Verendrye's travels as he guided himself by referring to prominent landmarks such as Turtle Mountain and Dogden Butte.

President Jefferson, in 1803, writing of the difficulty of communication in the land to be explored by Lewis and Clark, noted: "many of the present establishments are separated from each other by immense and trackless deserts." At the time Jefferson was writing, the term "desert" referred to any treeless and uninhabited land, unsuitable for agriculture, regardless of whether it was arid or not.

Early explorers' perceptions of the area that is now North Dakota were not necessarily favorable. In 1823, the Stephen Long military expedition followed the Minnesota River upstream to the headwaters of the Red River of the North, and then down the river to near the Canadian boundary, before heading east. They saw little of what is now North Dakota, except the extreme eastern edge, but they mentioned how dry it was. Recent studies verify that when the Long expedition passed through the area, the northern Plains were experiencing a time of severe drought.

William Keating, a naturalist who accompanied the 1823 Long expedition, took note of the dry conditions and described the country:

> a rather dry argillaceous ground, which, within a few miles of the river and its tributaries, yields good grass, but at a distance from it presents but a scanty growth . . . It is probable, that the fires, which annually overrun these prairies, destroy all the vegetable matter, and tend to keep the ground in an impoverished state.

However, Keating understood that fire was important in maintaining the prairie vegetation. He stated: "We may . . . consider fire as the cause of the continuance, if not the original existence of prairies."

Following his expedition, Long produced a map labeling the area the Great American Desert. In his report, Long, like Keating, used the term "argillaceous" (the word means "rich in clay") to describe the soil. He stated:

> the dwarfish appearance of the herbage it supports, especially on the upper [southern] portion of the [Red River] valley, indicates either a want of fecundity or the admixture of salts or other ingredients not congenial to vegetation.

The party's geographer, Edwin James, wrote of the region:

> I do not hesitate in giving the opinion, that it is almost wholly unfit for cultivation, and of course, uninhabitable by a people depending upon agriculture for their subsistence. Although tracts of fertile land considerably extensive are occasionally to be met with, yet the scarcity of wood and water, almost uniformly prevalent, will prove an insuperable obstacle in the way of settling the country.

The sentiment was repeated by dozens of explorers. In 1837, Peter Garrioch said of the Fargo area that it was "one continuous stretch of wild and barren plains." He referred to the area as a "desert." Nine years later, Robert Clouston, while in the Fargo area, described the land as a "waste," "treeless and lifeless." "Despite the charm of the scenery," he wrote, "all seems desolation, and a painful feeling of solitude casts a shade over one's mind as he wanders through those seldom trodden wilds." Other explorers during the same period apparently did not necessarily share Clouston's sense of oppression about the Red River Valley, but they passed over the prairie as if nothing between rivers and woods existed. For them, it was merely an expanse to cross, featureless as a sea, and it scarcely merited a mention, either for what it was or for what it might become (Gilman, 1978).

In short, many early North Dakota explorers were spectacularly wrong in their evaluation of the natural environment.

Later, when the area that was to become Dakota Territory began to be settled, attitudes concerning the area changed (but the change was probably due more to economic self-interests than to any new understanding of the true character of the land). By the 1860s, steamboat pilots on the Missouri River used lignite as a fuel, and in the 1860s and 1870s, personnel from Forts Rice, Buford, Stevenson and Lincoln used lignite. During the Settlement Period of the 1870s and 1880s, the railroads, especially the Northern Pacific, encouraged people to settle in Dakota Territory by publishing articles and pamphlets describing the exceptional attributes of Dakota, such as the rich soils, moderate climate and extensive coal resources. A promotional blurb from an 1882 NP brochure reads:

> As there are no large bodies of timber in Dakota, and no beds of the anthracite and bituminous coal found in the East, it is supposed by many that this country is entirely destitute of fuel, and that its only means of supply is by purchasing coal in the East at the most exorbitant rates. This is an error. Beds of LIGNITE coal are found in this locality in almost inexhaustible quantities. This coal is entirely free from sulfur, stone, and other foreign elements. It burns up as clean as wood, without waste, and without clinkers or cinders.

The territorial government provided promotional materials. This from 1887:

> This lignite, or brown coal, is of soft variety, excellent for heating purposes, and . . . superior to almost any coal discovered on the continent . . . lignite burns readily and furnishes the settlers of a prairie country with that inestimable boon, cheap, domestic fuel.

Newspapers had a lot to say in promoting settlement. From the March 24, 1884, edition of the *Bismarck Tribune*, regarding McLean County, the "heaven-favored territory":

> The inexhaustible beds of lignite coal, which underlies so much of this locality, are of untold value to the settler. It makes excellent fuel, and can be supplied at a reasonably low cost. This coal is being largely used in lieu of wood, and for both heating and cooking purposes it meets every requirement . . . These immense deposits of lignite stretch far away on every side, and their value to the country can never be estimated.

Neither did local newspapers characterize winter weather accurately. Rather, they played it down. Newspaper editors wanted to dispel what they considered "myths" regarding Dakota Territory weather, as they were concerned that potential settlers might be dissuaded from coming. We read such phrases as "has been just cold enough to retain the snow and make it the most delightful winter," or "ours is a dry cold, with but little wind." In any case, as we all know, weather perception is in the mind of the observer. Keep that in mind when you shovel snow next January!

Potential settlers had to be assured that western-most North Dakota, on the eastern edge of the "Great American Desert," had enough precipitation to grow crops. Along with other experts, Professor Alfred Atkinson, president of Montana State Agricultural College from 1919 to 1937, sounded a reassuring note, emphasizing a popular saying of the day: "Rain follows the plough." In 1917, Atkinson wrote:

> It seems to be a matter of common observation that rainfall in a new country increases with settlement, cultivation and tree planting. Louis Agassiz in 1867 predicted that this increase in moisture would come about by the disturbance of electrical currents caused by the building of the railroads and the settlement of the country.[66]

Literature like that I quoted above, fueled by what passed for science at the time, published and circulated by government officials, the railroads and the newspapers, helped convince thousands of Europeans and eastern Americans to come to Dakota Territory, resulting in an influx known as the Great Dakota Boom. People emigrated from the British Isles, Germany, Russia, Norway and a host of other European countries. Many immigrants came individually; others joined large colonization groups. Several villages and towns in western North Dakota, such as New Salem, Glen Ullin, Richardton, Gladstone and Dickinson, were direct results of these colonization projects.

The area of a homestead was designated at 160 acres. We know now that that amount of land is insufficient for either a farm or ranch, but the 160 acres suited the railroads and their plan for settling the area. John Wesley Powell, a 19th century geologist who led several scientific expeditions in the western states, in 1878 proposed allocations of a full four sections in semiarid regions. Such large areas were controversial because they threatened to reduce the total number of settlers who could obtain land on the frontier. Powell was dealing primarily with areas to the west of Dakota Territory, but his ideas applied to what is now North Dakota as well. Powell wrote:

> The grass is so scanty that the herdsman must have a large area for the support of his stock. In general a quarter section of land alone is of no value to him; the pasturage it affords is entirely inadequate to the wants of a herd that the poorest man needs for his support.
>
> Four square miles may be considered as the minimum amount necessary for a pasturage farm, and a still greater amount is necessary for the larger part of the lands; that is, pasturage farms, to be of any practicable value, must be at least 2,560 acres and in many districts they must be much larger.

Powell was one of the first to realize that the limited availability of water would be an important limiting factor in the American west—hence his large-land comment.

Periodic severe flooding along the rivers was another factor that influenced the distribution of the settlers, particularly in the Red River Valley. Pembina, established before 1800, was the earliest permanent settlement in what is now North Dakota. It was a Métis settlement, but other colonists, including Swiss and Scottish people, settled there in 1812. As a result of the deep snows of 1825–26, the Red River flooded so badly that the Swiss people left and moved to Illinois. In the mid-1800s, Pembina was not at all like people today expect a village to be. It had a population of 1,000 in 1849. Most people, both whites and Métis, lived in skin or bark lodges because the spring floods washed everything away almost every year. Finally, in 1853, most of the inhabitants of Pembina moved to higher ground at St. Joseph (now Walhalla) because of the floods of the early 1850s, especially the one in 1852, which was probably the highest since 1800.

Much of the Red River Valley was settled earlier than most of the rest of the state due to its proximity to Lord Selkirk's Red River Settlement (Winnipeg) in the Lower Fort Garry area. At the time of the Red River cart trails, before the railroad was completed east-west across Canada, the Red River Settlement depended on north-south trade to St. Paul. Commerce between Winnipeg and St. Paul followed two main routes in North Dakota and several in Minnesota.

Some of the first attempts to establish permanent settlements in what is now North Dakota were made by traders of the Hudson's Bay Company, which began operations in the Red River Valley in 1793. In 1797, C. J. B. Chaboillez, a French trader with the North-West Fur Company, built a trading post on the southern bank of the Pembina River, near its mouth, but this was soon abandoned. Three years later, Alexander Henry built two trading posts in what is now North Dakota, one on the western bank of the Red River near the Park River, where he lived until 1808. In 1801, John Cameron erected a trading post for the North-West Fur Company on a site where Grand Forks is now. All of these early settlements were situated at strategic locations along the Red River, and all of them were subject to periodic flooding.

Among the first St. Paul-Winnipeg travel routes in North Dakota, the Ridge Trail followed the Lake Agassiz beaches along the edge of the lake plain because the land there was drier and sandier (therefore less mud), and it had fewer mosquitoes. It offered a better view, too, and the lack of woods reduced the chance of ambush by hostile Indians.

Another travel route was the West-Bank River Trail, which was used in dry years in autumn and winter. It followed the river, encountering many marshes, sinkholes and muddy areas. Near the Red River, tributaries such as the Park, Forest and Turtle rivers are quite large, and crossing them presented problems. The water in these rivers was so salty that it could not be used for drinking or cooking. It was not until 1871 that the stage route between St. Paul and Winnipeg was in operation and, with it, bridges and daily mail service.

For those interested in the distribution and history of the various ethnic groups found in North Dakota, I suggest a 1983 book titled, *Prairie Mosaic: An Ethnic Atlas of Rural North Dakota,* by Father William C. Sherman. Sherman's book provides a comprehensive analysis of the surprisingly large number of ethnic groups that came, first to Dakota Territory, and then to early North Dakota. His book includes maps showing where these groups were located as of 1965, and it presents something of both the times of settlement and the various social institutions around which the national communities were known to have rallied.

Sherman's intent was not necessarily to tie the settlement patterns to the geology or natural environment, but he cites numerous examples of how the early settlers were influenced by the environment. I will mention just a few examples.

Wherever the Anglo-Americans, people of British Isles background, settled in North Dakota, they tended to choose the rougher areas, becoming ranchers; many of their descendants are still ranchers. The Anglo-Americans entered southwestern North Dakota from Montana and South Dakota to populate the rugged badlands area of southwestern North Dakota. These people belong to the western cattle culture, the cowboys. A natural affinity seems to exist between cattle-raising and a British Isles background. Many Anglo-Americans came with the trail herds that moved up the eastern slopes of the Rocky Mountains. They preceded the dirt-farmers, and chose to settle in the rougher areas.

The earliest Germans from Russia arrived in 1884 and settled in the south-central part of the state. They came to North Dakota as a result of railroad advertising and favorable reports of advance scouts. If the scout's reports were promising, the German-Russians arrived in force with people and goods. They were farmers from the steppe of Russia, with a tradition of farms, vineyards and gardens. For this reason, they avoided rough country. One of their quotations says: "Every hill is a thorn."

Life on the steppe had made the Germans from Russia suspicious of mountains and comfortable on the plains of North Dakota, even though, prior to moving to Russia, they had come from the hills and forests of Germany. They knew how to build an adobe home, lay out farmsteads, build a root cellar and dig a well. This knowledge came from their previous Russian experience.

Usually, the German-Russian practice of choosing lands similar to those with which they were familiar was successful, but sometimes it led to serious mistakes. In Pierce and McHenry counties, for example, some of them chose the sand hills (collapsed outwash and lake-plain dunes) because the sand areas in Russia where they came from were good farmland in that climate. In North Dakota, however, where it is usually drier, sand hill farming was often a disaster. After the topsoil was disturbed, it began to blow away.

The German-Russians coming to North Dakota expanded northward, out of South Dakota, across the more customary east-west settlement pattern. German-Russian people living in north-central North Dakota have been quoted as saying "there were too many rocks down in southern North Dakota" and that's why they moved north. This comment does not necessarily make sense geologically because, overall, the two areas are essentially similar in that respect, although one of the most bouldery areas I know of is in the Zeeland-Venturia vicinity, in southern McIntosh County.

Icelanders, who had settled near Gimli, Manitoba, in 1875, moved south in 1878 to the Mountain, Hallson and Gardar areas (all in northwestern Pembina County). They chose the edge of the flat Red River Valley floor, where it meets the Pembina Escarpment, ostensibly because from there they had a view to the east of the flat lake plain, which

▲ This view is eastward from the Pembina Escarpment in Cavalier County. The town of Mountain is four miles to the east, in Pembina County. The broad and level land in the distance is the glacial Lake Agassiz plain, the Red River Valley. At the time Icelandic settlers moved south from Canada to settle in this area, no shelterbelts existed, and the land bore a closer resemblance to Iceland than it does now. JPB: Photo: 6/29/2009 (it was said) resembled Iceland. To the west, the land is hilly, sandy and somewhat rocky along the former shoreline of glacial Lake Agassiz. I have been in eastern Iceland, from where many of the people emigrated, and I am not entirely convinced there is any real resemblance between the treeless prairie there and the view eastward from the town of Mountain.

The first Norwegian settlers, arriving before 1870, came to the Wild Rice River area of Cass County and the Fort Abercrombie area of Richland County. Their farms commonly extended from the rivers onto the nearby prairies. The Norwegians usually chose areas near the river valleys, and the old observation that Norwegians liked the valley seems to hold true in North Dakota, especially in the Sheyenne and Goose River valleys. Perhaps the valleys reminded them of the fjords of Norway, or maybe it was just sensible to have access to the water in the rivers, but the presence of rivers and creeks definitely influenced how the Norwegians dispersed throughout the state. They followed the Sheyenne and Goose rivers upstream and, in fact, the Sheyenne River Valley is an almost exclusively Norwegian river from southern Cass County for 200 miles upstream.

In north-central North Dakota, the Norwegians settled first along the Souris River in Bottineau County, and then went on to fill up the treeless prairie. They settled along the southern side of Turtle Mountain where water and wood were abundant. The earliest Norwegians in Bottineau County were country people who did not like towns and villages. For example, in 1886, of the 100 people living in Dunseith, only one was a Norwegian. He ran a saloon.

People of Belgian ancestry came to parts of northwestern North Dakota from Minnesota in the early 1900s. They originally came to work in the coal mines, but eventually became farmers. Finnish settlers preferred timber and hillside areas, and many of them ended up with poor-quality land. Some Polish immigrants, along with some Bohemians, settled in southwestern North Dakota and became coal miners, due in part to their mining experience in Europe. Those who remained eventually became farmers or moved to towns. In eastern North Dakota, in the Warsaw area, Polish immigrants came to farm on the level land of the old Glacial Lake Agassiz. Much of central and northern Poland is quite flat. Perhaps they felt comfortable on the lake plain.

The French settlers, who arrived before 1870, chose land that included a combination of water, woodland and prairie. A French farm was a "long lot," extending from the water line, through the trees, and

▲ When the Germans from Russia settled in North Dakota they brought their distinctive building methods with them, using locally available materials at relatively low cost. The Hutmacher Farmstead, 10 miles southwest of Killdeer in Dunn County, is an excellent example of the use of a variety of geologic materials by settlers. Built of local resources for less than $100 between 1928 and 1964, the buildings are excellent examples of building techniques used by Germans from Russia.

The Frank Hutmacher farmstead incorporates a traditional, ethnic architectural style that originated in the Black Sea region of Russia and the Ukraine. The farmstead features five stone-slab buildings with roofs of clay and straw supported by timbers and branches. The buildings include a farmhouse, livestock barn, garage, granary, outhouse, oil shed and poultry barn (as well as a "cluck house" for baby chicks). The exterior and interior wall surfaces of the farmhouse were covered with a clay and straw mixture to finish the walls and provide an additional barrier from wind and water.

The buildings, constructed of sandstone, clay, brush and straw, used only small amounts of lumber and other salvaged or purchased materials. The walls are built of sandstone slabs (Sentinel Butte Formation) gathered from nearby hilltops. The stones were stacked about 16 inches thick and held in place with clay/mud mortar. This stone slab construction technique was used in North Dakota by both Ukrainian and German-Russian immigrants. JPB Photo 8/6/2009 out onto the prairie. French settlers were not aggressive in their acquisition of land and, although they arrived early and presumably had their choice of the best and as much land as they might have wanted, they kept their holdings small.

The first Ukrainians homesteaded in the Belfield area in 1896. They had lived on the steppe, and most of them came initially to farm, choosing the better homestead lands in the area between Belfield and the badlands. Some of them came to work in the coal mines in the early 1900s. The Ukrainians built the well-known adobe-type houses and barns found in southwestern North Dakota. The Grassy Butte "sod" post office is really a conventional Ukrainian log building, covered in adobe, with a sod roof.

Ukrainians came to Wilton to work in the coal mines, too. Along with some Austrians and Hungarians, they mined the lignite at Wilton. The hills to the south of the periphery of the town of Wilton are spoils of the old Truax-Traer Mine, which was a surface mine. However, underground mining in the area dates to 1901. In 1907, the Wilton underground mine was billed as the largest underground lignite mine in the world. At times, the mine had more than 900 men on the payroll, which was as much as $90,000 a month.

Summary

The most obvious link between geology/natural environment and settlement patterns by the various ethnic groups relates to their choice of landscape. Many settlers chose to settle in familiar areas, similar, in appearance at least, to their Old World homeland, sometimes in spite of the availability of better farmland nearby. Their choices were not necessarily wise ones. Some ethnic groups chose places to settle based on their background. Some came as miners. This was true in other states as well. In Montana, the Welsh and Irish came to Butte to work in the copper mines. My wife, Mary, grew up near Streator, Illinois, where a large proportion of the settlers came from Slovakia to mine coal (but Mary's ancestors came from Ireland, to farm). In North Dakota, some settlers preferred ranching and chose the rough badlands in the west. Wherever they went, immigrants to North Dakota were influenced by their impression of the landscape, which is a manifestation of the geology.

CHAPTER NINE

NORTH DAKOTA'S ENERGY RESOURCES

In addition to geology's role in shaping North Dakota's landscape, geologic processes have been of vital importance in developing our state's energy resources:

- Lignite coal, produced in large surface mines, is used to provide fuel for electrical generating plants and to produce an array of chemical products.
- Garrison Dam harnesses the Missouri River's hydroelectric power.
- Geothermal energy from the Earth heats schools, municipal buildings and homes.
- Wind turbines perched on the edge of the Missouri Escarpment and other points of geologic prominence produce electricity.
- Solar energy shows great promise in North Dakota, where sunshine is abundant.
- Oil and gas are pumped from the deeply buried geologic formations of the Williston Basin.

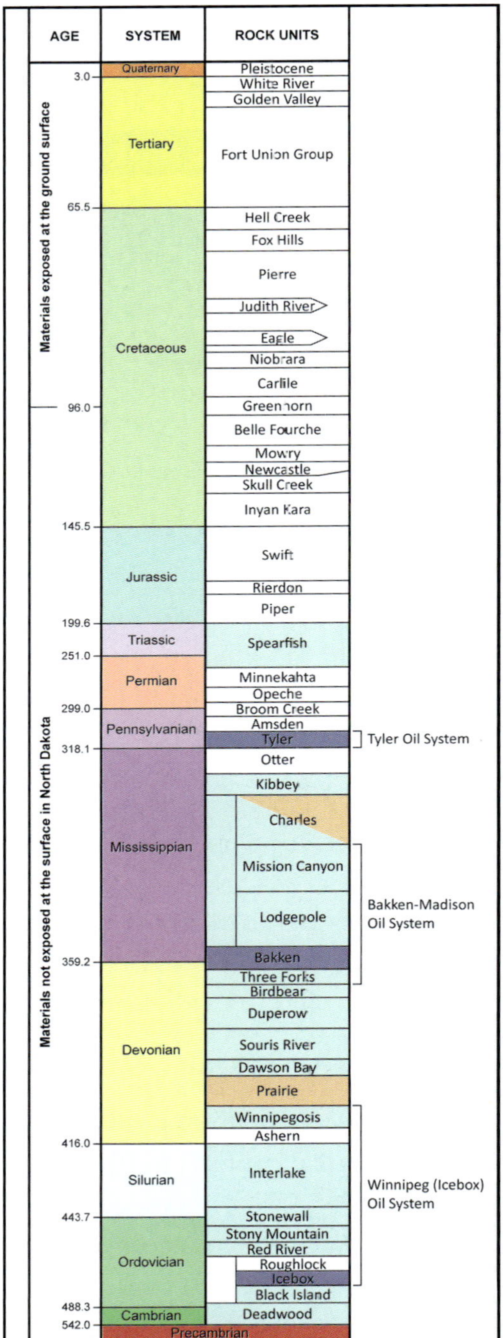

Geologic Formations in the Williston Basin

Lignite

Geology

Lignite coal underlies much of the western two-thirds of North Dakota. It occurs in geologic formations exposed at the ground surface, or just beneath the surface, buried under a few tens of feet of glacial deposits or bedrock (sandstone, shale, etc.). The Paleocene Bullion Creek and Sentinel Butte formations, which contain most of our lignite, reach a maximum depth of 1,800 feet in the center of the Williston Basin near Watford City; however, we utilize only the uppermost few hundred feet, as deeper lignite seams are beyond the reach of surface mining operations.

◀ This illustration shows the geologic formations that occur in the Williston Basin of North Dakota (**Rock Units** column), along with the geologic systems in which they occur (Cambrian, Silurian, Tertiary, etc.). The illustration represents the rock units that might be encountered by a hole drilled in the thickest part of the Williston Basin.

Systems are age units, which are determined by radiometric, fossil, relative position and other geologic evidence. The younger geologic formations, nearer the top of the stratigraphic column, are exposed at the ground surface. The colors used to differentiate the systems from one another are largely those recommended by the 2009 International Commission on Stratigraphy.

The left-most side of the stratigraphic column (**Age**) shows the ages of some of the units in the **Rock Unit** column. Ages are given as millions of years ago. Only the very uppermost rocks of Precambrian age are represented. To portray the entire 4.6 billion years of the Earth's history would require the column to be nine times as long as is shown.

In addition to formation names, the **Rock Unit** column shows the main recognized oil-producing formations, as well as formations that contain salt. Formations colored light blue (Interlake, Duperow, Lodgepole, etc.) have produced oil in North Dakota. The dark blue formations (Icebox, Bakken and Tyler) serve as source beds from which oil is thought to have migrated to other formations—and they have themselves produced oil.

Salmon-colored formations (Charles and Prairie formations) in the **Rock Unit** column contain salt, mainly potash.

This illustration is based, in part, on information included in an article by Julie A. LeFever in the July 2014 issue of *Geo News*, vol. 41, No. 2, p. 17–22.

Lignite, at 55 million to 60 million years old, is geologically younger than coals in most other states, where it may be more than 300 million years old. It is quite soft because it was never subjected to extensive alteration by heat and pressure through geologic time. Lignite is a low-rank coal, meaning that it contains less carbon and, with a heating value of approximately 7,000 Btus per pound, it has less energy per pound than higher-rank coals. Even though it provides less energy than coals of higher rank, such as subbituminous and bituminous coals, the sheer volume and easy accessibility of North Dakota's lignite make it a vital source of energy.

North Dakota has a total lignite resource of 1.27 trillion tons, but of this, only 2 percent (25 billion tons) can be economically recovered by surface mining. At the current rate of production, North Dakota's recoverable lignite resource would last 825 years. The state's lignite reserve is the second largest in the world (only Australia has more).

Production History

The American Indians probably used lignite for heating for centuries. Lewis and Clark understood the importance of lignite in an area where fuel was otherwise scarce. While traveling through today's McKenzie County, Meriwether Lewis discussed outcrops of lignite along the Missouri River in his journal entry for April 16, 1805. In 1867, while at Fort Stevenson, General Philippe Regis deTrobriand conducted tests on lignite to determine its heating qualities.[67]

The earliest written account of lignite mining dates to 1873, when small mines were situated along the main transportation routes in western North Dakota.[68] Early settlers mined lignite by tunneling short distances into the outcrops, or simply by removing shallow overburden and digging out the lignite, a procedure similar to modern strip mining. By 1900, at least 73 mines were operating in the state. Most were small, seasonal mines in which coal was removed from the face of the outcrop. These mines were referred to as wagon mines because farmers and ranchers brought their wagons to the site and filled them with coal. Some underground workings extended horizontally for short distances into outcropping lignite beds.

Prior to 1902, half of North Dakota's lignite output was from small strip pits along outcrops, where overburden was thin. Before

▲ **The lignite seam exposed in this road cut 10 miles west of Twin Buttes (14 miles northeast of Dunn Center) is 7 feet thick. This exposure is fresh (recently cut during road construction), so the entire thickness of the lignite seam can be seen. Most natural exposures of lignite tend to be slumped and weathered so that the actual thickness cannot be determined without digging into it to find the top and bottom of the lignite seam.** JPB Photo: 9/25/2009

power stripping equipment was developed, 10 feet of overburden was the practical maximum to allow mining a 6-foot-thick seam. Between 1902 and 1919, underground mining developed on a larger scale and, for a time, underground production overtook the stripping method. By the early 1920s, approximately 250 mines operated in the state, with equal numbers of underground and surface strip mines.

In 1918, the Whittier Coal Company and Truax Brothers began stripping lignite with horse-drawn elevating graders and dump wagons at Kincaid, in Burke County. The wagons were used both to dispose of overburden and to haul lignite to a tipple, a structure used for loading coal into railroad cars. In 1919, Truax introduced the first power equipment, a 1½-cubic-yard steam shovel, which replaced the elevating grader for loading overburden into dump wagons. The same shovel was used to load lignite into narrow-gauge pit cars that were hauled to the tipple by steam locomotives.

Due to the increasing use of natural gas and fuel oil, the number of underground lignite mines in North Dakota dropped from 320 in 1940 to zero in 1966, when the last underground mine closed.[69] Since then, several large electrical generating stations, consuming lignite from nearby surface mines, have operated in the state. Because water is required to generate electricity, these plants are clustered near the plentiful water supply of Lake Sakakawea in west-central North Dakota.

In a modern surface-mining operation, the overburden is removed, without explosives, using draglines with bucket capacities of more than 100 cubic yards. The lignite is broken up with explosives to facilitate loading by power shovels, which are more effective than front-end loaders for digging in the thick coal seams.

Uranium Contained in Lignite

Several areas of radioactive mineral deposits occur in western North Dakota, along with lignite resources. Investigations done during the late 1940s to the late 1970s identified areas in Billings, Bowman, Golden Valley, Slope and Stark counties where uranium and other radioactive elements were found associated with beds of lignite.

Beginning in 1956, small amounts of uraniferous lignite were shipped to processing mills in other states, but the mills, which were designed to process uraniferous sandstone, could not efficiently process the low-grade-ore lignite. This problem was remedied by burning the lignite in open pits at the local mine sites, sometimes burning the lignite bed in place after the overburden had been removed. The incineration process took up to two months. Diesel fuel and old tires mixed with the lignite helped it burn sufficiently. The resulting ash from the mines was further reduced by burning in kilns.

The kiln-burned ash was shipped to mills in other states. At least seven uraniferous lignite mines operated in North Dakota. Record-keeping was poor for a variety of reasons, including national security during the Cold War period. Most of the mines were short-lived operations, and mining was discontinued in 1967. Renewed interest in uraniferous lignite in the mid-1970s resulted in a number of additional lignite/uranium investigations, but the accident at the Three Mile Is-

▲ This dragline five miles north of Washburn is removing overburden on top of the lignite seam being mined at Falkirk Mine in McLean County. The machine does not handle lignite. Rather, it removes the material overlying the lignite, making it possible to access the lignite and load it into trucks for transport to the nearby Coal Creek Station electrical generating plant.

The 13.5 million-pound dragline, named Chief Ironsides, has a bucket that holds up to 125 cubic yards of overburden. The mine has two of these draglines; the other, the Prairie Rose, is the same size. The base of each dragline covers an area the size of a baseball infield. The dragline boom is 215 feet from the top of the boom to the ground and the boom is 360 feet long. The bucket (20 by 16 by 10 feet) weighs 244,000 pounds (122 tons) empty and holds 160 tons of material. The dragline uses enough electricity to supply a city of 1,775 people.

Chief Ironsides and Prairie Rose are "walking draglines." They walk on 16- by 80-foot shoes at a rate of 7 feet per step, moving 1/10th of a mile per hour. On January 26, 2010, Chief Ironside traveled across U.S. Highway 83. To keep from damaging the highway, the company built a 10-foot-thick soil road on top of the highway for the dragline to walk on. Chief Ironsides will remain east of the highway for 20 years.

Each year, Falkirk mines between 300 and 400 acres of land and reclaims the same amount, delivering 7 million tons of lignite to Coal Creek Station. At the current rate of production, the 75,000-acre mine has sufficient lignite reserves for 100 years of production (650 million tons of coal). JPE Photo: 6/9/2013 land nuclear plant in 1979 effectively halted all uranium exploration in the United States.

In recent years, the uranium deposits associated with lignite have been investigated as health hazards rather than potential mineable products. Concern is expressed for human and livestock consumption of uranium contained in ground water in some parts of southwestern North Dakota. Elevated radon levels near the deposits are a concern, too.

North Dakota's lignite/uranium mines operated without any reclamation laws, and the open pits were left as they were when the sites were abandoned. In the 1980s and 1990s, the North Dakota Public Service Commission reclaimed the sites for health reasons by burying the most radioactive material in the bottom of the pits and leveling the surface.

The Current Lignite Industry in North Dakota

The production and use of lignite is one of North Dakota's largest industries, along with agriculture, oil and gas production, tourism and manufacturing. Currently, lignite is produced at four large mines: the Beulah Mine, Center Mine, Falkirk Mine and Freedom Mine. A fifth mine, the Coyote Creek Mine, proposed by North American Coal Corporation, is planned to be operational in 2016. In recent years, around 30 million tons of lignite have been mined annually in North Dakota. In 2010:

- 79 percent was used to generate electricity for more than 2 million people;
- 13.5 percent was used to generate synthetic natural gas for distribution to 400,000 homes and businesses in the eastern United States;
- 7.5 percent was used to produce fertilizer and other products.

Electricity is generated at six lignite-fueled power plants in western North Dakota and one in eastern Montana. These power plants provide most of the electricity in the north-central U.S. region at considerably lower costs to consumers than in much of the rest of the United States.[70] The average price of electricity generated by our lignite power plants is less than half what residents of states on the East and West coasts pay.

The gasification of North Dakota's lignite began in 1984 with the completion of the Great Plains Synfuels Plant, near Beulah. The plant is the nation's only coal-to-(synthetic) natural gas (syngas) plant. It produces 153 million cubic feet of gas daily, most of which is piped to Ventura, Iowa, for distribution in the eastern United States. The plant produces a variety of agricultural fertilizer products such as anhydrous ammonia and ammonium sulfate, as well as other chemicals, such as liquid nitrogen (used in food processing), naphtha (used in solvents), phenol (as a resin), and krypton and xenon gases (used in lighting and thermopane windows). It captures 8,500 tons of carbon dioxide daily, which is transported, via pipeline, to oil fields near Weyburn, Saskatchewan, where it is stored, underground, while helping maintain pressure in the old, partially depleted, oil fields, increasing and prolonging oil production in those fields.

Reclamation

North Dakota enacted its first mined-land reclamation law in 1969, and made major revisions to it in 1973 and 1975. Following the federal "Surface Mining Control and Reclamation Act of 1977" (SMCRA), the North Dakota Legislature in 1979 adopted a new law to comply with the federal law. Today, ongoing mining operations incorporate concurrent reclamation of the mined land as mining progresses.

Lignite mining in North Dakota disturbs approximately 2,000 acres of land surface each year, and the same amount of land is reclaimed. The state requires, at minimum, the restoration of mined land to its original fertility and productivity. Some land has been reclaimed for other purposes. For example, the town of Underwood turned 150 acres of mined land into a golf course east of town. Post-mine topography usually has gentler slopes than pre-mine land because post-mine topography is determined, in part, by the landowner's preference for how the land should be reclaimed. Land that is reclaimed into a smooth surface configuration tends to result in more productive cropland.

Broad expanses of crops growing on formerly mined land testify to the success of reclamation. If you travel along U. S. Highway 83 near Underwood, or near any large lignite-mining operation, you can watch huge draglines carrying out mining operations. Not far from the mining activities, you may see fields of grain growing on land recently reclaimed. Except for small signs posted along the perimeter of reclaimed fields, it is not evident that the land has been mined.

Reclamation notwithstanding, scarred land remains in some areas mined prior to enactment of the 1969 reclamation law. The Truax-Traer Coal Company operated the Custer Mine six miles east of Garrison, at the junction of U. S. Highway 83 and State Highway 37, from 1947 until 1964. The mined area includes the underground Hummel Mine, which operated during the 1930s. Although the land was left unreclaimed when the mine closed, local residents began planting trees on mine spoil banks while the mine was still in operation. Tree-planting "bees" were initiated by the Garrison Sportsmen's Club in May 1949. Youth groups such as the Boy Scouts, Girl Scouts, 4-H and school groups helped with the task. The spoil banks today are heavily vegetated and support abundant wildlife, thanks to the thousands of trees planted there.

Another area, mined between 1920 and 1940 in the Noonan-Columbus area, Divide and Burke counties, was never reclaimed. Reclamation was considered, but a study of the ground water conditions and the natural salts within the spoil piles in the early 1980s recommended that the spoil piles be left as they are because, the study concluded, leveling them might harm the quality of ground water in the area. However, the Public Service Commission in January 2012 announced that 350 acres of the mine spoils south of Columbus would be reclaimed and new vegetation planted on the leveled land.

In some places where underground mining took place, mine shafts have collapsed when the timbers rotted, so that the land surface above is pitted. When an underground mine was abandoned, miners would remove ("rob") lignite from the supporting pillars. Because of the inherent instability of the unconsolidated materials overlying underground mine voids, the surface above the mine has collapsed in many places, resulting in sinkholes. The North Dakota Public Service Commission has largely remedied the sinkhole problem by digging out the old mine workings where they are close to the surface, and by pumping grout into the deeper workings to stabilize them. Part of the coal severance tax is used to fund the reclamation of old underground mines.

In closing this discussion of reclamation, it seems appropriate to quote North Dakota Governor Arthur A. Link's "When the Landscape is Quiet Again" statement, made on October 11, 1973:

> We do not want to halt progress. We do not plan to be selfish and say 'North Dakota will not share its energy resource.' No, we simply want to insure the most efficient and environmentally sound method of utilizing our precious coal and water resources for the benefit of the broadest number of people possible. And when we are through with that and the landscape is quiet again, when the draglines, the blasting rigs, the power shovels and the huge gondolas cease to rip and roar! And when the last bulldozer has pushed the last spoil pile into place, and the last patch of barren earth has been seeded to grass or grain, let those who follow and repopulate the land be able to say, "Our grandparents did their job well. The land is as good and, in some cases, better than before." Only if they can say this will we be worthy of the rich heritage of our land and its resources.

The Future of Lignite

New technologies and processes continue to improve the efficiency of North Dakota's lignite as an energy source. New and proposed projects utilizing lignite include an electrical generating plant with a capacity of 99 MW (megawatts) at Spiritwood, Stutsman County. The Spiritwood Station plant supplies steam to a nearby malting plant and to an ethanol plant. The lignite used in the plant is dried and refined at Great River Energy's Coal Creek Station near Underwood before being shipped to Spiritwood. The Spiritwood Station power plant was completed in 2011 and became operational in October 2014.

Another proposed mine, planned near South Heart in Stark County, would convert lignite to synthetic gas and hydrogen, which would be used to power turbines generating up to 175 MW of electricity. The plant's carbon dioxide production would be sold to oil companies for pressurizing oil fields.

Much of North Dakota's lignite occurs at depths much too deep to use conventional strip-mining methods. In places too deep to surface mine, underground coal gasification methods might be used to produce synthetic natural gas (syngas) in place, without mining it. Underground gasification involves converting the lignite to gas by injecting air (oxygen) and steam into the lignite at high pressure and temperature to ignite and fuel underground combustion. High-pressure combustion then takes place at temperatures as high as 2,700°F, decomposing the lignite into gases, which are brought to the surface and processed into usable synfuel products.

Using underground gasification methods, the amount of North Dakota's usable lignite resource could be increased by many times. Because it would take place at depths up to 2,000 feet, underground gasification would not disturb the land surface. Syngas can be processed into a variety of products, including methane, hydrogen, urea and liquid fuels. Because the process takes place underground, air pollution is essentially eliminated.

Hydroelectric Power

The Missouri River provides a reliable flow of water, the result of the melt-off of annual snowfall in the mountains of Montana and Wyoming, and the system of large dams that have been built on the river. North Dakota has one major hydroelectric plant, at Garrison Dam, which spans the Missouri River Valley between Riverdale in McLean County and Pick City in Mercer County. One of five large dams on the Missouri River, Garrison Dam holds back Lake Sakakawea, which floods 480 square miles and holds 24 million acre-feet of water, the third largest reservoir in the United States, by volume. Lake Sakakawea extends a distance of 200 miles upstream from Riverdale to near Williston and has 1,240 miles of shoreline. Constructed between 1947 and 1954, Garrison Dam is 210 feet high and 2½ miles long.

The five turbines of the Garrison Dam hydroelectric generating facility have a maximum capacity of 583 megawatts (MW), with an average output of 240 MW, enough power to supply electricity for 350,000 people (a megawatt hour is enough electricity to provide an hour's worth of power to more than 800 homes). The plant is operated by the U. S. Army Corps of Engineers, with five hydropower generating units in the powerhouse.

Lake Oahe, the reservoir stored behind Oahe Dam, at Pierre, South Dakota, extends northward 230 miles to near Bismarck. Lake Oahe and its dam are comparable in size to Garrison Dam and Lake

Sakakawea. The lake floods 580 square miles, holds 23.5 million acre feet of water, and has a shoreline extending a total of 2,250 miles. Oahe Dam has a maximum hydroelectric power capacity of 786 MW.

Geothermal Energy

The temperature of the Earth increases with depth. At its core, the temperature is approximately 12,000° Fahrenheit, the combined result of heat produced due to the pressure of the overlying rocks, radioactive decay of isotopes and the residual heat from planetary accretion, the original heat that accompanied the formation of the planet 4.6 billion years ago. The Earth's natural heat provides geothermal energy. The amount of that heat reaching the surface totals twice the energy humans consume from all sources. Heat flows constantly from the Earth's interior, but it reaches the surface in greatly varying amounts in different places. In Yellowstone Park, for example, the heat is sufficient to cause water near the surface to boil and geysers to erupt. In the summer of 2014, it even melted some asphalt roads in part of the park. In areas where it is most abundant, geothermal heat can be used for direct heating purposes: desalination, spas and generating electricity. Approximately 0.3 percent of the global demand for electricity in the United States is currently provided by geothermal heat.

Western North Dakota has potential for geothermal heat from deep sources (deep enhanced geothermal systems – EGS). EGS technology uses heat from deep in the ground to turn water into steam, which drives turbine generators to produce electricity. However, the geothermal energy currently used in North Dakota does not utilize geothermal heat in the sense that many people think of it; no hot springs, geysers, deep, molten magma or ultra-hot rocks are involved. Such EGS technology sources are restricted to countries such as China, Iceland, Japan and New Zealand, as well as a few locations in the western United States, in California, Nevada, Wyoming (Yellowstone Park area), and similar potential in Canada.

In North Dakota, the upward flow of heat keeps the ground water below a certain depth, below the frost zone, at a near-constant temperature. The geothermal heat pumps in North Dakota extract heat from ground water. Since the mid-1980s, almost all geothermal installations in North Dakota have been closed-loop, water/antifreeze systems. In such systems, a U-shaped loop of pipe is placed in a vertical drill hole 150 to 200 feet deep, and the hole is back-filled with clay. Some closed-loops are placed horizontally in 5- to 10-foot-deep trenches or in lakes or ponds. Heat pumps add or remove heat from the circulated solution, depending on whether heating or cooling is desired.

No ground water is consumed by a closed-loop geothermal system. The temperature in the shallow ground water zone, or in the ground itself if water is absent, ranges from 43° to 50° Fahrenheit in North Dakota. A heat pump operates on the same principle as a home refrigerator or air conditioner, both of which are actually one-way heat pumps. A geothermal heat pump moves heat in either direction: out of the ground in winter for heating, into the ground, from the home or building in summer, for cooling. Water for home use can be heated and

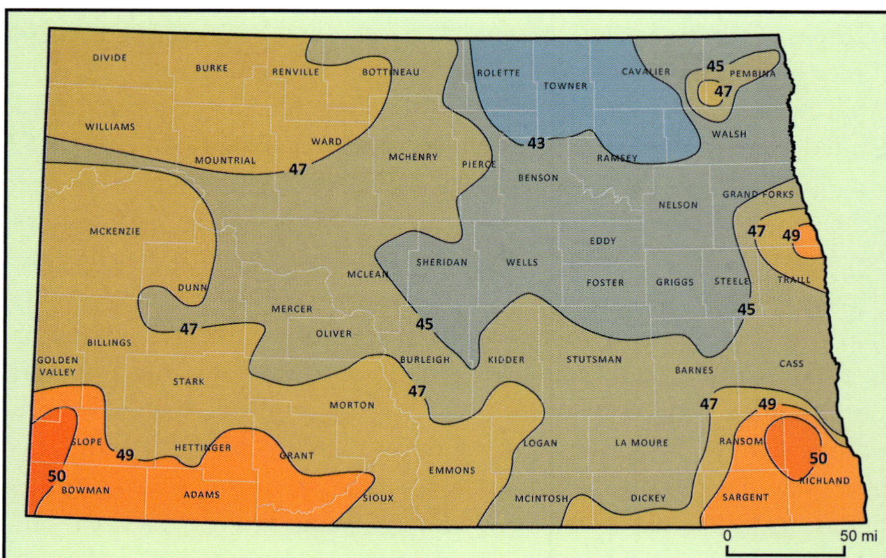

▲ This map of North Dakota shows the expected ground water temperature in degrees Fahrenheit at a depth of 100 feet. Cooler temperatures tend to be in the east, warmer in the southwest and southeast.

Map is modified from one by Harris, K.L., Howell, F.L., Wartman, B.L., and Anderson, S.B., 1982 (see references).

stored in a separate hot-water heater.

Through 2013, a total of 1,393 permits had been issued in North Dakota for new geothermal system installations. Most commercial-size systems are located in the main urban areas. The Davies High School in south Fargo utilizes the largest geothermal system in the state (as of 2014). The 276,000-square-foot complex is heated and cooled by 928 wells, each drilled to a depth of 180 feet. In Bismarck, a geothermal system heats and cools a 97,000-square-foot addition to the North Dakota Heritage Center on the State Capitol grounds. At Minot State University, an even larger system is planned to heat and cool the entire campus.

In recent years, North Dakota has ranked either first or second in the nation in the number of geothermal heat pump systems per capita. Such systems may become the most popular installation in new homes, as well as in older homes when conventional climate-control systems need to be replaced. Currently, the most economical way to heat your home in North Dakota is with a ground-water heat pump. The costs of installing geothermal systems are higher than conventional gas furnace/electric air conditioner systems, but because operating costs tend to be lower than natural gas, fuel oil, propane or electric heat, a geothermal system can pay for itself in a few years.

Wind Energy

David Thompson, a Canadian explorer, surveyor and map maker, was one of the first explorers in North Dakota to comment on wind. While he was traveling down the Souris River from the Assiniboine River in 1797, the winds were so strong that he surmised they were responsible for bison distribution and herding behavior, as well as the lack of trees on uplands. In December 1797, he commented: "I am utterly at a loss, to account for such violent winds on this part of the Plains, and this may account for the few Bison we have seen, and the smallness of the herds, which rarely exceed twenty." He later commented: "It is in these Bottoms [along the Souris River] that Trees grow, and are sheltered from the Storms; for on the level of the Plain, it is not possible a tree can grow."

We know now that climate and weather result in wind energy. The wind results from the interaction of two separate forces:

- Pressure gradients, produced by uneven solar heating (the air moves between atmospheric highs and lows)
- The rotation of the Earth, which produces a separate, additional, movement of the air

Wind speed and density both determine wind power class. Wind density is a measure of energy, in watts per square meter. The denser the air, the more energy it provides to a turbine. Air density is greater at lower elevations, at colder temperatures and with decreased humidity. Cold, dry air near sea level—moving fast—has the potential to power a wind turbine most efficiently.

The wind turbines that have been installed in North Dakota in recent years are 262 feet tall (the blades reach that high as they turn).

▲ Wind turbines (not yet operational at the time this photo was taken in August of 2009) stand on the upturned edges of ice-thrust blocks in an area of ice-thrust topography 10 miles north of Rugby in Pierce County. JPB Photo: 8/20/2009

This is 20 feet higher than the State Capitol building. The wind is usually strongest at 125 feet above the ground.

Wind speed varies according to time of day and season, and from year to year. In North Dakota, the average daily maximum wind speed occurs between 11:00 a.m. and 6:00 p.m. The peak speed is more pronounced during the summer months and less noticeable during the winter months. Minimum daily wind speed is usually lowest during the early morning hours. The time of year has an effect on wind speed and duration, too. The highest mean monthly wind speeds are in May and the lowest are in July, with the highest speeds occurring from December through May. Short-term weather conditions are often a factor.

The configuration of the terrain influences the wind speed and direction. Differences in elevation affect the wind resource, but overall, the state has a class 4 wind regime, averaging 14 to 15 miles per hour or greater. High-potential areas, with class 5 wind speeds, 16 to 18 miles per hour, include the Missouri Escarpment, which rises 600 feet above the surrounding terrain, the Pembina Escarpment, the upturned edges of an array of ice-thrust blocks in northern Pierce County, and Turtle Mountain, where elevations range from 600 to 700 feet above neighboring topography. The highest wind speeds occur over smooth terrain, such as elevated plateaus or lake areas, where friction is low (rough terrain causes friction, which decreases wind speed).

Wind power potential is classified into seven categories. North Dakota has the most area of any state with wind power class 4 and above, and our wind energy potential ranks among the top (windiest) states.[71] Because the power produced by a wind turbine increases in proportion to the cube of the wind speed, small increases in wind speed can result in substantial differences in the power provided by a wind turbine. As of December 31, 2014, North Dakota had 1,886 MW of installed wind energy capacity, ranking it 11th in the nation. North Dakota ranks first among the states in wind-generated electricity per capita. The number one state, Texas, had 12,214 MW of installed capacity. North Dakota, as this is written, accounts for 3.5 percent of the electricity generated in the United States. Completion of wind-energy projects proposed or under construction in the next two years will result in the state having 2,000 MW of wind-energy capacity.

Solar Energy

North Dakota's long summer days and dry climate, particularly in the western part of the state, provide us with more sunlight than much of the rest of the United States, more than any other state along the Canadian border.

The energy from the sun's radiation striking the Earth's land surface amounts to the equivalent of 26,000 terrawatts (1 terrawatt equals 1 trillion watts), or 1,600 times the current worldwide energy used by humans. Enough solar energy falls on the surface of the Earth every 40 minutes to meet 100 percent of the entire world's energy needs for a full year. Currently, most of that solar energy goes to waste (at least in a strictly marketable sense; it does, however, grow crops, control weather

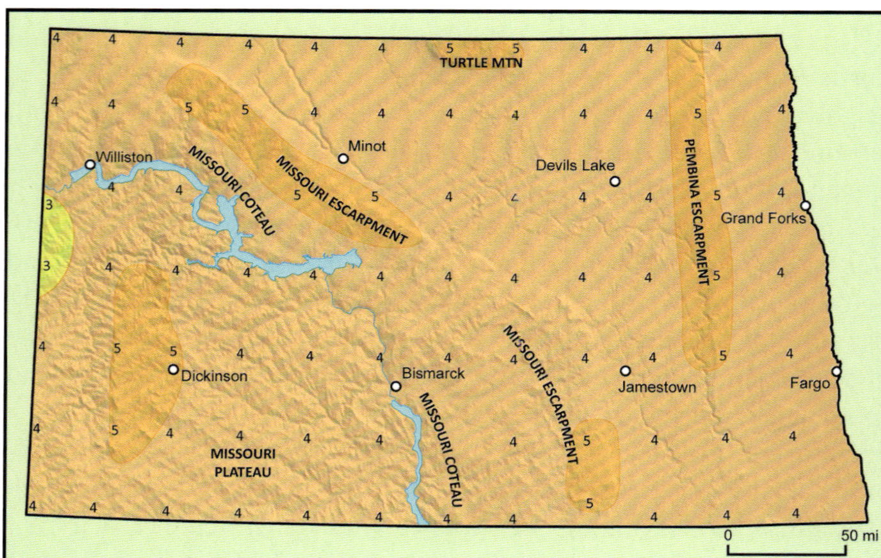

▲ **North Dakota, as a whole, is considered, overall, to have a Class 4 wind regime, meaning that wind speeds average at least 12 mph at a height of 30 feet off the ground. Some areas are designated Class 5 (average wind speeds about 14 mph). Wind speed and wind density are used to determine wind-power class. Seven wind-power classes have been designated. North Dakota has the most area of any state, with an average wind-power class of 4 and above.** Map was compiled by the U.S. Dept. of Energy.

and give suntans, among other things). North Dakota has an average solar energy density of 4 to 5 kwh per square meter per day, enough sunlight to derive sizeable amounts of solar energy. Even so, very few high-capacity solar installations currently exist in the state. The Energy Information Administration reports shipments of photovoltaic cells to North Dakota totaling only 31 kilowatts in 2008 and 2009, and the installation of 3,622 square feet of solar thermal collectors in the same period.

Solar energy technology can be classified into two main types: photovoltaics (PV) and concentrated solar power (CSP). CSP uses mirrors or lenses to concentrate large areas of sunlight into small areas, thus creating heat, which can be used to drive a heat or steam engine. PV power uses solar radiation to create direct current electricity.

Solar is an economical energy source for many locations where no electric power exists. You may have noticed the small solar panels that provide power to operate such things as railroad crossing signals. Solar

▲ **The solar panel, on right, generates electricity during the day to power the night-time lights on either side of the sign at the entrance to Cross Ranch State Park in Oliver County.** JPB Photo: 8/10/2014

energy is an economical choice vs. windmills and engine-driven generators for many locations where farmers and ranchers want to power remote pasture wells used primarily during the dry summer months.

In the Verendrye-Minot area, Verendrye Electric Cooperative is installing hundreds of solar arrays at stock wells in the six-county area it serves around Minot. The $800 solar panels are far more cost-effective than paying to extend transmission lines to remote sites.

Solar power will continue to grow in importance in the future, both in the United States and in North Dakota. In 2014, combined solar and wind power capacity in the United States overtook hydropower for the first time. The United States can expect to add 351 gigawatts of energy capacity in the next 25 years. Of that, most will be from gas-fired power plants, but more than 10 percent of the new capacity will be from solar power.

▲ This photovoltaic pasture well apparatus is located about eight miles south-southwest of McClusky in Sheridan County in an area of dead-ice moraine. Three solar panels provide 100 volts of electricity to provide power for an eleven-gallon-per-minute pump, which draws water through a 75-foot well. The water is pumped into two storage tanks, each with a capacity of about 2,000 gallons. The tanks can be accessed separately. When I was there, an electric fence, which runs through the pasture and between the tanks, restricted the cattle to the south part of the pasture. The rancher can balance pasture usage by moving the cattle between the north and south pastures. The system was installed in the spring of 2015.

I thank Tom Jesperson, Verendrye Electric Cooperative, who suggested this site to illustrate solar-powered pasture wells. JPB Photo: 7/17/2015

Oil and Gas

Early Oil and Gas Development in North Dakota

By the middle of the 19th century, basic knowledge of petroleum geology had progressed to the point that geologists understood that certain geologic structures are important in the search for oil. Geologists knew, for example, that oil tended to concentrate along the crests of anticlines. The existence of the Cedar Creek Anticline in southwestern North Dakota was already known in the late 1800s, and geologists recognized its importance to oil production.

The first hydrocarbon used commercially in North Dakota was natural gas, which was contained in the water flowing from artesian wells. Gas production began in southeastern North Dakota as early as 1892 from several artesian wells extending southward from Jamestown to Merricourt (north of Ellendale in Dickey County). The gas, known around the turn of the 20th century as marsh gas, was collected in tanks and was used for a time at Edgeley for lighting, cooking and heating. Eventually, when the artesian pressure head was lowered so that it stood below the land surface, gas production ended.

In Bottineau County, from 1907 until 1910, enough gas was produced near Lansford and Westhope to provide heat and light for several homes. The North Dakota Gas Company drilled eight gas wells, costing 13.6 cents a foot to drill. It built a 20-mile-long pipeline to Westhope and sold the gas to customers for 30 cents per 1,000 cubic feet in summer, 40 cents in winter. Local farmers used the gas to heat barns and other structures for several years before 1910. These gas wells, and others in north-central North Dakota near Mohall and Maxbass, were drilled with the ***intention*** of producing gas. Until that time, gas produced with water in North Dakota was an incidental byproduct of water wells.[72]

In Bowman County, in the southwestern corner of the state, gas was discovered on the Cedar Creek Anticline in 1914, and two gas fields produced there for many years. The gas was produced from Cretaceous sand formations, well above and unrelated to much deeper beds in the same area that later became important oil-producing formations.

A. J. Collier identified the Nesson Anticline in 1918 and, in 1920,

A. G. Leonard, North Dakota's third state geologist, described several more anticlines in North Dakota. The first actual attempt to drill an oil well in North Dakota (not a gas well or a water well that incidentally produced gas) was made in 1916, when the Pioneer Oil & Gas Company began drilling the Pioneer Oil & Gas No. 1 well near Williston.

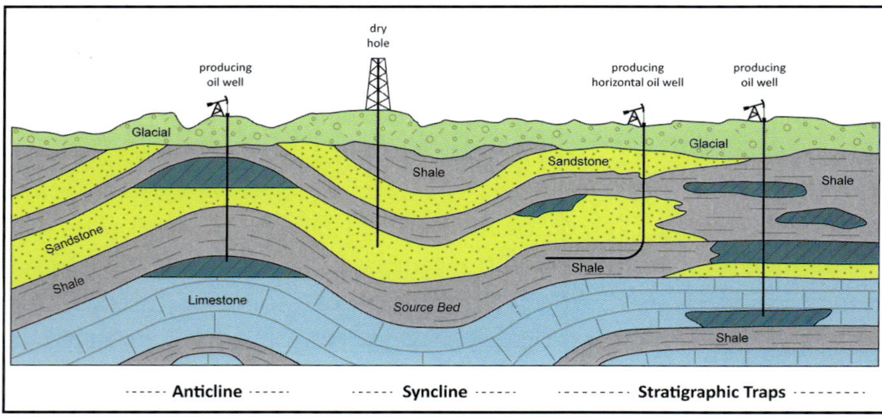

▲ This cross section—a vertical slice though part of the Earth's crust—shows an anticline on the left; a syncline in the center; and some stratigraphic oil traps on the right. The colors indicate the kind of geologic material: light blue is carbonate, gray is shale, yellow is sandstone and green (on the surface) is glacial sediment.

Anticlines, which are upward folds in layers of rocks, usually form during lateral compression. As two regions or tectonic plates are pushed toward each other, something has to give. Sometimes the rocks are crushed, and/or they may be pushed up to form mountains or, as shown here, they may bend, much like pages of a soft-backed book bend if you push on the binding and the side opposite the binding.

A syncline is a downward fold in layers of rock. It may form when lateral pressure produced downward folds. Synclines do not typically produce oil, as anticlines may.

Petroleum (oil and/or gas, shown as darker blue), which may form in oil-producing shale beds, can collect and become concentrated deep underground on the crests of anticlines, or in stratigraphic traps in some kinds of rocks, such as carbonates, or it may remain in the shale in which it formed.

The weight of overlying rocks deposited on top of the shale (and the oil it contains) may force the oil upward and/or horizontally into traps, such as anticlines. Or, the oil—all or some of it—may remain in the shale, as it has in the Bakken and certain other shale formations.

▲ **This oil pumper is located six miles south of Medora in Billings County.**
JPB Photo: 9/9/2009

The well produced only water and was abandoned at a depth of 2,107 feet in 1920.

The same time the Pioneer well was being drilled, a North Dakota corporation calling itself the Des Lacs Western Oil Company was organized in Minot. The company attempted to drill several wells in northwestern North Dakota, but it is difficult to be sure just what was going on much of the time, as the various press releases and reports that were released were a blend of facts, hearsay and promotional statements. Coincidentally, one Des Lacs Western hole was located approximately two miles from modern production in today's Lone Tree Field in Ward County. Wells in Lone Tree Field produce from the Sherwood Sand, part of the Mississippian-Madison Formation, at depths of 6,500 feet. The choice of a location was fortuitous, and if the Des Lacs well could have been drilled 4,000 feet deeper, it might have been a success. However, such a depth was well out of reach at the time.

A couple of other attempts to find oil involved test wells drilled on the Nesson Anticline in locations recommended by State Geologist A. G. Leonard. The Big Viking Oil Company No. 1, drilled in 1936, and the California Company Nels Kamp No. 1, drilled in 1939, were dry holes drilled within yards of each other. The Amerada-Hess No. 3-31 eventually found oil in Silurian rocks in 1985 at that same location. Either of the two early wells might have been North Dakota's discovery well had they been drilled deeper.

On April 4, 1951, Amerada Petroleum Corporation's No.1 Clarence Iverson well discovered oil 2½ miles southwest of Tioga in Williams County. Only a matter of months separated the discovery of oil in the Williston Basin in Manitoba (January 1951), North Dakota (April 1951), and Montana (July 1951). Oil was discovered in Saskatchewan in 1953.

Following its discovery, oil production climbed until 1966, when 27 million barrels of oil were produced – 74,000 barrels of oil per day (bopd). Production then declined, but in 1984, as a result of higher prices and therefore increased drilling, production peaked at 52.7 million barrels (144,000 bopd). That remained North Dakota's peak annual production until 2008, when the Bakken production pushed past the 1984 high, and North Dakota's production continued to increase through 2014. Early in 2014, North Dakota's oil production surpassed a million barrels a day.

For those interested in more detail, I wrote a brief history in 2001 of the first 50 years of North Dakota's oil industry, stressing the events relating to the discovery in 1951 ("The 50th Anniversary of the Discovery of Oil in North Dakota").

The Williston Basin

Large amounts of oil and gas occur in the thick sedimentary rock sequence of the Williston Basin, a large down warp in the Earth's crust in which more than 16,000 feet of sedimentary rocks have accumulated.[73] The Williston Basin underlies the western half of North Dakota, as well as adjacent areas in eastern Montana, northwestern South Dakota, southern Saskatchewan and southwestern Manitoba. The thickest rock section is near Watford City in McKenzie County. The rock formations in the Williston Basin represent 540 million years of the Earth's history and include a nearly complete section of geologic units with rocks from all geologic periods represented. Such a complete geologic section is uncommon in a sedimentary basin. Hydrocarbons in the Williston Basin in North Dakota occur in economic amounts in the lowermost Mesozoic and most of the Paleozoic formations (rock units older than 250 million years), and in smaller amounts in some of the younger formations.

The Bakken Formation

GEOLOGIC STUDIES OF THE BAKKEN

In recent years, hundreds of petroleum geologists and engineers have devoted hundreds of thousands of hours to studies of the Bakken Formation. They have published technical papers and determined how to best produce the oil in the formation. Surely, though, Julie LeFever, a petroleum geologist with the North Dakota Geological Survey, deserves more credit than anyone else for our current understanding of the Bakken. Since she joined the North Dakota Geological Survey in 1981, Julie has been studying the Bakken Formation; she has researched the geology of the Bakken rock cores for virtually her entire career with the Survey. Julie, director of the Wilson M. Laird Core and Sample Library on the University of North Dakota campus since 1989, has published dozens of technical reports on the Bakken. Her work has added immensely to our knowledge of the formation and the oil resource it contains. Much of the recent success of the Bakken oil play can be attributed to her perseverance through the years, and the excellence of her research.

In 1993, the North Dakota Geological Survey and its counterpart, the Saskatchewan Energy and Mines Branch, began sponsoring annual meetings to foster the exchange of information among geologists in industry and government across the U.S-Canadian boundary. Initially, the meetings emphasized horizontal drilling technology, but the range of geologic and engineering topics soon grew to include many kinds of information, including Bakken Formation geology. The annual meeting alternates between North Dakota and Saskatchewan and is now known as the Williston Basin Petroleum Conference.

DESCRIPTION OF THE BAKKEN

The Mississippian-Devonian age Bakken Formation (359 million to 360 million years old), underlies western North Dakota and contains the largest known reservoir of sweet crude oil yet discovered

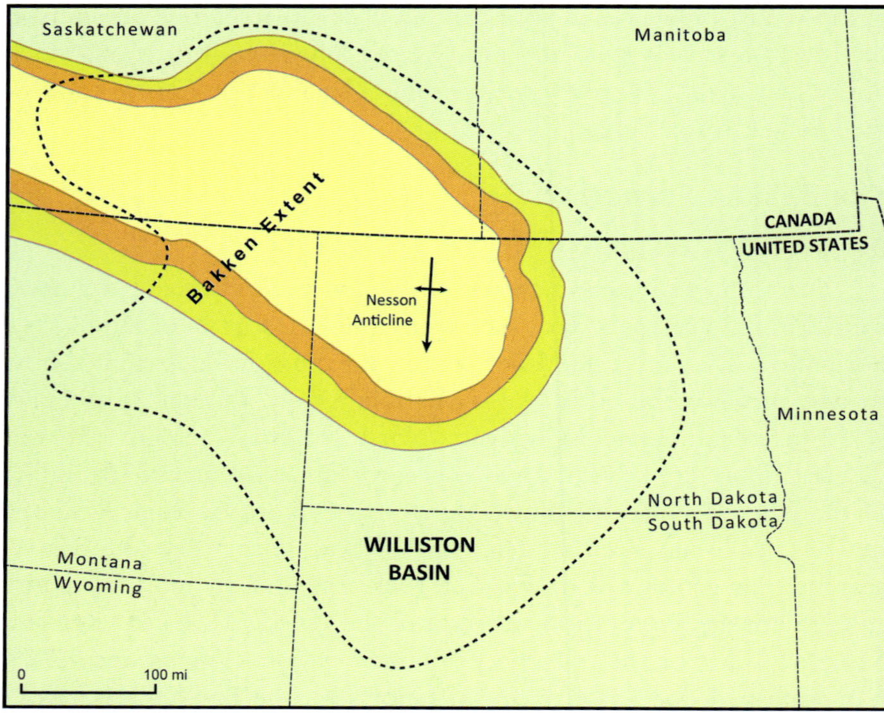

▲ The Bakken Formation underlies most of northwestern North Dakota and extends into surrounding states and provinces. The approximate limit of the Williston Basin is shown by the dotted line.

The Bakken Formation has three members, which vary in thickness and character from place to place. The characteristics of the Bakken are (very generally!) as follows: The lowermost shale member, deposited first, extends throughout the entire area in which the Bakken occurs. It is about 30 feet thick. The Bakken Sea shrank as deposition continued, so that the middle member, a dolomitic sandstone, ranging from 30 to 50 feet thick, extends over a smaller area than the lower member. The sea continued to shrink, so the upper shale member is even less extensive. It is also thinner than the other two members: about 20 feet thick. Most of the oil production, up to now, has come from the middle member of the Bakken. This map shows the upper member (light yellow) partially covering the middle member (orange), which overlies the lower member (darker yellow).

A variety of similar interpretations, all differing slightly from one another, are available. I adapted this version from one provided by the United States Geological Survey.

in North America.[74] Unlike oil produced from structural geologic features, such as anticlines and stratigraphic traps, the oil in the Bakken is contained in a widespread layer that is continuous over much of the area I mentioned.

The Bakken Formation consists of three members. The upper and lower members are black, organic-rich shales that serve as seals on the top and bottom of the middle, oil-bearing member of the Bakken and the top of the underlying Three Forks Formation, which is also a shale and a major oil resource. The average thickness of the Bakken is about 75 feet, but it reaches 145 to 150 feet thick southeast of Tioga, along the eastern base of the Nesson Anticline.

The Devonian-age Three Forks Formation (370 million years old) directly underlies the Bakken, farther east and south in North Dakota. The Three Forks exists over a larger area and extends beyond the limit of the Bakken in most places. It is as thick as 250 feet in eastern McKenzie County. Where the Three Forks is present, the two formations, Bakken and Three Forks, are often produced as a single unit.

In some places, a silty sandstone zone, referred to as the Sanish Sand, occurs at the base of the Lower Bakken. A similar (or identical?) zone, known as the Pronghorn, lies immediately below the base of the Lower Bakken in some places. The Pronghorn has been defined as a member of the Bakken and can be considered to include the Sanish, although some geologists assign one or both of the two zones to the Three Forks.

The Bakken Formation is overlain by the Mississippian-age Lodgepole Formation. The Lodgepole was deposited beginning 359 million years ago and continuing for several million more years. The Lodgepole consists of dense limestone and calcareous shale. Although it served as an oil-generating formation, too, and produces oil in North Dakota, the Lodgepole is less rich in hydrocarbons than is the Bakken. The Lodgepole Formation has a maximum thickness of 900 feet in eastern McKenzie County.

FORMATION OF BAKKEN AND OTHER WILLISTON BASIN SHALES

When the Bakken Formation rocks were deposited, the land mass that includes North Dakota was situated just north of the equator, and the climate was tropical. Shallow, marine sediments of the lower member of the Bakken were deposited as the Devonian was ending. Depositional conditions gradually changed so that terrestrial sediments were deposited in early Mississippian time (marine sediments are deposited in sea water, whereas terrestrial sediments are deposited on land, by rivers and streams).

Major tectonic uplift and erosion of the Earth's crust occurred along the margins of the Williston Basin sometime during the transition from Devonian to Mississippian time. This caused the Bakken Sea to shrink as elevations rose. As a result of the uplift, by the end of Devonian time, the sea covered a smaller area, approximating the current extent of the upper part of the Bakken shale. The earliest (lowermost) member of the Bakken extends over the largest area, the middle member extends over less area, and the youngest (uppermost) member covers the least area of the three Bakken members. When the earliest Bakken shales were deposited, the sea environment was still rich in oxygen. The later, overlying shales were deposited in a sea that was lacking in oxygen.

The increasingly anoxic (oxygen-poor) conditions in the Bakken Sea allowed plants and animals growing in the sea to sink to the sea floor when they died and accumulate as an organic-rich layer of debris. Bacteria that otherwise would have decomposed and destroyed the dead plant and animal matter could not do so without oxygen. This anoxic environment resulted in the formation of large quantities of materials that generated the hydrocarbons being produced today.

Anoxic conditions are necessary for source rocks to generate oil and gas such as the Bakken shale. Source rocks are rich in extra-fine organic materials, consisting of compounds made primarily of carbon and hydrogen. Their high organic content consists of decayed phytoplankton (plant material consisting of microscopic algae) and decayed zooplankton, tiny animals that feed on phytoplankton. Such tiny plants and animals live in oceans, seas and bodies of fresh water. In contrast, coal is formed of ancient plant life (larger, not microscopic-size plants) that lived in swamps on land.

No analog for the Bakken Formation exists in the world today. The Black Sea may be an example of a somewhat similar modern environment, but black shales are no longer being deposited in the Black

Sea. Furthermore, the Black Sea exists at a higher latitude and in a different depositional and geomorphic regime than did the tropical Bakken Sea. Other examples of modern anoxic lakes or seas include the Red Sea, Persian Gulf and Lake Tanganyika (in southeast Africa).

SOURCE ROCKS

The oil in the Bakken, Three Forks and other Williston Basin shales is tightly held in the rock because shales have low permeability, which means that pore spaces are poorly connected. The oil cannot flow easily through the rock to a wellbore. In addition to the Bakken and Three Forks shales, the Lodgepole, Tyler and Winnipeg are other oil-generating shales in the Williston Basin.

RESERVOIR ROCKS

Non-shale formations that have produced oil in North Dakota consist of reservoir rocks, sandstones and carbonates. Reservoir rocks have good permeability, meaning that fluids are able to move through them, and to the wellbore. Oil and gas form in source rocks and then, through geologic time, migrate from those source rocks to reservoir rocks. Until the Bakken play, with its greatly improved horizontal drilling and hydraulic-fracturing technology, all of the oil produced in North Dakota was through vertical wells drilled into reservoir formations.

Considerations Relating to Oil and Gas Production in North Dakota

REGULATION

The development of North Dakota's petroleum resource has been an orderly, deliberate process. In 1941, State Geologist Wilson M. Laird convinced legislators to pass a regulatory bill for oil and gas development and, because he did so, North Dakota already had a sound conservation law in place when oil was discovered in 1951. Today, the State Industrial Commission regulates oil and gas exploration, drilling and production.

By specifying the placement and number of wells that can be drilled, production can be maximized with the fewest necessary wells. Deeper drilling usually requires fewer wells and wider spacing between wells because pressures increase with depth, and high pressures help force the oil into the wellbore. For this reason, fewer wells are needed to efficiently extract the oil. Other factors that need to be considered to determine spacing include the type of oil, its viscosity, how hot it is and its gravity.[75] Horizontal wellbores drain larger volumes of rock than do vertical wells drilled to the same depth. The legally designated space a well is presumed to drain (the "spacing unit") varies from as small as 40 acres for some shallow wells to several sections for deeper wells. Spacing units for deeper wells and horizontal wells, such as those drilled in the Bakken, are larger: 1,280 acres (two sections) and more.

HORIZONTAL DRILLING

Horizontal drilling for oil has fundamental advantages vs. vertical drilling. More of the oil-bearing rock comes in contact with the wellbore, so more oil can be extracted. A vertical well can penetrate no more than the total thickness of the oil-rich horizon, which for the Bakken might be 75 feet. A well drilled horizontally in the Bakken might penetrate several thousand feet of potentially productive rock—rock that will come in contact with the drill bore. In addition, horizontal drilling makes it possible to produce oil and gas from otherwise inaccessible areas such as beneath Lake Sakakawea or from beneath public lands, towns or parks, where drilling operations are difficult or otherwise restricted.

The drilling of a horizontal oil well begins by drilling a vertical hole to a depth just under the deepest potable water zone ("potable" water is drinkable). In the part of western North Dakota where oil is being produced, the deepest potable water is contained in the Cretaceous Fox Hills Formation aquifer, which directly overlies the Pierre Formation shale. The vertical hole is usually drilled through the Fox Hills and 50 feet into the Pierre, using a 13.5-inch bit. Then, a 9.6-inch casing is installed to the drilled depth and cement is pumped down the casing, through the hole to the base of the casing. Setting the base of the casing into the Pierre ensures that the entire Fox Hills Formation will be protected from contamination. The pump pressure causes the cement to rise from the bottom of the casing, up along the outside of the casing, sealing it off from the fresh-water-bearing zone. Up to this point, the procedure is essentially the same as it has been for all of the vertical oil wells that have been drilled in North Dakota for the past several decades.

If the well being drilled is a horizontal one, after the cement has cured, the drill is again lowered to the bottom of the casing and the drilling resumes. The hole is then drilled down to the kick-off point, 500 feet above the target zone in the Bakken. The kick-off point is the

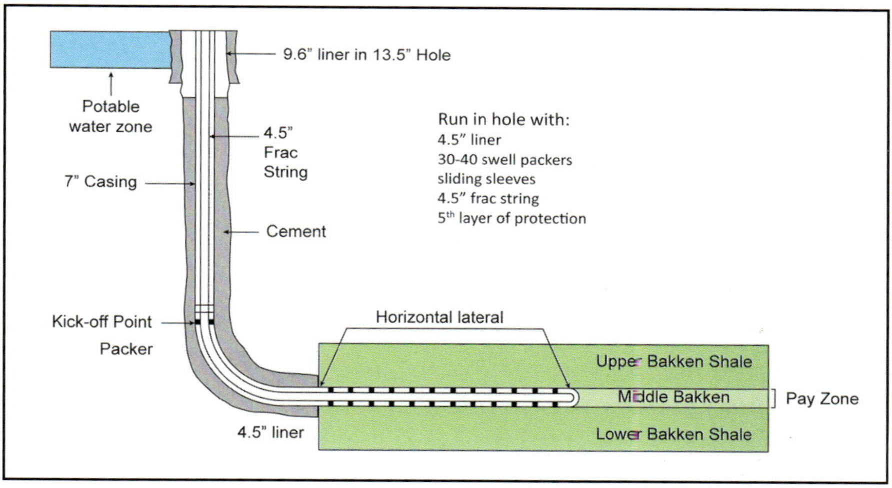

▲ This schematic diagram shows some of the main elements of a horizontal oil well. In North Dakota, the vertical part of the wellbore may be drilled 10,000 feet before turning at the kick-off point, which might be 500 feet above the pay zone (the rock layer that contains oil). At that point, the wellbore begins its curve so it ends up being horizontal through the pay zone, the Middle Member of the Bakken Formation.

The vertical part of the well is drilled by an apparatus powered by a motor located on the drill rig on the land surface. The horizontal part of the well is drilled by a bit powered by a down-hole motor in the hole, not by the motor on the above-ground drilling rig. The horizontal-drilling apparatus is referred to as MWD (measuring while drilling). As the horizontal leg is drilled, the master driller, on the surface, makes constant adjustments to keep the drill in the pay zone.

After the horizontal leg of the well is completed, packers are placed in the desired positions within the leg and then expanded to seal off the wellbore at that point The packers (shown here as the black bars within the horizontal lateral) subdivide the horizontal leg of the well into shorter segments, allowing a procedure called "stage fracking," in which the operator can concentrate the hydraulic fracturing applications to parts of the wellbore where it is likely to be most effective.

depth at which the drill stem needs to begin its bend to become horizontal in the planned pay zone. A 7-inch casing is then installed and cemented in the hole, sealing it from the surrounding sediments.

The machinery used in the wellbore to drill horizontally includes a forward unit consisting of three components. In the lead is the drill bit, which can be maneuvered or adjusted as drilling proceeds. The second component, behind the drill bit, is a down-hole motor, which provides power to the bit. Behind the motor is the instrument casing, which is referred to as "measuring while drilling" or MWD. The MWD unit sends information to the surface so the master driller can make adjustments on the drill bit, as necessary.

The drill rig keeps pressure on the drill stem and the motor/drill bit unit. The motors on the above-ground drill rig no longer power the drill stem, as they did when the vertical section of the well was being drilled. Rather, the down-hole motor/MWD unit drives the bit through the horizontal leg in the pay zone.

LINEATIONS

Western North Dakota's land surface is marked by a pattern of surface lineations ("linears" or "joints") that can be seen on high-altitude aerial or satellite photographs. The lineations are apparent only in places not covered by glacial deposits. Two sets of lines are apparent, one trending roughly northwest to southeast, and the other northeast to southwest. The lineations are the surface expression of fractures that extend upward from the basement rocks (granite, etc.) through all of the overlying formations, except the glacial sediments, which were deposited so recently that the joints have not had time to propagate upward through them. Knowledge of the lineations was tied into oil exploration several years ago, when geologists defined the overall orientation patterns for the lineations, which form a criss-crossing pattern. In certain situations, the fractures can act as conduits through which oil can travel to horizontal wellbores drilled to intersect them.

HYDRAULIC FRACTURING

Because any naturally occurring lineations/fractures that might exist are normally closed tight in the subsurface, no fluids can flow along them unless they are forced open. Hydraulic fracturing, coupled with horizontal drilling and MWD, makes successful drilling in the Bakken possible. Hydraulic fracturing is not new; it has been used for more than 65 years. What is new is the ability to drill directionally by steering the drill bit.

Hydraulic fracturing is a process that uses water, chemicals and proppants. Proppants may be sand that has been sieved so that all the grains are essentially the same size, or they may consist of manufactured ceramic beads, all the same size (the size of a pin head). A mixture of proppants and fluids is injected, under pressure, into any open existing fractures in the oil-bearing rock and into new fractures created during the hydraulic fracturing procedure. A horizontal well is hydrofractured at strategic intervals along the horizontal length of the wellbore. After being forced open, the fractures are held open by the proppants, thus creating a pathway to allow oil to flow to the wellbore. Natural pressure in the formation forces the oil to flow through the pathways to the wellbore so it can either flow or be pumped to the surface through the wellbore.[76]

In early horizontal wells, the hydraulic fracturing procedure involved fracturing the whole lateral leg of a well at one time. Because the fracturing fluid took the path of least resistance, parts of the horizontal leg were fractured, while other parts were not, and results were spotty. Current methods involve a procedure referred to as "stage fracking," which allows operators to subdivide the horizontal leg of a well into shorter segments by using packers. Using stage fracking, operators are able to concentrate on specific sections of the horizontal well leg and the parts of the formation that are considered to have the greatest potential to benefit from receiving the most intense treatment.

Because the oil-producing portion of the Bakken Formation in North Dakota is two miles deep, it would be practically impossible for the fracturing fluids used during drilling operations to escape from the Bakken to the surface, and pollution resulting from hydraulic fracturing should be a rarity. Even so, if the vertical portion of a wellbore is not sound, or if the casing is not properly cemented, contamination can occur.

In addition to hydrocarbons, oil and gas wells produce considerable amounts of salty formation water (sometimes more water than oil) that requires disposal. This water is usually injected back into the ground. In North Dakota, most of the waste water is injected into the Dakota Formation, which is deeper than 6,000 feet in the area of oil production.

In some states, small earthquakes have been noted in areas of hydraulic fracturing activity and it has been suggested that the earthquakes are a result of fracturing procedures. It is probably more likely that most such earthquakes are related to wastewater injection procedures. Injection wells have been suspected of causing tremors in Ohio, Arkansas, Texas and Oklahoma, where fracturing and injection commonly take place at much shallower depths (less than 2,000 feet). In North Dakota, where both hydraulic fracturing and waste-water injection take place at much greater depths, no such problems have been noted.

PIPELINES

In much of the area where Bakken oil is being produced, the crude oil still has to be trucked to its destination, greatly increasing rural road traffic and road maintenance costs. Pipelines are the safest and most cost-effective way to transport oil, but current pipeline capacity is insufficient to handle the amount of oil being produced, or projected to be produced. The relative efficiency of moving crude oil through a pipeline can be illustrated by considering the 77-mile-long Four Bears Pipeline, completed in January 2012. This pipeline carries oil through Dunn and McKenzie counties to the Butte Pipeline, in Baker, Montana, and to the Bakken Oil Express Railroad Terminal near Dickinson. The Four Bears Pipeline transports 105,000 barrels of oil per day. In contrast, the average load for an oil-tanker truck is about 220 barrels of oil per load. The pipeline saves 50,000 to 75,000 truck miles per day. Still another recently approved 16-inch pipeline between the Bakken Oil Express Railroad Terminal and a pumping facility near Killdeer will have a capacity of 165,000 barrels a day. This 38-mile-long pipeline will eliminate the need for 825 tanker trucks a day between those two points.

NATURAL GAS DESCRIPTION AND PROCESSING

The Bakken and Three Forks formations are best known for their immense oil resource, but the two formations contain associated natu-

▲ This Bakken well site is located 10 miles north of New Town in the Little Knife River Valley in Mountrail County. The pumper is on the right and oil-storage tanks are to the left of the pumper. The flared gas is to the left of the tanks. At the time I took this picture, in 2010, natural gas produced with the oil had to be burned or used on site because no well-site-to-market pipeline had been built.
JPB Photo: 7/16/2010

ral gas as well—gas that is produced along with the oil. With the development of the Bakken and Three Forks, natural gas production has risen dramatically.

As you travel through a recently developed oil field, you are likely to see flares, which are flames resulting from burning the gas that is produced with the oil at many well sites. As I write this (2014), gas pipelines have not yet been built to transport the gas to market in many places where newly producing Bakken wells have been drilled. The gas either has to be used at the well site or it must be flared. Currently, 25 percent of the natural gas coming to the surface with recently developed Bakken oil is flared because natural gas cannot be allowed to vent to the atmosphere. Methane (CH_4), the main constituent of gas that would escape if it were not flared, is a potent greenhouse gas, much more so than carbon dioxide, the main combustion product of flaring. The amount of gas flared should drop as pipelines are built to transport the gas to processing plants and as other new technologies for dealing with the gas are perfected.

Natural gas may be classified as either "wet" or "dry." Dry natural gas is methane, with a very low content of liquid hydrocarbons. Wet gas contains liquid hydrocarbons, usually less than 85 percent methane, but with more butane, ethane and other complex hydrocarbons. The natural gas produced from the Bakken and Three Forks formations is considered to be wet, meaning it is high in natural gas liquids, such as propane, butane and natural gasoline, which are held in the produced natural gas stream and must be separated at a natural gas processing plant, using a process known as fractionation.

Apart from associated gas being produced with oil from the Bakken and other wells, potential gas-producing strata, beyond the area of the Williston Basin in North Dakota, include much shallower formations such as the Dakota Group, the Newcastle (Muddy) Sandstone, and the Eagle and Judith River sands (both part of the Pierre Formation). Gas is present in some organic-rich glacial sediments, too. The Niobrara and Greenhorn formations, both of which are found near the surface in eastern North Dakota, produce gas in other states. Studies by the North Dakota Geological Survey have confirmed that sizeable shallow gas deposits exist in many parts of the state, well beyond areas that currently produce hydrocarbons.

Summary

By the end of the first decade of the 21st century, an estimated 1 trillion barrels of oil had been produced worldwide. Some experts theorize that another trillion producible barrels still exist. The first trillion barrels were considered to be "easy" oil, easy-to-find oil, close to the surface or concentrated in large, easily accessed reservoirs. Such oil is readily produced from reservoirs that have sufficient pressure, porosity and permeability to flow freely. The rest of the existing oil resource then, is "tough" oil, buried off shore, deep underground, tied up in impermeable shale or scattered in small, hard-to-find reservoirs. Tough oil is sometimes referred to as "unconventional" oil—oil which does not flow freely or requires special technologies to extract, so it is more expensive to produce. Twenty years ago, North Dakota's Bakken oil was considered to be part of the second (tough/unconventional oil) category, but now it is being produced.

The development of any large oil production play, from the initial discovery to the last well being abandoned, usually follows a pattern. For the Bakken oil play, the discovery phase does not really apply as we have known ever since oil was discovered in North Dakota that oil occurs in the Bakken. For this reason, the discovery phase might more aptly be considered the point at which technology made the Bakken oil economically feasible to produce. That point might be in 2007, when production from the Parshall Field proved that the Bakken oil could be produced economically.

Following the discovery of the Parshall Field, the next phase was the drilling done to define the geographical extent and limits of the Bakken Formation oil. That phase is essentially complete. The next phase involves the development drilling, when oil companies secure their leases of Bakken acreage. That is currently under way, along with the following phase, infill-drilling, which is in progress. During the infill-drilling phase, the optimum number and placement of necessary wells for the most efficient production is determined and accomplished.

The last phase involves secondary and tertiary drilling. As oil wells mature and their production rates decline, production can often be maintained and prolonged by injecting water into the oil-producing formation (a process known as secondary recovery) or by injecting carbon dioxide, nitrogen or some other substance into the formation (tertiary recovery). Such methods prolong production by adding pressure to the oil-producing formation. Secondary and tertiary recovery methods have been used in North Dakota's older producing formations for many years already.

The Bakken oil play has resulted in a great increase in oil activity and production in North Dakota.[77] If estimates of reserves are close to accurate, much more oil will eventually be taken from the Bakken than has already been produced from all of the formations in North Dakota since oil was discovered in 1951. The Bakken Formation contains a resource, estimated by the United States Geological Survey in 2012, at 400 billion barrels of oil. The USGS estimates that 7.4 billion barrels of that resource is *recoverable*. However, any official estimates of recoverable oil from the Bakken are likely to be too low because, as technology improves, the definition of "recoverable" will change to include a larger fraction of the total resource.

If we were to stop drilling new Bakken wells, the growth in production could end abruptly, and overall production could decline rapidly. The steep and rapid declines in production of these tight-oil wells require an ever-increasing number of new wells just to keep the overall production rate constant. In the short term, though, the extensive drilling we have experienced for the past few years is sufficient to more than compensate for decline in individual wells. According to one estimate, maintaining the Bakken production requires 100 new producing wells per month. It is kind of like the dilemma described by the Red Queen in *Alice in Wonderland*: "It takes all the running you can do just to stay in the same place."

It remains to be seen how long the increase in oil production in North Dakota will be maintained, and how quickly it will fall when drilling slows down. As I write this, in 2014, daily production is more than a million barrels a day. Other factors remaining stable, production will likely continue to increase, at least until all the wells considered necessary to produce the Bakken oil have been drilled.

Future production figures are speculation, best left to petroleum engineers and economists. Nevertheless, as a geologist, it seems to me that, with a resource base as large as that found in the Bakken/Three Forks, it is likely that North Dakota can expect several more years of drilling and development, followed by decades of continued, healthy oil and gas production.

EPILOGUE

In this book, I have dealt with one small part of the Earth that is North Dakota. I have explained why and how the state became as it is and how the land influences us. My goal has been to help readers better understand this place. Geology is a dynamic science that demands an open, objective and imaginative mind. Geologists tease the past out of the clues the Earth provides in its rocks and landscapes.

The land that is North Dakota is built on its geological inheritance, one that has been developing for eons. We, as individuals, can observe only small natural changes in the landscape during our lifetimes. We experience only a brief moment, a scratch on the surface, of our Earth's long and dynamic history. Even so, as enough small events, along with occasional larger ones, continue long enough, they all combine to transform the landscape.

The landscape is the end result of the story told by the geologic forces, whether visible or buried, that have shaped it. Our landforms are the bold, unmistakable inscriptions marking the surface on which we live. The story of our land may be likened to a history book with many of the pages torn. The book records the natural events and, in the final chapter, the human experiences, that have shaped the Earth and our lives. If we read between the lines, interpret the endnotes and piece in as many of the missing pages as we can, we will better understand the deeper, hidden layers of much older, buried landscapes and seascapes. Ancient sea floors, coral reefs, shore lines, coastal swamps, tropical river systems, melting ice sheets and a variety of other environments combine to form the framework of our North Dakota landscape.

The land attests to the underlying geology. A change in surface terrain often indicates a geologic change beneath the ground. Most of what we see: elevation, drainage, surface materials and the climate—even snow, rain, wind and warmth—are products of geologic activities. The landscape is where all human activity takes place, and learning to live in harmony with it is essential. If we understand the building blocks and mechanisms—bedrock, glacial materials, soil, water and air—and their inherent relationships, we will be better able to protect our future heritage.

Our landscape is geologically young. Much of it dates to the Ice Age, only yesterday in terms of geologic time. Many of the landforms found in states like North Carolina, Wyoming or Arizona are hundreds of times older than ours. We tend to think of the landscape as permanent, but our hillsides, broad valleys and ravines, buttes, badlands, rolling prairies, moraines, lake plains and river flood plains are dynamic, always changing. People can sometimes be a major force in hastening the change. Humankind has, in fact, become a geologic force: consider Lakes Oahe and Sakakawea, reservoirs behind massive dams. We have developed the technology, the equipment, the power and the will to shape our environment—whether subtly or dramatically, purposefully or unintentionally, for better or worse.

We ask much of our land—and the land, in turn, challenges us. We farm the fields, drink the water, build our homes and businesses, pump the oil and gas, mine the coal and sand and gravel, bury the wastes and play in the parks. Our land provides the basis for all we are. City people depend on it just as much as those who live in rural areas. Each of us needs clean water, fresh air, open space and the chance to experience the natural marvels surrounding us.

As we look at today's North Dakota, we strive to appreciate yesterday and anticipate tomorrow. Both our past and our future are rooted in our land, and in its ability to work for us and with us. If we understand the geological history and natural processes that have shaped the landscape, we can better apply that knowledge to conserve and renew it. As we constantly mold and remodel the land to suit our needs, it is important that we adapt to the changes we impose upon it. However we choose to treat our land—gently and with respect, or harshly and callously—we are shaping the legacy we leave our children.

ACKNOWLEDGMENTS

When I retired, my wife, Mary, suggested I write a book about the features people see as they travel around North Dakota. Beginning in 2008, we dedicated four North Dakota summers to researching and taking photos for this book. We toured every county and most of the towns as we drove tens of thousands of miles on gravel roads and trails, rediscovering North Dakota and reacquainting ourselves with the state. It was an enjoyable journey. As we drove along, we described the landscape and I took thousands of photos of the geology.

I want to thank some of the many people who helped me along the way. They include (in alphabetical order):

Eric Brevik, Ph.D., professor of geology and soils and chair of the Department of Natural Sciences at Dickinson State University, who thoroughly reviewed the science of my manuscript, assisting me with a variety of topics, such as soils, glacial mechanics, and crustal rebound. I also want to thank Eric for writing the foreword to this book.

Alex Chaky, retired Williston petroleum engineer, who knows far more about petroleum geology than, well, anyone. Alex clarified parts of my chapter on energy, ensuring I was accurate in my discussions of the Bakken Formation, horizontal drilling, hydraulic fracturing, and any number of oil- and gas-related topics.

Doreen Chaky, geologist, historian, and author of a recent award-winning book, *Terrible Justice,* read one of the earliest versions of my book, as well as several subsequent drafts. Doreen's helpful advice enabled me to improve my writing style. She encouraged me to write in an active voice. Her experience and advice on writing and publishing were uniquely valuable.

Reinhard Mandl, renowned photographer with the State Museum in Vienna, Austria, provided me with on-going ideas and photo advice. We first met Reinhard and Christine while they were in North Dakota preparing a pictorial history of Sitting Bull. Reinhard's photo-biography of the Dakota chief was subsequently displayed in many of Europe's leading museums, and now has a permanent home in the *Museum für Völkerkunde* (Museum of Ethnology) in Vienna.

Lorraine Manz, Ph.D., geologist and editor with the North Dakota Geological Survey, found references, photos and diagrams of features I otherwise could not have included. I am especially grateful for her help with my discussion of geothermal resources. Lorraine and her engineer husband, Brad, share my own appreciation for single-malt Scotch whisky, the geology of Scotch whisky and the leading role early Scots had in the development of geology.

Edward C. Murphy, M.S., state geologist and friend for more than 30 years, helped me clarify many of my thoughts. Ed's knowledge of western North Dakota stratigraphy, clays, landslides, geologic history and any number of other geologic concerns is practically unbounded.

Mike Renner, president of "Up and Running Design.net" and a talented Bismarck website developer, provided valuable advice, reassurance and practical knowledge about the various kinds of software we used.

Donald C. Schwert, Ph.D., professor of geology and chair of the Department of Geosciences at North Dakota State University, read the manuscript, suggested topics to include and advised me on ways to approach them. Don's advice as I wrote my chapter on Lake Agassiz and the Red River Valley was especially valuable.

Robert W. Seabloom, Ph.D., professor emeritus of biology at the University of North Dakota and friend for more than 40 years, provided me with unique perspectives stemming from his life time of mammalian research. Bob wrote the book *Mammals of North Dakota,* which was published in 2011 by the North Dakota Institute for Regional Studies.

C. Thomas Shay, Ph.D., University of Manitoba professor emeritus of biology, read early drafts of my book and advised me about a variety of topics relating to anthropology, biology and soils. I had the opportunity to read parts of a book Tom is planning to publish on the natural history of the prairie states and provinces. His book provided me with background and ideas for my own work.

Many others provided help and encouragement as I wrote this book. Some who come to mind include long-time Grand Forks friends Fr. William Sherman, who gave me early advice about what North Dakotans want to know, and Ken Carlson, who suggested ways I might word difficult passages. Paul Picha, Claudia Berg, and Eric Holland and others at the North Dakota Heritage Center in Bismarck, and David Borlaug at the Lewis & Clark Interpretive Center in Washburn, offered ongoing encouragement. Steve Van Dyke, with the North Dakota Lignite Energy Council, provided current information about North Dakota's lignite industry. Jeff Jennings and Robert Post Johnson, consulting geologists with Harris, Brown and Klemmer Firm of Consulting Geologists, provided data and discussion relating to oil production, Bakken decline curves and a variety of other topics. Robin McQuinn, geologist, read the manuscript and made suggestions.

I gained valuable insights and many friendships during the years I mapped geology in North Dakota as I visited with farmers and ranchers while traveling around the state. They know and understand their land, are always curious about the geology and ask questions to satisfy their curiosity. Often their questions made me think about things that hadn't occurred to me. Their questions helped me identify topics people are curious about and what they want to know about the geology of North Dakota's landscape.

I was fortunate to have the chance to teach a class on the geology of North Dakota for the Osher Lifelong Learning Institute (OLLI) in 2013. The topics I covered were essentially the same as the material in this book, and the new perspectives I gained from the questions students in the class asked were valuable background as I prepared this book.

I thank Dean Kent Sandstrom and Dr. Ross Collins and his staff at North Dakota Institute for Regional Studies. I especially appreciated Ross's friendly, low-key approach as he responded to my questions and concerns, and his patience as I did my best to provide the actions and materials needed to expedite the publication of this book.

Dr. Suzzanne Kelley, Editor in Chief for the North Dakota State University Press, has guided this book through its final stages, leading to publication. I appreciate and am deeply grateful for all she has done. Suzzanne's experience and knowledge of the publishing business, coupled with her attention to detail, ensured the technical quality of this book. She helped me meet deadlines with good-natured encouragement and humor. Working with Suzzanne has been a pleasure.

I want to offer special thanks to Deb Tanner for her expert layout and design of my book. Arranging an array of photos and diagrams, some of them quite technical in nature, and all so beautifully embedded within a fairly complicated text, was a considerable challenge. I also appreciate Deb's choice of the castellate badlands scene for the cover. It speaks geology, which is what I had hoped for.

Finally, and most importantly:

Our son, Bill Bluemle, an anthropologist and archaeologist, has worked for more than 20 years on the northern Plains. Bill periodically read and edited drafts of my manuscript. Our daughter, Irene Bluemle, suggested the title for my book. Their early ideas helped me formulate a plan that I followed from the beginning.

Valerie Barbie-Bluemle, Bill's wife and also an anthropologist, provided ongoing encouragement with her stimulating questions and insightful observations.

Mary E. Bluemle, Ed.D., fellow geologist and my wife, served as initial editor and overall coordinator through the many revisions of this book. Mary handled the necessary accounting and travel planning, and kept our office systems and software working smoothly. Without her involvement, the book would never have gotten to the publisher.

APPENDIX A
ICE AGE CHRONOLOGY

The Earth's Major Glacial Periods

During its 4.6 billion-year history, the Earth has undergone at least five periods of long duration, tens to hundreds of millions of years, during which repeated episodes of glaciation were common.

Beginning with the oldest, the five major periods are as follows:

1. **Huronian Glaciation**, from 2.4 billion to 2.1 billion years ago (early Proterozoic time). Deposits of this glacial period have been found in the Lake Huron area of Ontario, in Wyoming and in Africa. The glaciation was likely initiated by the so-called Great Oxidation Event, a time when an oxygen-rich atmosphere was replacing the previous carbon dioxide-rich atmosphere. The loss of the carbon dioxide resulted in cooler climates because less greenhouse gases were available. The time of the Huronian Glaciation is sometimes referred to as First Snowball Earth, implying that glaciers covered the entire surface of the Earth.
2. **Cryogenian Glaciation**, from 950 million to 580 million years ago (late Proterozoic time). "Cryogenian" is from the Greek *cryos*, "cold," and *genesis*, "birth"; the Cryogenian is known also as the Varangian. This time is often referred to as the Second Snowball Earth, again implying that glaciers covered the entire Earth's surface. Although glaciation was widespread in both the Huronian and Cryogenian, it is unlikely the Earth was ever entirely covered by ice.

However, the Earth's continental plates were fast-moving during Proterozoic time and, when they periodically slid into positions over both polar regions, the result was repeated episodes of glaciation. Glacial deposits dating to this period have been found on all continents except Antarctica.

3. **Andean/Saharan Glaciation**, from 460 million to 430 million years ago (late Ordovician time). Glacial deposits of this age are found in the Andes and Brazil in South America, in Africa and on the Arabian Peninsula. The super-continent, Gondwana, was positioned over the South Polar Region during the Andean/Saharan Glaciation. Gondwana included most of the landmasses in today's Southern Hemisphere: Antarctica, South America, Africa, Madagascar and Australia, as well as the Arabian Peninsula and India, both of which have since moved into the Northern Hemisphere.
4. **Karoo Glaciation**, from 360 million to 260 million years ago (Devonian to Permian time). Deposits of the Karoo Glaciation are found in South America, Africa, India, Australia and Antarctica. "Karoo" is an Afrikaans term for a region in South Africa; it means "hard and dry." As during the Andian/Saharan Glaciation, the super-continent, Gondwana was positioned over the South Polar Region during the Karoo Glaciation.
5. **Holarctic-Antarctic Glaciation** probably began 30 million years ago and it continues today. Deposits of this glaciation are found on all of the Earth's continents. The Holarctic-Antarctic Glaciation intensified in late Pliocene time, 3.5 million years ago, and evidence of continental-scale glaciation has been found in Ontario dating to that time. It was probably not until the early Pleistocene that glaciation directly affected North Dakota. Since the Holarctic-Antarctic Glaciation began, the Earth has experienced several episodes of advancing glaciers, each lasting a few tens of thousands of years. The most recent such episode was the Wisconsinan Stage. Glacial ice remains over the South Polar Region, Greenland and Iceland. Until the Antarctic continent moves away from the south polar area, we can expect recurring

episodes of glaciation.

Deposits of the Holarctic-Antarctic Glaciation are the only ones known in North Dakota. If deposits of any of the earlier glaciations exist in the state, they occur in the deeply buried rocks of the Williston Basin.

The Holarctic-Antarctic Glacial Period

Although one of the prerequisites for glaciation seems to be the presence of one or more land masses over a polar region, other factors exert varying degrees of control over how often glacial conditions occur, how severe they are and how long they last. Global climate, since mid-Pleistocene time a million years ago, has varied through a cycle of approximately 100,000 years. During each complete 100,000-year cycle, the glacial part lasts from 85,000 to 90,000 years. This is followed by a warm, interglacial part, which may last from 10,000 to 15,000 years. Afterwards, the ice masses on the continents begin to grow again, lowering sea level up to 500 feet, compared to the peak levels in the interglacial cycle.

The 100,000-year cycle I just mentioned is known as the Milankovitch cycle, named after the Serbian scientist Milutin Milankovitch. This and other cycles identified by Milankovitch determine variability in the Earth's orbit around the sun and variations in the Earth's rotation. Several cycles combine to determine Earth's climate. One of these is the 105,000-year variation in the shape of the Earth's elliptical orbit. Another is the 41,000-year cycle in the tilt of the Earth's axis, and a third is a 26,000-year cycle in the placement of the day at which the Earth is nearest to the sun as it moves along on its orbit. The last of these is referred to as the axial precession, a cyclic wobble of the Earth's axis of rotation (with respect to the solar orbital plane). This causes the spring equinox to shift through the months from January, through February, on to March and so on, back around to January again at the end of 26,000 years.

At the higher latitudes, North Dakota and northward, the cumulative effect of all these cycles is to determine the degree of variation between summer and winter temperatures. At one extreme, northern areas will have warm summers and cold winters. At the other extreme, the same areas will have cool summers and warmer winters. The summer temperature is the most important factor. Glaciers develop when cool summers do not allow all of the snow that fell the previous winter to melt. This occurred most recently during the Little Ice Age between 1350 and 1850 A.D., making Greenland uninhabitable and greatly expanding the glaciers in the Rocky Mountains.

We do not know for sure how many times North Dakota was glaciated during the current Ice Age. Projecting the cycles I just described into the past, we see that as many as 20 periods of glacial conditions *could* have existed during the past 3 million years, the Pleistocene Epoch. We know that Pleistocene glaciers advanced into the state several times prior to the Wisconsinan. A glacier that advanced as far south as Missouri 2.4 million years ago very likely flowed over much of North Dakota. We know that a glacier covered the state 500,000 years ago. It is a certainty that additional glaciations occurred between those dates (2.4 million and 0.5 million years ago). At least two glaciations, the Illinoian and Wisconsinan, occurred since 0.5 million years ago.

All of the glacial landforms we see today in North Dakota were deposited during the most recent, Wisconsinan Glacial Stage, which began 90,000 years ago and ended 11,500 years ago. Early glacial deposits can be seen at the surface in a few places, but they are not widely preserved as landforms where they can be easily observed. Rather, they fill some low areas, such as old river valleys or holes dug by early glaciers during ice-thrusting events.

Some of the best exposures of till in North Dakota are located two miles north of Riverdale, along the east shore of Lake Sakakawea in McLean County, and at Beulah Bay, 15 miles north of the town of Beulah, along the south shore of the lake. In both places, three distinct layers of till can be seen: the youngest, at the top, is Late Wisconsinan in age; the middle till is probably Early Wisconsinan, and the lower-most till is pre-Wisconsinan. Prior to construction of Garrison Dam and the flooding of Lake Sakakawea, the deeper tills were buried and remained undiscovered because they were buried in topographically low areas (ancient river valleys) beneath layers of younger glacial sediments. I first saw the wave-cut exposures near Riverdale in the

1960s when Lake Sakakawea was only a few years old and wave-cut exposures were fresh. Wave-erosion since then has modified the shoreline, making the relationships less obvious, but all three till units can still be observed.

▲ This wave-cut cliff of till is two miles north of Riverdale on the east shore of Lake Sakakawea. This till is probably Early Wisconsinan in age, more than 70,000 years old. It is oxidized throughout its total thickness of 30 feet. The till is mainly a grayish shade of tan, and it contains a high percentage of clay, along with numerous limestone pebbles, but few boulder-size rocks. Vertical joints have formed in the till so that, when erosion by waves removes material from the face of the cliff, large chunks of till fall away from the main body of the cliff face, separating from the main body of till along the joint lines. The lithologies of rocks included in the till indicate that it was deposited by glacial ice flowing from the north or northeast.

The fact that the 30-foot-thick till is oxidized throughout reflects the 35,000-year duration of the Middle Wisconsinan deglaciation, which provided sufficient time for a climate comparable to that which we enjoy today to weather the entire thickness of the till, prior to the advance of the Late Wisconsinan glacier.

The top surface of the underlying, much older, lighter-colored layer of till can be seen in the lower left part of the photo (see following photo). This older till is much harder than the upper till, and wave erosion has formed a flat bench on its upper surface. JPB Photo: 10/9/2012

The lower-most till has a lighter coloration than the two upper tills and is much harder and more indurated than the two tills overlying it. It is deeply oxidized to depths exceeding 50 feet, indicating that it was exposed to weathering for a prolonged period of time after it was deposited, likely for tens of thousands of years. Our modern land surface has been exposed to weathering for the 10,000 or so years since the last glaciers melted. I named the lower-most till the Dead Man Till because it occurs in Dead Man Coulee. This till is the oldest I have found exposed in North Dakota. It may be several hundred thousand years old.

The Dead Man Till at Dead Man Coulee is sandy and contains abundant chunks of lignite, suggesting that it was deposited by a glacier coming from the northwest, an area where lignite-bearing bedrock formations occur. In some places where the Dead Man Till is exposed along the shore of Lake Sakakawea, an iron-cemented gravel (the Four-Bears Gravel) overlies the Dead Man Till, separating it from the overlying till unit. The two tills overlying the Dead Man Till contain more clay and lack lignite fragments. The ice that deposited them must have come from the north or northeast, over bedrock formations that do not contain lignite.

Elsewhere in the state, deposits of older glaciations are covered by younger glacial deposits and are therefore known only through test-hole drilling.[78] At least seven buried till units, totaling more than 600 feet thick, have been identified near Devils Lake beneath the Late Wisconsinan deposits at the surface. These layers of till, which were penetrated during test drilling, are separated by oxidized zones and paleosols, or they are otherwise distinguishable from one another in some way (mineral content, texture, etc.). The ages of most of the buried tills that have been found in North Dakota during test-hole drilling are not known.

The only place where the age (or at least the minimum age) of deeply buried tills has been determined in North Dakota is in the valley of Cherry Creek in McKenzie County. Cherry Creek flows in a largely buried valley that was once, before the badlands began to form, the route of the Little Missouri River into Canada. The two buried tills are separated by a layer of volcanic ash known as the Lava Creek B ash, which has been radiometrically dated at 640,000 years old. The ash

▲ Two till layers are being exhumed two miles north of Riverdale on the east shore of Lake Sakakawea. The till in the foreground is pre-Wisconsinan in age, more than 130,000 years old. This till, the Dead Man Till, is light tan in color, very pebbly to sandy, with abundant lignite chunks. Notice the quartzite boulder embedded in the till on the lower right.

The abundant lignite included in the till suggests that the till was deposited by a glacier flowing from the northwest over mainly Tertiary bedrock formations.

A lag of abundant boulders lies on the eroded surface of the Dead Man Till in front of the cliffs in the background. The till in the background is Early Wisconsinan age. JPB Photo: 10/9/2012 traveled to North Dakota from the area that is now Yellowstone Park when the Yellowstone Caldera erupted. This ash occurs above a till unit that fills—blocks—the old valley of the Little Missouri River. The 640,000-year date may be the minimum age for the glacial event that diverted the Little Missouri River eastward from its northerly route into Canada because the glacial till it overlies has to be older than the ash, but we do not know how much older. In any case, at least two buried tills exist in the Cherry Creek valley. One of the till layers is older than 640,000 years and the other is younger.

Southwest of the Missouri River in North Dakota, indirect evidence exists—on the land surface—for at least three glacial advances, early or pre-Wisconsinan in age. The relative ages of the advances are based on the average numbers of glacial erratic boulders found within the outer boundary of each glaciated area, each successive glaciation being less extensive, but more bouldery, than the prior ones; that is, the more recently glaciated areas are covered by more boulders. Radiometric age determinations of the length of time glacial erratics have been exposed to the elements on the land surface help us understand the ages of past glacial events.

North Dakota's glacial deposits may be classified as follows (ybp = years before present):

Wisconsinan Glacial Stage (11,500–90,000 ybp)
Sangamon Interglacial Stage (90,000–130,000 ybp)
Illinoian Glacial Stage (130,000–190,000 ybp)
Several Pre-Illinoian Glacial Stages (all before 190,000 ybp).[79]

The Wisconsinan Glacial Stage, which began 90,000 years ago, can be divided into three discrete parts: early, middle and late. The early part, a major glacial advance, lasted from 90,000 until 70,000 years ago. The middle part of the Wisconsinan, between 70,000 and 35,000 years ago, was a time of alternating cooler and warmer periods in North Dakota, but it was ice-free, and the climate was much like we have today. The middle Wisconsinan deglaciation was followed by a second major glacial advance, the Late Wisconsinan, which reached its maximum southern extent between 25,000 and 16,000 years ago and ended 11,500 years ago. During the advancing, and most intense, phase of the Late Wisconsinan glaciation, between 25,000 and 16,000 years ago, the unglaciated area beyond the glacier margin was intensely cold, resulting in widespread permafrost everywhere southwest of the Missouri River in North Dakota. The Late Wisconsinan glacier began to melt 16,000 years ago.[80]

The Wisconsinan Stage can be broken down as follows:

Holocene deglaciation: 11,500–present time
Late Wisconsinan glaciation: 35,000–11,500 ybp
Middle Wisconsinan deglaciation: 70,000–35,000 ybp
Early Wisconsinan glaciation: 90,000–70,000 ybp

Holocene Time

For the past 11,500 years, since the end of the Late Wisconsinan Glaciation, we have been experiencing an interglacial stage characterized by a climate that alternates from dry to wet and cool to warm. Geologically, this period is called the Holocene (from Greek: "entirely recent"). Initially, beginning 14,000 years ago, and soon after the glaciers had melted sufficiently to allow some ice-free areas, North Dakota summers were slightly colder than those today, but winters were still considerably colder. Precipitation was less than today because the jet stream hugged the southern margin of the main glacier, deflecting moist air to the south.

The immediate post-glacial condition was not tundra, as one might expect. Tundra did develop farther north (and still exists) in places like Yukon and Nunavut, as the glaciers receded. In North Dakota, though, because the recession of the ice margin lagged behind the quickly warming climate, tundra did not develop. The environment that the Paleo-Indians lived in, at least in eastern North Dakota, was an open, white spruce parkland (parkland is open land with scattered groups of trees). Spruce trees, along with shrubs and some deciduous trees such as aspen, ash and birch, lined the shores of lakes (Lake Agassiz as well as numerous smaller lakes). Prairie grasses covered the uplands.

I have found mammoth teeth in the Herman Beach (the uppermost and oldest shoreline of Lake Agassiz). Mammoth bones and teeth (not the ones I found) recovered from the Herman shoreline have been dated at 11,500 ^{14}C yr BP (Harington and Ashworth, 1986). It is likely that Paleo-Indians hunted these animals, along with other game.

The Altithermal was once considered to be a prolonged drought, but we know now that the climate was quite variable. From 10,000 until 6,000 years ago, it alternated between severe aridity and moist intervals. As early as 9,000 ^{14}C yr BP (equivalent to 11,000 calendar years), summer temperatures were 3° to 7°F warmer than modern (pre-industrial level) temperatures; insolation (the amount of radiation received from the sun) was at its post-glacial maximum. Winters were cooler than modern (again pre-industrial) by 3° to 10°F. The time from 8,000 ^{14}C yr BP until 7,700 ^{14}C yr BP was dry; between 7,700 and 6,000 ^{14}C yr BP, conditions were more moist (and likely quite cold). The driest part of the Holocene was between 6,000 and 5,000 ^{14}C yr BP. This variability is reflected in the way Devils Lake has ranged between overflowing to the Sheyenne River and being a dry lake bed.

Summary

Glacial events occur in a cyclical pattern and with a strong periodicity, and they have tended to last longer than have the intervening periods of warmer climate. The major glacial and interglacial periods are excellent examples of the natural climate change that has occurred through geologic time because they have taken place, with consistent regularity, every 100,000 years during the last several million years, and each change has followed the same pattern of slow, steady cooling, followed by abrupt warming, returning to conditions similar to today's. The evidence for this periodicity is found in the chemical records preserved in polar ice, mainly the East Antarctic ice cap, in which the glacial record matches that of ocean sediments.

During post-glacial Holocene time, North Dakota's climate has been variable, but tending to be warmer and drier than it is today.

APPENDIX B
GLOSSARY

All of the terms in this glossary are terms I have used in this book. My definitions are based, in part, on the Glossary of Geology, Fourth Edition *(1997), published by the American Geological Institute, but I have written new definitions in some instances. I have taken care that all definitions apply to the usage in this book.*

Many terms used in the science of geology owe their origin to names that were applied by early (mainly 18th and 19th century) geologists, most of whom were educated in classical studies. They often drew upon Greek, Latin, Old French or Old English terminology to construct a word describing a feature or process. A knowledge of the roots of various geologic terms can often provide insight into the idea that the geologist naming the feature was trying to convey and it can help to more clearly understand the meaning of the term. For this reason, I have included brief statements of the etymology of some of the terms in this glossary.

ablation. Process by which a glacier loses snow and ice; melting, evaporation and sublimation resulting in the separation of rock material, the formation of residual deposits, and their removal through wind action or the washing away of loose and soluble materials; from the Latin *ablationem*: a taking away.

abrasion. Process of mechanical wearing, grinding, scraping or rubbing of rock surfaces by friction and impact; the tools of abrasion include wind, ice, waves, running water and gravity, all of which transport solid rock particles away; from the Latin *abradere:* to scrape away, shave off.

acre-foot. Volume of a 1-foot-deep layer over a 1-acre area (43,560 square feet; an acre-foot amounts to 43,560 cubic feet of material). Acre-feet are commonly used in measuring volumes of water or reservoir storage space, or the volume of reservoir rock in an oil or gas field.

aerial photograph. An image taken from the air, such as a photograph of a part of the Earth's surface. Technical geomorphic studies utilize vertical air photos, as well as satellite images.

agate. A translucent variety of microcrystalline quartz; a variegated form of chalcedony (a translucent form of quartz) commonly mixed or alternating with opal, and characterized by colors arranged in alternating stripes or bands, or in moss-like forms. Agates may be any color. They are common in Pliocene gravel beds found on the surface in western-most North Dakota; agate is from the Greek *akhates*: the name of a river in Sicily where the stones were found—Pliny.

age/stage. (1) **Age** is the formal geochronologic (time) unit of lowest rank, below epoch, during which the rocks of the corresponding **stage** were formed; (2) the term used informally to designate the length of geologic time during which the rocks of any stratigraphic unit were formed; (3) a geologic division of time of unspecified duration in the history of the Earth, characterized by a dominant or important type of life form, e.g., the "age of mammals"; (4) the time during which a particular geologic event or series of events occurred or was marked by special physical conditions, e.g., the "Ice Age"; (5) relative position of anything in the geologic time scale, e.g., "the rocks of Miocene age." "Age" is often expressed as ybp ("years before present" or B.C.E. ("before the common era" or "before the Christian era").

alluvium/alluvial. Sediment eroded from adjacent areas and deposited by running water in and along rivers and streams. Alluvial materials were deposited by running water; from the Latin *alluvius*, meaning "flood."

Altithermal Interval. The dry postglacial time (from 7,500 to 4,000 years ago) during which temperatures were considerably warmer than today; corresponds to the Atlantic Interval in Europe, known as the thermal maximum; the middle part of the Hypsithermal Period.

anoxic. Without oxygen or short in oxygen.

anticline. Layers of rock that have been laterally compressed to form a fold, which is convex upward and whose core contains the stratigraphically older rocks. The Nesson and Cedar Creek anticlines are examples in North Dakota.

aquifer. Saturated sediment or rock, sufficiently porous and permeable, so it is capable of providing a sustainable yield of suitable quantities of ground water; a water-bearing stratum; from the Latin *aqui-* meaning "water" and *fer* meaning "bearing"; **please note:** the "fer" at the end of the word "aquifer" should be pronounced like "fur," not "fire."

archaeology. The study of human cultures through the recovery and analysis of their material relics; also spelled archeology; from the Greek *arkhaiologia,* "the study of ancient things."

areal. Refers to the geographic area over which a rock or sedimentary unit occurs, either buried or exposed at the surface; as "areal distribution" or "areal extent."

artesian. An adjective referring to ground water confined under hydrostatic pressure; an artesian aquifer is a confined aquifer; from the French *puits artésien* "wells of Artois," a French province where such wells were first bored in the 18th century.

badlands. Intricate, stream-dissected topography characterized by a fine drainage network with high drainage densities and short, steep slopes; often with little or no vegetative cover; American English translation of the French *mauvaises terres a traverser,* meaning "bad land to travel across." The American translation "badlands" is a shortened version of the French translation.

basement. (1) Undifferentiated complex of rocks that underlies the rocks of interest in an area; (2) the crust of the Earth below sedimentary deposits, extending downward to the Mohorovicic Discontinuity (the boundary between the Earth's crust and the underlying mantle, aka MOHO); in many places, the basement rocks are considered to be those that are igneous and metamorphic, of Precambrian age.

bed. A layer of sediment or sedimentary rock; *see* **stratum** or **strata.**

bedrock. Solid rock underlying the soil or other unconsolidated, unlithified surficial material; the rock or sedimentary material beneath the covering of glacial sediment.

bentonite. Soft, plastic, porous, light-colored rock composed of clay minerals of the montmorillonite (smectite) group, plus colloidal silica, and produced by devitrification (the conversion of glass to crystalline material) and accompanying chemical alteration of a glassy igneous material, usually volcanic ash or tuff (cemented volcanic ash); often, a clay that will swell in volume when it becomes wet; *see* **smectite;** probably named for Fort Benton, Montana.

"big bang" hypothesis. Hypothesis that the observed expansion of the universe may be extrapolated back to a primeval cosmic event; the "beginning of time." Current calculations place the big bang at 13.81 billion years ago.

bog. (1) Waterlogged, spongy ground consisting primarily of mosses and containing acidic, decaying vegetation that may develop into peat; (2) vegetation characteristic of this environment, especially sphagnum, sedges and heaths; also: fen, marsh or swamp; from Irish-Gaelic *bogach*, meaning "soft, moist."

boreal. (1) "Northern" or pertaining to the north, or located in northern regions; (2) pertaining to the northern biotic area (or Boreal Region) characterized by tundra and taiga and dominated by coniferous forests; (3) pertaining to the Boreal postglacial period, characterized by a cool climate like that of the present Boreal Region; also said of the climate of such a period; (4) pertaining to a boreal climatic zone or to the climate of such a zone; from Greek *Boreas*, name of the god of the north wind. The southern counterpart to boreal is "austral."

boulder pavement. (1) Accumulation of cobbles and boulders once contained within a layer of glacier sediment and remaining in the position in which they were set down after the finer material was removed by waves, currents, running water or other forms of erosion; (2) a smooth surface strewn with striated and polished boulders abraded to flatness by the movement of an overriding glacier.

butte. Conspicuous, isolated, usually flat-topped hill with steep slopes, usually capped with a resistant layer of rock; buttes represent erosional remnants carved from flat-lying rocks; the summit area of a butte (a French word meaning "hillock" or "rising ground") is smaller in extent than that of a mesa.

calcareous. Containing calcium carbonate; from Latin *calcarius* "of lime."

cannonball. A large, spherical concretion resembling a cannonball.

caprock. Hard and erosion-resistant rock layer, usually sandstone or limestone, overlying softer, more easily eroded materials; top of a butte or mesa (this definition refers to North Dakota buttes—the term has other definitions, as well, relating to hydrocarbons, coal and others).

carbonate rock. Rock consisting chiefly of carbonate components, such as limestone or dolomite; a sedimentary rock composed of more than 50 percent by weight of carbonate minerals; carbonate: from French *carbonate,* salt of carbonic acid.

cast. Secondary rock or mineral material that fills a natural mold; a replica or reproduction of the external details (size, shape, surface features) of a fossil shell, skeleton or other organic structure produced by the filling of a cavity formed by the decay or dissolution of some or all of the original hard parts of which the organism consisted; in a geologic sense: from Swedish *kasta,* meaning "to form in a mold."

castellate. A physiographic feature such as a cliff or butte that has a towering or battlement structure resembling a castle; caused by vertical flow of running water causing rills.

cementation. The diagenetic (chemically changed) process by which coarse clastic sediments become lithified or consolidated into hard, compact rocks, usually through deposition or precipitation of minerals in the spaces among the individual grains of the sediment. Cementation y occur simultaneously with sedimentation, or at a later time.

channel. (1) Stream bed where a natural body of surface water flows, a natural passageway or depression of perceptible extent containing continuously or periodically flowing water, or forming a connecting link between two bodies of water; a watercourse; (2) deepest or central part of the bed of a stream (aka the thalweg) containing the main current, and occupied more or less continuously by water; (3) abandoned or buried watercourse represented by stream deposits of gravel and sand; (4) artificial waterway, such as an open conduit, an irrigation ditch or canal, or a flood way.

clastic. Rock or sediment composed principally of broken fragments derived from pre-existing rocks or minerals that have been transported some distance from their places of origin; also said of the texture of such a rock. The term is usually used as a noun in the plural, e.g., the commonest clastics are sandstone and shale; from the Greek *klastos*, broken in pieces.

chert. A form of quartz, very finely crystalline; may form irregular nodules in other sediments.

clay. Very fine-grained sediment defined as having particles less than 4 microns in diameter.

climate change. Change in overall climatic conditions, usually gradual, due to natural, cyclic changes in a variety of factors, primarily solar output and changes in the Earth's geometry relative to the sun. The Earth's climate is constantly changing on both geologic and shorter-term human-awareness scales.

clinker. Sediment that has been baked by heat from buried, burning coal (lignite, in North Dakota). Clinker may range from rock that 1) has completely melted to assume a glassy texture, 2) to brick, 3) to material that is only slightly colored and hardened. Commonly referred to as "scoria" in North Dakota (but true scoria is of volcanic origin).

coal. A combustible organic sedimentary rock formed of compressed decomposed plant remains.

concretion. (1) A hard, compact mass or aggregate of mineral matter, often subspherical, but may be oblate, log-shaped, disk-shaped or irregular; formed by precipitation from aqueous solution about a nucleus or center, such as a leaf, shell, bone or fossil; a hard inclusion in a sedimentary rock, usually of a different composition than that of the rock that surrounds it and from which it is sharply separated; most concretions formed during diagenesis (chemical change) after burial. (2) a collective term applied loosely to various primary and secondary mineral segregations of diverse origin, including irregular nodules, spherulites, crystalline aggregates, geodes, septaria and various related mineral bodies.

contact. A planar or irregular surface between two types or ages of rock; a contact usually marks some kind of obvious change in material, texture, color, etc. A contact may or may not represent a break in time.

continental deposits. Formed on land rather than in the sea; continental deposits may be of lake, swamp, wind, stream, glacial or volcanic origin; compare with marine deposits.

continental displacement (continental drift). A general term, relating to a theory originally propounded by Alfred Wegener, a German geophysicist, in 1912; the theory states that thin plates of crust move freely around on the Earth's surface. The original theory stated that crustal masses, composed of both continental and oceanic crust, are propelled over the Earth's surface and move in various directions independently of each other. The theory eventually evolved into the theory of plate tectonics.

correlate. To show a correspondence in character and stratigraphic position between geologic formations in two or more separated areas.

cosmogenic. Changes caused by the action of cosmic rays; cosmogenic nuclides are isotopes that are produced by interaction of cosmic rays with the nucleus of the atom; analysis of such nuclides on glacial erratic boulders can help determine how long they have been exposed to sunlight.

coteau. The word "coteau" can refer to any of a variety of features: a range or sharp ridge of hills; a high plateau; a hilly upland, including the divide between two valleys, esp. a prominent and dissected escarpment forming the edge of a plateau (as the Missouri Coteau in North Dakota); Canadian French: slope of a hill, hillock, little hill).

creep. The slow, continuous downslope movement of rock and soil particles due to gravity; many types of creep have been described using criteria relating to material properties, stress level, stage and rate of deformation, mechanics of failure, geometric patterns and cause of deformation.

cross section. A two-dimensional diagram or drawing that shows the sequence of rock and sediment layers as they occur in a vertical plane; commonly drawn from the ground surface down to some selected depth, such as the top of the bedrock; a side view of a vertical plane cut through part of the Earth's crust.

crude oil. Petroleum in its natural state as it emerges from a well, or after passing through a gas-oil separator, but before refining or distillation.

crust. The outermost, solid layer of the Earth, separated from the underlying mantle by the so-called Mohorovicic Discontinuity.

crustal rebound. Return of the Earth's crust to the position it was before the weight of glaciers depressed it. Syn. isostatic rebound; glacial isostatic adjustment. The term may apply to rebound of the crust in response to the removal of earthen material due to erosion.

cutbank. A term used in the western United States for a steep, bare slope formed by a stream eroding laterally.

dead ice. (1) Ice that is not actively flowing forward and is not receiving material from an accumulation area; the ice of a stagnant glacier. (2) Detached blocks of ice left behind by a retreating glacier. The ice blocks are commonly buried beneath glacial sediment and, when this happens, they melt slowly, without the production of large quantities of water, and, as the overlying materials collapse, the result is a variety of irregularities in the topography.

dead-ice moraine. Areas of glacial deposits that were once underlain by stagnant glacial ice; numerous hummocks and depressions, wetlands (ponds and sloughs) and a variety of associated landforms such as "doughnuts," collapsed lake plains or perched lake plains; dead-ice moraine is poorly drained with few streams; known as hummocky moraine.

deglaciation. The uncovering of a land area from beneath a glacier or ice sheet by the wasting of ice as a result of shrinkage due to melting.

dimension stone. Building stone that is quarried and prepared in regularly shaped blocks according to specifications.

disintegration ridge. A ridge, which may be straight or curved in any way, formed when stagnant glacial ice melts and cracks or intervals between ice bodies become filled with glacial or glaciofluvial sediment; the sediment remains as a ridge when all of the ice has melted.

doughnut. Small, circular, doughnut-shaped closed ridge of glacial origin; associated with areas of glacial stagnation and dead-ice moraine.

drift/glacial drift. A general term applied to all rock materials transported by a glacier and deposited by or from the ice, or by running water originating from a melting glacier. The term "drift" was introduced by Roderick Murchison in 1839 for material (then called diluvium) that he regarded as having drifted in ocean currents and accumulated under the sea in comparatively recent times; this material is now known to be a product of glacial activity.

drumlin. A low, smoothly rounded, elongate to oval hill, mound or ridge of compact till or, less commonly, other sediment of glacial origin (sand, sandy till, banded clay), built under the margin of the ice and shaped by its flow, or carved from an older moraine by re-advancing ice; the longer axis parallels the direction of movement of the ice. The nose of a drumlin is usually blunt and pointing in the direction from which the ice approached; a gentler slope tapers in the direction the glacier was flowing. In North Dakota, many drumlins have extreme length to width ratios, and are much longer than drumlins in other areas; from Irish-Gaelic *druim*, "back, ridge."

Dryas. The Younger Dryas was a cold period, a return to glacial conditions after they had been warming for several thousand years. Sometimes referred to as the "Big Freeze," it began 12,800 years before present (BP) and ended 11,300 BP. The name "dryas" comes from an alpine flower, the Dryas octopetela. The Younger Dryas caused rapid re-advance of glaciers in North Dakota, resulting in the formation of the Devils Lake Basin, and the long drumlins near Verendrye. The Older Dryas, which occurred between 18,000 and 15,000 BP, did not have a big impact in North Dakota, most of which remained covered by glaciers throughout that time.

earthflow. A mass-movement landform and process characterized by downslope sliding of soil and weathered rock over a discrete basal shear surface within well-defined lateral boundaries; grades into landslides and mudflows through a continuous range in morphology associated with increasing fluidity.

earthquake. A sometimes disastrous, sudden motion or trembling in the Earth caused by the abrupt release of slowly accumulated strain, usually associated with moving continental plates; tectonic activity that may result from isostatic rebound of the Earth's crust associated with deglaciation or from volcanic activity.

end moraine. A ridge of material formed by the accumulation of glacial sediment at the edge of a glacier. A moraine deposit at the lower or outer edge of a glacier that marks its farthest extent.

eolian. Wind-deposited sediments, including deposits of loess and wind-blown dune sand, as well as sand-blasted surfaces (from the Latin *Æolus*, "god of the winds," and the Greek *Aiolos*, "of the wind"; the ancient district of *Aiolis* in Asia Minor, named for the wind god Aiolis).

erosion. The general process or group of processes whereby the materials of the Earth's crust are loosened, dissolved or worn away, and simultaneously moved from one place to another by natural agencies, which include weathering, solution, corrasion and transportation; the mechanical destruction of the land and the removal of material (such as soil) by precipitation runoff, waves and currents, moving ice or wind; from Latin *erosionem* or *erosio*, meaning "a gnawing away."

erratic. Boulder or other rock fragment transported by glacial ice from its place of origin to an area where the material on which it lies is different; a rock or boulder that is out of place; from an Old Swiss/French word meaning "wandering, straying, roving" or "ground that has wandered."

escarpment. A long, more or less continuous cliff or steep slope facing in one general direction, breaking the continuity of the land by separating separate level or gently sloping surfaces; produced by erosion or by faulting. The term is often used synonymously with scarp, but the term "escarpment" more often refers to a cliff formed by differential erosion; an escarpment implies a more extensive feature than does "scarp," which commonly refers to a more local feature; from French *escarpment*, meaning "make into a steep slope."

esker. A narrow, sinuous ridge of (mainly) sand and gravel deposited by a meltwater stream flowing upon, within, or beneath a melting glacier. Eskers range from a few hundred yards to several miles in length; from Irish *eiscir*, "ridge of gravel."

fault. A fracture or zone of fractures in Earth materials along which displacement of the sides relative to one another, up or down or sideways, parallel to the fracture, has occurred. It has been hypothesized that the Missouri Escarpment is due, in part, to a fault in the underlying bedrock (beneath the glacial deposits), but this idea is inconclusive (used in a geologic sense: from Old French *faute*, meaning "opening, gap, failure").

firn. Material between snow and ice; older and denser than snow but not yet transformed into glacier ice.

flint. A microcrystalline variety of chert (quartz); used as artifacts (tools, including spear points and arrowheads); an example in North Dakota is Knife River Flint; "flint" is an Old English word meaning "hard, rock that splits."

flood plain. Surface or strip of smooth land adjacent to a river channel, constructed by the present river in its existing regimen and covered with water when the river overflows its banks; specifically, the flat land along a river that periodically floods. The extent of the area that a lake may flood or has flooded due to fluctuations in the water level.

fluvial. (1) Pertaining to a river or rivers; (2) existing, growing or living in or near a stream or river; (3) sediments or landforms produced by the action of a stream or river; from Latin *fluvius*, meaning "river."

fool's gold. Iron pyrite, usually in cubic crystals that have the appearance (color) of gold.

formation. The basic geologic unit of lithostratigraphic (rock-strata) classification. A formation must be identifiable on the basis of easily recognized physical properties and widespread enough to be mapped at a regional scale; formations may be divided into smaller units (called members) where these, too, are recognizable and mappable. Several formations may be combined into groups when useful and appropriate.

fossil. Trace or imprint of a plant or animal preserved in the Earth's crust since some past geologic or prehistoric time; loosely, any remains or evidence of past life; said of any object that existed in the geologic past and of which evidence still exists. Originally an adjective meaning "ancient;" now used as either a noun or an adjective; from Latin *fossilis*, "dug up."

friable. Property of a rock that can be disintegrated by the slightest pressure, such as that between two fingers; from Latin *friabilis* "easily crumbled or broken."

geographic center. Point on which an area on the Earth's surface would balance if it were a plate of uniform thickness (i.e., the center of gravity of such a plate). The geographic center of the conterminous U.S. (excluding Alaska and Hawaii) is located in the eastern part of Smith County, Kansas (lat. 39°50'N, long. 98°35'W); the geographic center of North America is located six miles west of Balta, Pierce County, N.D. (lat. 48°10'N, long. 100°10'W). The geographic center of North Dakota is located 2.8 miles southwest of McClusky, in Sheridan County (lat. 47° 26' 46.824" N, long. 100° 28' 07.896" W). The geographic center of the U.S., *including* Alaska and Hawaii, is near Belle Fourche, South Dakota (lat. 44°58'N, long. 103°46'W).

geologic age. (1) The age of a fossil organism or of a particular geologic event or feature referred to the geologic time scale and expressed in terms of either years before present (absolute age) or comparison with the immediate surroundings (relative age); (2) an age datable by geologic methods; (3) a term used to emphasize the long-past periods of time in geologic history, as distinct from present-day or historic time.

geologic hazard. A naturally occurring or human-made geologic condition or phenomenon that presents a risk or is a potential danger to life and property. Examples include landsliding, flooding, volcanism, earthquakes, ground subsidence, coastal and beach erosion, and faulting. Human-caused events, such as dam leakage and failure, mining disasters, pollution and waste disposal, are sometimes considered to be geologic hazards.

geologic map. A diagram or drawing on a horizontal plane of part of the Earth's surface showing, by means of lines, colors, symbols and orientation, the distribution of selected features such as particular surface or subsurface units of rocks or sediments; a map on which is recorded geologic information, such as the distribution, nature and age relationships of rock units (surficial deposits may or may not be mapped separately), and the occurrence of structural features (folds, strike/dip, faults, joints), landforms, mineral deposits and fossil localities.

geologist. A person who is trained in and works in any of the geological sciences.

geology. The study of the solid part of a planet; the term is broadly applied, covering topics such as the moon and other planets; locating economic minerals and fuels; identifying stable sites for construction; predicting potential dangers associated with our active planet; understanding how the landscape formed; and other uses; from Modern Latin *geologia*, "the study of the Earth."

geomorphic/geomorphology. The science that treats the configuration of the Earth's surface; the study of the classification, description, nature, origin and development of present landforms and their relationships to underlying geologic structures, and of the history of geologic changes as recorded by these surface features.

glacial. (1) Relating to the presence and activities of ice or glaciers, as glacial erosion; (2) distinctive features and materials produced by or derived from glaciers and ice sheets, as glacial lakes; (3) pertaining to an Ice Age or region of glaciation. A glacial age, or glacial stage of a glacial epoch, esp. of the Pleistocene Epoch; e.g., the Wisconsinan glacial age or stage; from Latin *glacialis,* "icy, frozen, full of ice." The rate at which something happens, e.g., very slow.

glacial lake. (1) A lake that derives much or all of its water from the melting of glacier ice, e.g., fed by meltwater, or lying on glacier ice and due to differential melting; (2) a lake occupying a basin produced by glacial deposition, or one held in by glacial ice or by a morainal dam and containing both meltwater and water from precipitation run-off.

glacial lake plain. A flat area underlain by fine-grained sediment deposited in a lake formed by ponded meltwater from a glacier.

glacial striations. Long, straight and parallel scratches or furrows on a bedrock surface caused by the dragging and scraping of rock fragments embedded in the base of an overriding glacier; striations are usually oriented parallel to the direction of ice movement.

glaciation. (1) A collective term for the geologic processes of glacial activity, including erosion and deposition, and the resulting effects of such action on the Earth's surface; (2) the formation, movement and recession of glaciers; (3) the covering of large land areas or mountainous areas by glaciers or ice sheets; (4) the geographic distribution of glaciers; (5) any of several parts of geologic time during which glaciers were more extensive than at present; a glacial epoch, or a glacial stage; (6) a climatic episode during which extensive glaciers developed, attained a maximum extent and receded.

glacier outburst flood. A sudden release of meltwater from a glacier or glacier-dammed lake, sometimes resulting in a catastrophic flood; formed by melting of ice blocking a drainage channel or buoyant lifting of ice by water or by subglacial volcanic activity. Syn: *Jökulhlaup* (an Icelandic term for such a flood).

gneiss. A metamorphic rock, altered from another rock and with bands of granular material; may consist of any of several minerals;

from Middle High German *gneist*, "spark"; so called because the rock glitters.

ground moraine. (1) An accumulation of glacial sediment after it has been deposited or released from the ice during ablation, to form an extensive area of low relief; may lack pronounced transverse or longitudinal linear elements or may be marked by washboards and drumlins; (2) a gently rolling ground surface underlain by till deposited beneath a glacier.

ground water. (1) That part of the subsurface water in the zone of saturation. (2) Loosely, all subsurface water as distinct from surface water. Also spelled as one word: groundwater.

hazardous waste. Waste that poses a present or potential danger to human beings or other organisms because it is toxic, flammable, radioactive or explosive, or has some other property that produces substantial risk to life or safety.

hoodoo. A column, pinnacle or pillar of rock produced in a region of sporadic heavy rainfall by differential weathering or erosion of horizontal strata, facilitated layers of varying hardness, and occurring in varied and often eccentric or grotesque forms; found in areas of badlands topography (American English: literally "something that causes or brings bad luck," but this does not adequately explain its local usage).

hummocky. Uneven topography, most common in areas of dead-ice moraine.

hydrodynamic blowout. Forceful release of water from an artesian aquifer upon the release of the mechanism holding the water in place; large amounts of sediment (usually sand) may accompany the water as it rushes to the surface.

hydrocarbon. Any organic compound, gaseous, liquid or solid, consisting of carbon and hydrogen, including natural gas and crude oil; crude oil is a complex mixture of hydrocarbons.

Hypsithermal. Same as Climatic Optimum and thermal maximum; the part of the Holocene when mean annual temperatures exceeded those of the present. Extends from 9,000 years ago to 2,500 years ago.

Ice Age. A loosely used synonym for glacial epoch, or time of extensive glacial activity; the latest of the glacial epochs, known as the Pleistocene Epoch.

igneous rock. One of the three basic categories into which rocks are classified, of which the other two are sedimentary and metamorphic. Igneous rocks are formed by the cooling of molten rock called magma, deep within the Earth's crust and rock that is extruded at the surface, such as a lava flow (igneous: from Latin *igneus*, "of fire, fiery").

impermeable. The condition of a rock, sediment or soil that renders it incapable of transmitting fluids under pressure; lacking connection between pores. Syn: imperviousness.

indurated. Rock or soil hardened or consolidated by pressure, cementation or heat (from Latin *induratus*, "to make hard, harden").

interglacial. Pertaining to or formed during the time interval between two successive periods of glaciation. The term implies both the melting of ice sheets to their present extent and the maintenance of a warm climate for a sufficient length of time to permit a change in vegetational regime.

internal drainage. Surface drainage insufficient to allow precipitation that falls to reach the ocean; drainage only toward the lowermost or central part of an interior basin. May be local, as in an area of dead-ice moraine on Turtle Mountain or the Missouri Coteau, where glacial deposits have been haphazardly distributed and streams have not had time to develop, or it may be regional, as in the Devils Lake Basin, which has internal drainage until the level of the lake rises high enough to flow out of the basin.

iron pyrite (or "pyrite"). Crystalline, metallic-looking iron sulfide, known as "fool's gold."

kame. A low hill or mound composed of poorly sorted sand and gravel that was deposited by glacial meltwater plunging into crevasses near the front of a glacier; a kame may be steep-sided or only a gentle rise. Through time, the term has undergone changes in meaning, but today it refers to a deposit of glaciofluvial and glaciolacustrine sand and gravel whose precise mode of formation may be uncertain (from an Old English word, "*camb*," meaning comb).

kettle. A steep-sided, basin- or bowl-shaped hole or depression, commonly without surface drainage, in glacial deposits, often containing a lake or swamp; formed by the melting of a large, detached block of stagnant ice (left behind by a retreating glacier) that had been wholly or partly buried in the glacial deposits. Kettles range in depth from 3 to tens of feet, and in diameter to as much as several miles.

lacustrine. (1) Pertaining to, produced by or formed in lakes, e.g., "lacustrine sands" deposited on the bottom of a lake, or a "lacustrine terrace" formed along a lake margin; (2) growing in or inhabiting lakes, e.g., a "lacustrine fauna."

lag. A lag deposit is the coarse material left behind after most of the finer material (silt, clay, sand or sometimes fine gravel) has been removed by wind, running water or wave action along a lake shore. In glaciated parts of North Dakota, a lag of boulders commonly covers the surface after erosion during a long period of time has removed the surrounding, finer materials.

lake plain. (1) The nearly level surface marking the floor of a former lake, underlain by well-sorted deposits from inflowing streams; (2) a flat lowland or extension of the lake bed bordering an existing lake.

landscape. The distinct association of landforms that can be seen in a single view.

landform. Any physical, recognizable form or feature of the Earth's surface having a characteristic shape, and produced by natural causes; major forms include plains, badlands, plateaus, canyons and mountains; minor forms include hills, valleys, slopes, eskers and dunes. Taken together, landforms make up the surface configuration of the Earth.

landslide. General term covering a variety of mass-movement landforms and processes involving the downslope transport, due to gravity, of soil and rock material en masse.

Laramide Orogeny. A time of deformation, resulting in the building of the Rocky Mountains from Late Cretaceous through Paleocene time (from French *orogénie*, "mountain forming").

lignite. Coal that still shows traces of the wood it once was; brownish-black coal, intermediate in coalification between peat and subbituminous coal; North Dakota lignite has a calorific value of 7,000 Btus per pound, on a moist, mineral-matter-free basis; known also as brown coal (esp. in Europe); from Latin, *lignum*, "wood."

limestone. Sedimentary rock consisting chiefly of calcium carbonate in the form of calcite; most limestone forms as a precipitate in seas.

lineation. Line visible on the land surface, viewed from a distance (satellite or aerial photography); the surface expression of fault or joint lines extending upward from the basement rocks. Lineations may form as a result of animals making paths and in various other ways.

lithified. Changed through time from loose sediment into a coherent and solid rock by means of cementation, compaction and compression; made into rock.

lithology/lithologic. A description of the major macroscopic features of a type of rock. The physical character of rocks or sediments, including mineralogical composition, grain size, color and structure; (lith + logy: study of rocks).

Little Ice Age. A cool, brief interval in an otherwise warm interglacial stage; specifically, as used in this book, the cool phase lasting from approximately the 14th into the 19th century, most notably in Europe.

loess. Windblown dust; a widespread, homogeneous, massive, unconsolidated but slightly coherent, fine-grained deposit that blankets much of today's North Dakota landscape and on which the soil is developed (from Swiss German *lösch*, "loose").

marine. Relating to oceans or seas; materials such as limestone and shale, deposited in a deep-water environment.

meander. One of a series of regular sinuous curves, bends, loops, turns or windings in the course of a stream. Meanders are produced by a mature stream swinging from side to side as it flows across its flood plain or shifts its course laterally. To wind or turn in a sinuous or intricate course; to form a meander (from Latin, *meander*, "a winding course," and Greek, *Maiandros*, name of a river in Turkey noted for its winding route).

meltwater channel. A channel-like valley of uniform depth and width that carried large volumes of glacial meltwater periodically issuing from proglacial lakes; such channels likely flowed full, repeatedly, but for brief periods of time, as massive amounts of meltwater were released; many meltwater channels do not carry modern streams. In North Dakota, the James, Sheyenne, Souris and Pembina river valleys are examples of large meltwater channels; innumerable smaller meltwater channels exist.

metamorphic rock. One of the three basic categories into which rocks are classified, of which the other two are sedimentary and igneous. Metamorphic rocks are those that have been changed in composition, mineral content, texture or structure by the application of heat or pressure; they were originally sedimentary, igneous or other metamorphic rocks; the word "metamorphic" was coined by the English geologist Charles Lyell in 1833.

methane. Colorless, odorless flammable gas, the simplest paraffin hydrocarbon, formula CH_4; the principal constituent of natural gas, found associated with crude oil and lignite (as coal-bed methane).

Mollisol. The main soil order in North Dakota, Mollisols form in semiarid to semi-humid areas, as a rule under a grassland cover. They are nutrient-enriched soils with deep, high-organic matter and a nutrient-enriched surface soil (a horizon; from the Latin *mollis*, "soft").

montmorillonite. A sheet silicate mineral found in some kinds of clay (most notably fuller's earth) and widely used for its absorptive properties. *See* **smectite**.

moraine. Mounds, ridges or other distinct accumulations of unsorted, unstratified glacial deposits, usually till, deposited chiefly by the direct action of glacier ice in a variety of topographic landforms; Swiss French word meaning "ridge of rock deposited by a glacier."

natural gas. Hydrocarbons that exist as a gas or vapor at ordinary pressures and temperatures. Methane is the most important constituent of natural gas, but ethane, propane and other gases may be present; nonflammable impurities include nitrogen, carbon dioxide and hydrogen sulfide. Natural gas may be associated with oil, or it may form as a byproduct of decay of organic material.

nodule. A small, irregularly rounded mass or lump of a mineral or mineral aggregate, sometimes having a warty or knobby surface and no internal structure, with a contrasting composition from the enclosing sediment or rock matrix in which it is embedded, e.g., a chert nodule in limestone or a pyrite nodule in lignite. Most nodules are secondary structures: in sedimentary rocks they are usually the result of post-depositional replacement of the host rock and are commonly elongated parallel to the bedding. Nodules can be separated as discrete masses from the host material (from Latin *nodulus*, "meaning small knot").

outcrop. Part of a glacial deposit or bedrock that is exposed (not covered by soil or vegetation) and visible at the Earth's surface.

outwash. Stratified sand and gravel washed out and away from a glacier by meltwater streams and deposited in front of or beyond the former margin of an active glacier. Fluvial material deposited on top of stagnant glacier ice that became pitted or collapsed when the buried ice eventually melted is referred to as collapsed outwash.

overburden. Any material that overlies bedrock; usually loose material.

oxidized/oxidation. Said of mineral deposits in which the sulfides have been altered to oxides and carbonates. Oxidation occurs in the materials that are in contact with oxygen in an environment in which soil-forming processes are active. In general, the thicker the zone of oxidation, the longer the land has been exposed to weathering processes.

paleontology. The study of plant and animal life in past geologic time; relationships of past life forms to existing plants, animals and environments. The main biologic tool used in characterizing the chronology of the Earth's history (Greek: literally, "ancient science or study of being").

paleosol. A soil formed on a landscape in the past, often with distinctive morphological features resulting from a soil-forming environment that no longer exists at the site; usually preserved by burial beneath younger sediments—in North Dakota, most commonly beneath or within glacial or alluvial deposits (Greek: "ancient soil").

patterned ground. Ground surface having relict polygonal or irregular patterns developed in poorly drained areas subject to intensive frost action and permafrost or by desiccation of mud flats, which are subsequently buried beneath fluvial or eolian sediment. Patterned ground is apparent when viewed from a height; it may not be apparent to a ground-based observer. The various patterns may be defined and accentuated by contrasting soils and vegetation.

pay or pay zone. The geologic structure or stratum that contains a mineral deposit. The interval of the stratigraphic section in an oil or gas field that yields oil or gas in economic quantities.

peat. An organic soil deposit that is flammable if dried; peat burns with a characteristic sweet smell. Peat is sometimes regarded as a primitive form of lignite (probably from Celtic root *pett* meaning "turf"; the earliest sense is not of the turf itself, but of the cut *piece* of the turf).

perched lake plain. One of many landforms associated with dead-ice moraine; a flat tract underlain by lake sediment and elevated above the surrounding area. Formed in a lake held in place by walls of ice, the lake floor was left standing in an elevated position after the ice melted. Also referred to as "elevated lake plains."

permafrost. Soil or rock at or near ground surface in arctic or subarctic regions; land that has been continuously frozen for a long time. Modern permafrost as deep as 3,000 feet underlies one-fifth of the world's land area. Each time North Dakota was glaciated, permafrost was widespread ahead of the advancing glaciers.

permeability. The capacity of a porous rock, sediment or soil to transmit a fluid; a measure of the relative ease of fluid flow under unequal pressure.

petrified. Turned to stone (from Middle French *pétrifier*, "to make or become stone").

petrified wood. Wood that has been lithified by impregnation with mineral matter of various chemical compositions (calcite, silica, etc.).

petroleum. A naturally occurring, complex, liquid hydrocarbon which, after distillation and removal of impurities, yields a range of combustible fuels, lubricants and petrochemicals; commonly referred to as "crude oil" prior to refining (from Latin *petra* "rock" + *oleum* "oil").

piping. Erosion by percolating water in a layer of subsoil, resulting in caving and in the formation of narrow, often vertical conduits, tunnels or "pipes" through which soluble or granular soil material is removed.

plateau. A broad, comparatively flat area of considerable extent and elevation; a plateau is usually higher and has more noticeable relief than a plain (from French *plateau*, "table-land," and Greek *platys*, "flat, wide, broad"), e.g., Missouri Plateau.

plate tectonics. A widely accepted theory of global tectonics in which the lithosphere (Earth's crust) is divided into a number of plates whose pattern of horizontal movement causes them to interact with one another at their boundaries, resulting in seismic and tectonic activity along these boundaries.

porosity. Pore space in rock; the percentage of the bulk volume of a rock or soil that is occupied by interstices within the solid material; if pores are connected, the material is permeable; pore space may be filled with air, water, hydrocarbons, etc.

pothole. A term sometimes applied to any depression occurring in a till or outwash plain. In North Dakota, the "prairie-pothole region" (Missouri Coteau) is an area of collapsed glacial topography with numerous potholes.

prairie. An extensive tract of level to rolling grassland, treeless, in the temperate latitudes of North America (from Old French *praerie* "meadow, pastureland"). Syn. steppe (but a steppe usually is considered to be more arid than a prairie).

proglacial. Immediately in front of or some distance beyond the outer edge of a glacier; proglacial refers to lake and stream deposits, and other features usually formed by or derived from water flowing away from melting glacial ice.

quicksand. A mass of fine sand consisting of smooth rounded grains and thoroughly saturated with water; most likely to occur at a river mouth or (in North Dakota) an area of ground water discharge (from Old English: *cwecesund* meaning "living sand").

radiometric age-dating. Methods of measuring geologic time; ways of calculating the age of a specimen by measuring the amount of a radioactive element remaining in it. Various methods of radiometric dating are based on the known, fixed rates of nuclear decay of natural elements such as carbon-14 (for short-term measurements—thousands of years) or potassium-40/argon-40, and certain other radioactive elements (for measurements of long duration—millions of years).

relief. A term used loosely to characterize the shape and configuration or unevenness of the Earth's surface; refers to the difference in elevation between hilltops and valleys of a given region. Places with little local variation in elevation have low relief; those with great variation have high relief.

reservoir rock. In petroleum geology: porous and permeable rock that yields oil and gas. Sandstone, limestone and dolomite are the most common reservoir rock materials.

riparian. Along banks of rivers and streams, as in riparian woodland (from Latin *riparius* "of a river bank"); "riverine."

rill erosion. The development of numerous, small and closely spaced channels, resulting from the uneven removal of surface soil by running water concentrated in streamlets. Rill erosion is an intermediate process between sheet erosion and gully erosion, most common in badlands areas.

sapping. As used in this book, the term refers to "spring sapping," which results when water rising to the land surface loosens the surface materials, making it possible for them to be removed as the water flows away.

sandstone. A sedimentary rock composed of sand-size particles of silicates bound together by a cementing material.

scarp. An abbreviation for "escarpment," but not so extensive a feature; a straight, cliff-like face or slope of considerable linear extent, breaking the general continuity of the land by separating surfaces laying at different levels. Some of the ancient shorelines of Glacial Lake Agassiz consist of scarps—sharp breaks in elevation—if no beach sand is present.

sediment. (1) Solid material, organic or inorganic, that has settled from suspension in a liquid, or (2) solid fragmented material transported and deposited by wind, water, or ice, chemically precipitated from solution, or secreted by organisms, that forms in layers in loose unconsolidated form; e.g., sand, mud, till (from Medieval French *sedere*, to sit, sink down).

sedimentary rock. One of the three basic categories into which rocks are classified, of which the other two are igneous and metamorphic. Sedimentary rocks result from the consolidation of loose sediment that has accumulated in layers; they form by the compaction and cementation of sediment or by the precipitation of dissolved minerals from salt or fresh water. Most sedimentary rocks were deposited from water, but they may be deposited by wind or glaciers as well.

seep. An area, commonly small, where water or oil percolates slowly to the land surface; refers to flows that are too small to be considered as springs. When used as a verb, seep means to move slowly through small openings in a porous material.

shale. A fine-grained detrital sedimentary rock formed by the consolidation, especially by compression of clay, silt or mud; shale is commonly characterized by a finely laminated structure, which imparts a fissility usually parallel to the bedding (origin of the word is unclear; possibly influenced by German *Schalstein*, "laminated limestone," and *Schalgebirge*, "layer of stone in stratified rock").

siltstone. A consolidated silt having the texture and composition of shale but lacking fine lamination or fissility; a massive mudstone in which silt predominates over clay; a rock whose texture is intermediate between that of sandstone and shale.

slopewash. Soil or rock material that has been transported down a slope by mass wasting, assisted by running water not confined to channels. Also slope wash.

slough. (1) A marshy tract contained in a swale or other local, shallow undrained depression on an area of dry land, as on the prairies of the Upper Midwest; a dry depression that becomes temporarily marshy or filled with water after precipitation or snowmelt; (2) a large wetland with characteristic, permanent vegetation. In North Dakota, sloughs are commonly habitat for waterfowl (from Old English *sloh*, soft, muddy ground).

smectite. The group name for the montmorillonite clay minerals; smectite minerals are derived from the alteration of volcanic glass and from the weathering of silicates. Smectite minerals are the chief constituents of bentonites and fuller's earth and are common in soils, sedimentary rocks, and some mineral deposits.

soil. (1) The unconsolidated mineral and/or organic material on the immediate land surface that serves as a natural medium for the growth of land plants; (2) the unconsolidated mineral or organic matter on the land surface that has been subjected to and shows the effect of genetic and environmental factors (from Latin *solum*, floor, base, ground, probably from *sedere*, to sit).

source rock. In petroleum geology, a sedimentary rock in which organic material under pressure, heat and time was transformed to liquid or gaseous hydrocarbons. Source rock is usually shale or limestone. The Bakken Formation shale is an example of a source rock.

stagnant glacier. Glacier ice that has become separated from the active glacier; stagnant glacier ice may become increasingly debris-covered as it melts. Syn: dead ice.

steppe. An extensive, treeless area of grassland similar to but considered to be drier than prairie. The term is used in southeastern Europe and Asia (from Russian *step*, of unknown origin).

strandline. (1) A beach or scarp, especially one higher than the present level of a sea or lake, or bordering a now-drained, former body of water; a former shoreline above the present water level; (2) the ephemeral line or level at which a body of standing water meets the land.

strata. Horizontal layers (plural of stratum). Single and distinct layers, or beds, of sediment or sedimentary rock easily distinguishable from layers above and below are strata. Stratigraphic units may be strata composed of a number of beds (from Modern Latin *stratum,* meaning thing spread out, coverlet, pavement).

stratigraphic, stratigraphy. The systematic definition and description of major and minor natural divisions of rocks and their arrangement according to composition, distribution, texture, correlation or other mutual relationships.

striations. Small marks and lines, frequently parallel, etched into a solid surface by some external force, such as glacial action; glacial striae (classical Latin: furrow, channel, flute of a column).

sweet spot. A place where a combination of factors results in a maximum response for a given amount of effort; in areas of oil production, a place where production is best.

talus. Rock fragments of any size or shape (but commonly coarse and angular) derived from above and lying on and at the base of a cliff or steep, rocky slope (from Latin *talutium*, a slope or outcrop of rock debris). Syn: scree.

tectonic. A reference to the formation of a major Earth structure, usually involving deformation or collision (colliding continental plates, etc.); tectonic forces include volcanism, earthquakes, sea-floor spreading and mountain building—forces that result in rearrangement or deformation of features of the Earth's crust (from Greek *tektonikos*, pertaining to building; used in a geologic sense since 1894).

terrace. A long, narrow, level surface, less broad than a plain, bounded along one edge by a steeper descending slope and along the other by a steeper ascending slope; a bench or steplike ledge breaking the continuity of a slope. A stream terrace marks a former level of the stream (from Old French *terrasse*, platform built on or supported by a mound of Earth).

texture. The size, shape and arrangement of the component particles of a sedimentary deposit such as sandstone or till, e.g., coarse-grained, fine, rounded, angular, etc. (from Latin *textura*, web, texture, structure).

till. Unsorted, unstratified rock debris composed of a wide range of particle sizes that was deposited directly by and underneath a glacier. Till is deposited mainly as mudflows (origin unknown, but a Scots Gaelic reference to boulder clay till is suggestive).

topography. The general shape of the land surface, including its slopes, relief and the configuration and position of its natural and man-made features (from Greek *topographia*, a description of a place).

tuff/tuffaceous. Consolidated or cemented volcanic ash, or containing these materials. In North Dakota, limestone or sandstone containing volcanic ash; commonly in butte-capping rocks.

tundra. A treeless, level to undulating plain, characteristic of arctic and subarctic regions; a marshy surface with a growth of mosses, lichens and low shrubs; tundra is underlain by dark, mucky soil and permafrost (from Lappish *tundar*, elevated wasteland).

unconformity. Surface of contact between two series of rocks in which the overlying rock differs substantially from the rock beneath; often implying a passage of time or other important change, and often representing the occurrence of major Earth movements or other changes between the two periods of deposition.

unlithified. Sediment that has not been changed into a coherent and solid rock; loose material.

varve. A sedimentary bed or lamina (thin layer) or sequence of laminae deposited in a body of still water during one year's time; a set of graded glaciolacustrine layers, seasonally deposited, usually by meltwater streams, in a lake. Counting and correlation of varves is a method to measure the ages, or longevity, of Pleistocene glacial-lacustrine deposits. *Rhythmites* are similar, and with laminae, but no annual connotation is implied (from Swedish *varv*, annual deposit of silt in a lake bed).

veblen. Mound or hill of fluvial material, kame-like in appearance, but formed as a result of ground water being blown out of the ground, carrying sediment to the surface as it exited the ground. Geologic features named for the town of Veblen, S.D., where they are common.

washboard moraine. One of several small, parallel, regularly spaced ridges that are oriented transverse to the glacier ice movement; viewed from the air, the successive ridges collectively resemble a washboard; abundant in the glaciated north-central U.S., including the plains of North Dakota and the western plains provinces of Canada.

weathering. The destructive process, or group of processes, by which earthy and rocky materials, on exposure to atmospheric agents at or near the Earth's surface, are oxidized and changed in color, texture, composition, firmness or form, with little or no transport of the loosened or altered material; specifically, the physical (mechanical) disintegration and/or chemical decomposition of rock that produces an in-situ (in place) mantle of waste, loosening and preparing sediments for transport.

well-sorted. A sorted sediment, commonly sand, that consists of particles all having approximately the same size.

wetland. General term commonly used by wildlife management specialists for an area of wet habitat; wetlands are a characteristic feature of certain kinds of geologic landforms such as dead-ice moraine and collapsed outwash; they include permanently wet and/or intermittently water-covered areas and characteristic flora and fauna. See: bog, fen, slough, lake, swamp, pond.

Wisconsinan. The name given to the most recent glaciation in North America. The European equivalent is the Würm Glaciation (named for the Würm River in Bavaria, Germany); the Wisconsinan and the Würm both are late Pleistocene in age.

NOTES

Preface and acknowledgments

1 Near Cooperstown, the North Dakota Historical Society has preserved examples of both a Launch Control Facility and a missile launch site (the Oscar-Zero Missile Alert Facility and the November-33 Launch Control Facility) at the Reagan Minuteman Missile State Historic Site. Visitors can tour both sites. Launch Control Facility sites were built to house personnel underground. The launch sites housed the actual missiles; no personnel were stationed there.

Some of the deeper excavations were unstable and dangerous when I inspected them. A vertical cut in shale or glacial sediment might give way without warning, sometimes almost filling the excavation. The Launch Control Facility excavations, which were as deep as 50 feet, were especially prone to collapse. I was never at an excavation when a wall slid, but I visited several sites where walls had collapsed before I got there.

Chapter 1: Geologic Origins of North Dakota's Landforms

2 These deeply buried rocks, and the mineral resources they contain, are described in the *Face of North Dakota,* published by the North Dakota Geological Survey in 2000.

3 An excellent discussion of North Dakota's fossil resources, written by Dr. John Hoganson, formerly with the North Dakota Geological Survey, is included in the 2011–2013 North Dakota Blue Book, published by the North Dakota secretary of state's office.

4 All of the exposed Cretaceous formations can be seen at least somewhere in North Dakota, but it is unlikely, unless you are particularly diligent, that you will see the one small exposure of Greenhorn shale. The Greenhorn is a dark-gray, limey shale that is exposed along the river bed of the Pembina River.

In 1967, a geologist with the U.S. Army Corps of Engineers showed me and two other North Dakota Geological Survey geologists the exact location of the proposed Pembilier Dam site, which was very near the point where the Pembina River crosses the Pembina-Cavalier county line (hence its name). The geologist was unsure of the character (age, formation, etc.) of shale he had found exposed at the site. He wondered whether it might be an exposure of Greenhorn

Formation shale, but the Greenhorn had never been identified at the surface in North Dakota.

The area of the proposed dam was heavily wooded. At the time we visited the site, the level of water in the Pembina River was low and we were able to see shale exposed near water level; the exposure was wet and slippery. Had the river level been higher, the exposure would have been submerged.

We verified that the shale exposure is Greenhorn. Had the dam been built, it would have covered the only exposure in the state (the dam itself would have covered it, not the floodwaters behind the dam). The Pembilier Dam was part of the Garrison Diversion project, which was eventually largely shut down. I do not recall the specific reason the Pembilier Dam was not built at that time (not because of the Greenhorn at that location). If the dam had been built, the resulting 20-mile-long reservoir would have inundated a large part of the wooded Pembina River Valley.

The obscure location of the exposure of Greenhorn shale ensures that only a very few people have seen it.

5 An asteroid is thought to have impacted the Yucatan Peninsula of Mexico at the end of Cretaceous time. Many scientists believe the impact was responsible for changing the climate, worldwide, killing much of the vegetation, and causing the extinction of much of the life on Earth, including the dinosaurs. Other scientists think that the dinosaurs were already declining and might have gone extinct at that time anyway.

6 The process of eroding large amounts of material from the Earth's surface, ultimately lowering elevations and carving out landforms, is known as denudation. Denudation is always going on, sometimes at greater rates than others. Recently I read that researchers calculated denudation rates from 990 river basins, worldwide, to estimate how much sediment is created annually as erosion wears down the land surface. Their conclusion was 5.5 billion metric tons globally, which translates, roughly, into one cubic mile of material. The steep slopes of mountains can produce sediments quickly, but low-sloping regions erode more steadily over a larger area and probably account for most of the world's sediments each year.

7 In the early 1990s, I had several conversations with a geologist from a major oil company. He was trying to determine how much glacial ice had covered parts of western North Dakota, especially in Dunn County where his company had holdings. He had become aware of a pattern of lineations in the area and wondered if glacial rebound, resulting from unloading of glacial ice, was responsible for the lineations. His hypothesis included the idea that the lineations, seen on air photos, may have formed in response to the release of overlying weight. He initially asked me how much glacial ice had covered the land, but after we ruled ice out for most of the area, his question became: "How much erosion of the land has taken place in the area?" He theorized that the lineations mark the locations of joints (joints are essentially faults along which no movement has occurred) that extend upward thousands of feet from the Precambrian basement rock to the land surface. He hypothesized that the joints might serve as conduits for the movement of hydrocarbons. Although his hypothesis was more involved than I have outlined, it eventually turned out to be plausible (except it did not involve glaciation). The idea eventually became an important consideration

relating to the horizontal drilling and hydraulic fracturing technologies being used to produce Bakken oil in North Dakota.

8 Erosion is always taking place on the landscape. Much of the time, eroded materials are simply moved from place to place, locally. In this section ("North Dakota's Modern Landscape"), I will be discussing a long period of time during which the main geologic process in North Dakota was erosion, which removed large amounts of material from the state so that the net effect has been a near-total re-shaping of the landscape.

9 John Steinbeck, 1962, *Travels with Charley*. I have always enjoyed Steinbeck's writing, including *Travels with Charley,* but much of what he wrote in the book has been questioned. He has been accused of fabricating some of the passages dealing with North Dakota. For example, his encounter with a Shakespearean actor at a campground near Alice (30 miles west of Fargo) seems unlikely. The actor mistakes Steinbeck for a fellow actor and greets him with a sweeping bow, saying, "I see you are of the profession." Some people have even questioned whether he really got to North Dakota. But it does not matter much to me whether parts of the book are fact or fiction. I still like it.

10 There is yet another possible explanation for the word *coteau.* The Métis fur traders, who traveled via ox cart between Winnipeg and St. Paul between 1820 and 1870, used the word *coteau* to describe the long, low ridges of land that bordered the Red River Valley. My impression is that, even though they were likely referring to the rise in elevation, apparent when viewed from the flat land of the Red River Valley, they probably were not referring to the beach ridges or other specific features. Their use of the word was derived from the French word *côte,* which meant, among other things, "coastline." The 19th century geologist George W. Featherstonhaugh (pronounced "Fanshaw"), who traveled through the area in 1837, referred to the edge of the Red River Valley as "an elevation like this 'côteau,' which stretches up and down the country, stands in the relation of a coast to the universal low level" of the plain.

Chapter 2: The Missouri Plateau

11 A 1984 study of prairie fires in North Dakota concluded that about six lightning-caused fires a year occur for every 10,000 square kilometers in the eastern part of the state, but 25 a year in the drier, western part of the state. – Ken Higgens, *Lightning Fires in North Dakota Grasslands and in Pine-Savanna Lands of South Dakota and Montana,* 1984.

12 The HT Butte clinker is named after the HT Ranch, a prominent horse ranch established in 1884 by the Little Missouri Horse Company. HT Butte is now known as Black Butte. The name "Black Butte" is probably derived from the dark color of the butte, compared with the white rocks of the nearby Chalky Buttes.

13 To put the age difference in perspective, the flat surface on top of the Killdeer Mountains is 2,000 ***times*** older than the flat surface between Fargo and Grand Forks.

14 An alternative explanation for the name of the Killdeer Mountains is that it refers to the call of a bird common in the area. Killdeer plovers are famous for attempting to lead one away from their nests by faking a broken wing. Their call approximates the words *Kill-dee-a!* or *Kill-dee-a!*

15 Slide Butte was so-named following a landslide of combined debris flows and rock falls on the north side of the butte in 1930 and 1932.

16 The names of North Dakota landforms owe their origins to a variety of sources. For example, the Métis fur traders, traveling along trails in the area of the beaches on the west side of the Red River Valley—the old shorelines of Glacial Lake Agassiz—imagined the broad grassland prairie to the east as a "sea" of grass. They referred to areas of wooded land in the sea as *ìles des bois* (islands of trees) and to the hills of the Pembina Escarpment to the west as *montagnes* (mountains).

Names such as these, applied by Canadian Métis, are examples of how people's perspectives rely on their past experience. The impression of the land that a Métis formed was much different than that of an American pioneer businessman, following the same trail northward from St. Paul in the 1840s. Geographer David Lowenthal, in 1961, wrote: "The surface of the earth is shaped for each person by refraction through cultural and personal lenses of custom and fancy. We are all artists and landscape architects, creating order and organizing space, time, and causality in accordance with our apperceptions and predilections." – David Lowenthal, *Geography, experience and imagination: towards a geographical epistemology,* 1961.

According to, Carolyn Gilman: "perceptions . . . differ according to the traveler's point of origin and destination, both intellectual and geographical . . . the geographical landmarks people identify and name can tell us much about the people themselves: how they lived and traveled—and, more important, how they thought." – Carolyn Gilman, *Perspectives of the prairie: cultural contrasts on the Red River trails,* 1978.

17 Regarding the name "Turtle Mountain," Patrick Gourneau, former tribal chairman of the Turtle Mountain Band of Chippewa Indians notes: "The Chippeway name is *Mekinauk Wudjiw* (Turtle Mountain). If it was Turtle Mountains, it would be *Mekinauk Wudjiw wum* (plural). The *Mechifs* referred to the hills as *la Montagne Torchue. La Montagne Torchue* is French meaning "Turtle Mountain." Gourneau further states: when the Chippeways first approached the wooded hills from a distance, the high butte . . . reminded them of a turtle because of its shape, like the back of a turtle." – Patrick Gourneau, *History of the Turtle Band of Chippewa Indians,* 1993.

Chapter 4: The Glaciated Plains and Missouri Coteau

18 T.C. Chamberlin, in 1895, named the "Wisconsin glaciation" (now referred to as "Wisconsinan") for excellent examples of glacial landforms in Wisconsin. Had Chamberlin done his work in North Dakota, he might well have coined the term "Dakota glaciation," as the landforms here are equally complex and representative.

19 At least five major, prolonged periods of repeated glaciation are documented during the 4.6 billion-year total history of the Earth. Each period lasted from tens to hundreds of millions of years, and each was marked by multiple expansions and contractions of glaciers, which individually lasted thousands of years. During these periods, one (or more) of the Earth's continental plates was situated over one (or both) of the Earth's polar regions. Such a situation is likely a requirement for widespread glaciation; that is, a land mass has to be located over a polar

region. This results in the potential for greater cooling because ocean currents cannot moderate worldwide temperatures as efficiently as they can when oceans cover the poles. We are currently [probably] in a contracting phase of the fifth of these major periods.

20 I have found small deposits of material exposed near Rhame that I have tentatively identified as till. Drill holes in southwestern North Dakota, beyond the known limit of glaciation, have penetrated pebbly clay that may be of glacial origin. However, no one has found conclusive evidence for glaciation in southwestern North Dakota, and I do not know when such early glaciations might have occurred.

21 Some geologists have suggested another term, "diamicton," instead of "till." Although diamicton refers to any material that may be essentially identical in content to till, it does not connote a glacial origin. The word "till" indicates that we know the material is of glacial origin. Thus, a mudflow that formed as a result of glacial processes might be a till, but an identical, modern mudflow might be referred to as a diamicton (I have tried to avoid becoming too involved in such semantical arguments).

22 Two of the best places to see several different layers of glacial deposits in North Dakota are near Riverdale, at the Wolf Creek inlet to Lake Sakakawea in McLean County and in Beulah Bay, 15 miles north of the town of Beulah in Mercer County. In both locations, three discrete till units, separated by cemented gravel layers or paleosols, are being eroded by waves along the lake.

23 The word *moraine* is a French/Swiss word, first used in 1789 by the Swiss scientist Horace de Saussure. It means simply "a heap of earth" (de Saussure did not realize the glacial origin). I do not know when the term was first used to refer specifically to glacial deposits, but the Swiss geologist Louis Agassiz (1807–1873) recognized alpine glacial deposits in Switzerland as early as 1837 and used the term "moraine" in his description of them.

The notion that the land had been glaciated was championed by Agassiz. Agassiz was himself a skeptic in the 1830s, but during the next few years, he became persuaded by fellow Swiss geologist Jean Charpentier that glaciers had once extended beyond their present limits in the Alps. As Agassiz, Charpentier and their associates accumulated evidence, it became clear that not only had the Swiss mountains been glaciated, but areas far beyond the alpine valleys as well.

Glaciation was the only way to account: 1) for the strange grooves and gouges on bedrock outcrops; 2) for boulders found far from their parent outcrops; or 3) for otherwise unexplained mounds of rock debris (moraines) that could not possibly be the work of local stream action. After years of fieldwork, Agassiz in 1840 published *Études sur les glaciers,* in which he argued that glaciers had not only extended across lowland Switzerland, but "great sheets of ice, resembling those now existing in Greenland, once covered all the countries in which unstratified gravel is found." Gradually, but grudgingly, the skeptics were won over. Based on more than 150 years of geological investigations, we now have a good picture of what happened during the Ice Ages.

By 1838, the English geologist William Buckland recognized that glaciers had covered Great Britain. The American geologist Edward Hitchcock, in his 1840 book, *Treatise on Geology*, discussed the glacial origin of moraines in Massachusetts. (I have a copy of Hitchcock's book [third edition, dated 1842], which was given to me while I was a college undergraduate student studying geology in the late 1950s. The elderly lady who gave it to me, a retired college professor, had used it as a text when she was a college student in the 1890s.)

24 The chances are good that a test hole drilled through a thick deposit of till will penetrate the combined accumulation of several separate glacial advances, layered one on top of another. Some of the till layers may be separated by old soil layers (paleosols), weathered zones that developed on an exposed till layer before a new layer of till was deposited on top of it by another glacier. Each paleosol records a long period of time during which the area was free of ice, as it is today, and the land surface was subjected to weathering processes. The thickness of the weathered zone may be approximately proportional to the length of time the land was exposed to weathering and other soil-forming processes, but in most instances, it cannot be determined with certainty how much elapsed time is represented by a weathered zone.

Although the formation of a soil is a complex process involving a variety of factors, one of the most easily recognized results is the oxidation of certain minerals. For example, iron-rich sediments are dark gray or black in their unoxidized state when the glacier deposits them, and they remain so while they are deeply buried beneath the water table. However, these minerals become brownish in appearance when they come in contact with oxygen above the water table. The oxidized zone is that thickness of material in which oxygen has altered many minerals to an oxidized state so they have assumed a tan or brownish appearance (just as iron becomes rusty when it weathers). The oxidation procedure can take many thousands of years; the longer weathering goes on, the deeper it affects the sediments.

The oxidized zone beneath our modern land surface in glaciated areas of eastern North Dakota is from 10 to 15 feet thick, the result of exposure to soil-forming processes for 12,000 years. Test holes drilled through thick layers of till (up to hundreds of feet thick in the Devils Lake area) may penetrate several separate buried oxidized soil zones. Some of these buried soil zones may be as thick as 30 feet or more, suggesting prolonged interglacial periods, much longer than the current interglacial period or resulting from warmer and wetter conditions leading to more intense weathering.

25 I had a chance to see washboard ridges as they were forming at the edge of a glacier in Lapland in Sweden several years ago. As the glacier advanced a short distance, it overrode its frozen-down outer edge, depositing ridges of material. I agreed with the Swedish geologists' conclusion that the ridges were not being "bulldozed" into place, nor were they simply small amounts of material that were being carried to the ice margin (like tiny "end moraines"). Neither were they annual accumulations, left behind as the glacier margin receded, stepwise, an explanation often offered for washboard ridges.

26 In Massachusetts, the term "hogbacks" is sometimes used to refer to drumlins. In the western part of the United States, the same term refers to bedrock ridges with sharp summits and steep slopes. The ridges in the western states are formed by erosion of the edges of slabs of resistant rock, such as sandstone; no glacial origin is involved.

27 During the summer of 2012, a glacier in Greenland advanced 10 miles, or 150 feet per day, the fastest speed ever recorded for a Greenland glacier.

28 This date, 12,500 ^{14}C yr BP, is a radiocarbon date. Years-before-present (ybp), when they relate to paleoenvironmental and archaeological data, (the last 30,000 years or so) usually use this time scale, rather than calibrated (calendar) chronology. This is because if a site has carbon (wood, charcoal, etc.), it is often possible to use carbon-14 dating techniques to determine its age. I will not comment on the mechanics of such dating techniques, as they are involved and not within the scope of this book.

Radiocarbon dates differ, systematically, from calendar dates; the older the date, the greater the difference. They result in a date that is slightly too young. Thus, a radiocarbon date of 20,000 years before present is the same as a calendar date of 24,000 years, a 4,000-year difference; a radiocarbon date of 15,000 years before present is the same as a calendar date of 18,000 years; a radiocarbon date of 10,200 years would coincide with a calendar date of 12,000 years, a difference of 1,800 years.

Because of the discrepancy, and because calendar years are more accurate and familiar to the average person, I have used calendar-year dates in this book whenever possible.

29 In the 1960s, when I first realized how hill-hole combinations like the one at Anamoose formed, I thought it would be appropriate to refer to such a feature as an "anamoose." However, the grammatical problems arising from that name were soon apparent. Are three such hills anameece? anamooses? anamoose? (as "I hit three moose standing on the road"). The town of Anamoose was founded by Romanian settlers who came to North Dakota from Saskatchewan. However, the name of the town, Anamoose, is an adaptation of a Chippewa word (uhnemoosh), meaning "female dog." Those people fascinated by unusual place names seem to be happy with Anamoose. In any event, the difficulties with the word soon led me to switch, simply, to "ice-thrust hills" or "ice-thrust topography."

30 Dogden Butte, so named because it had many prairie dog dens (some accounts say wolf dens), was originally named *Maison de Chien* (Dog's Home) by Pierre de la Verendrye in 1738. Explorer David Thompson mentioned the butte in his journal for 1797. The nearby town of Dogden was forced to change its name in 1927 to Butte because postal authorities complained the name was easily confused with Doyon, N.D. (the people in the town wanted to keep the name Dogden because it was unusual).

31 The first published references I have been able find describing ice-thrust features in North America and the glacial processes that formed them refer to features in Canada. Walter O. Kupsch, in a 1962 paper published in the *Journal of Geology,* described ice-thrust features in Saskatchewan and Alberta. The features had been described in Germany, but prior to Kupsch's work, they had not been documented in North America.

I first noted ice-thrust features in North Dakota in 1962, when I was studying excavations being made to house Minuteman missiles in eastern North Dakota (refer to my comments in the preface to this book). In 1967, I published a paper on ice-thrusting in Cavalier County, my first attempt at documenting the features in the scientific literature. That same year, I presented a paper on the topic at a Geological Society of America meeting. I recall that, during my talk, a geologist stood up and angrily berated me for my "idiotic comments" about the topic ("Glaciers can't do that!"). That's the only time anything like that ever happened to me. I eventually had an opportunity to visit with Dr. Kupsch about ice thrusting. Studies done through the years since then have greatly expanded our knowledge of the topic. I just looked up "glacial ice thrusting" on Google, and I see that it lists a total of 1,900,000 references to the topic.

32 The Dahlen Esker has been mentioned in various geologic reports dating to the late 19th century, but the best description and discussion of the feature was provided by Jack Kume for the North Dakota Geological Survey: *The Dahlen Esker of Grand Forks and Walsh Counties, North Dakota,* 1966.

33 Dr. Kenneth Harris, a North Dakota Geological Survey geologist, was the first person to recognize and describe veblens while he was mapping the geology along the South Dakota border in southeastern North Dakota in 1984. Dr. Harris named the features after the town of Veblen, South Dakota, located 15 miles southwest of Lidgerwood.

34 In 1989, Drs. Alan Kehew, Lee Clayton and Mark Lord described channels carved by catastrophic flows of water released from proglacial lakes, which carved valleys such as the Sheyenne and Souris in North Dakota.

35 Differentiating Devils Lake and Lake Minnewaukan is problematic. Lake Minnewaukan was dammed by glacial ice; Devils Lake was not. A logical way to distinguish the lakes may be to consider Lake Minnewaukan to have flooded land above the highest level reached by Devils Lake, approximately 1,460 feet. Lake sediment found above that elevation might, therefore, be considered to have been deposited in Lake Minnewaukan. All lake-related deposits and landforms below 1,460 feet could be considered to belong to Devils Lake.

Chapter 5: Other Geologic Features

36 Dunes need well-sorted sand to form. They cannot form on poorly sorted fluvial deposits (a mixture of sand, pebbles, cobbles, etc.) because as soon as the wind blows over the surface, it removes the finer material, leaving behind an armor of larger pieces (pebbles, etc.), thus preventing further wind erosion.

The theory of dune formation is complex. The shapes of dunes are the result of what is known as the "Gross bedform-normal transport rule," that is, "dunes orient in a direction maximizing the total sediment transport normal to their crest lines." Given a steady wind, dunes tend to form with their crest lines transverse to the prevailing wind. Their formation is primarily the result of a two-way interaction between sand and sediment. Moving air tends to pick up loose grains as it gains speed and it deposits them as it slows down. If the air-flow field around an obstruction (a ridge of sand or a bush, for example) is such that wind velocity peaks upstream of the obstruction's highest point, the wind's net effect is to transport sediment from the obstruction's base to its crest.

In some dune fields, the dunes have crests that are transverse to the prevailing wind direction, but in other fields, the crests are parallel to the wind direction; in still other dune fields, combinations of the two are found. It would be necessary to know the complete history of the wind activity that formed them to understand why dunes in any particular location are shaped as they are. That is, mature dunes retain—record—a history of the winds that shaped them, so it is difficult to separate the effects of more recent winds from those of earlier winds, and this detail of wind history is almost never available.

37 North Dakota is a prairie state, with less than 1 percent of the land forested, making it the least-forested state. However, the area of dunes near Denbigh is home to the Denbigh Dunes Experimental Forest. The 636-acre (one square mile) forest tract, established in 1931, consists of 60 percent tree plantings, with the remainder largely native prairie.

During the 1930s, the entire Great Plains region from Texas to North Dakota, and into Canada, was experiencing low precipitation, high summer temperatures and strong winds. The area of the dunes in McHenry County, with its thousands of acres of well-sorted sandy soil, was especially susceptible to damage due to these climatic conditions. Local people thought that forested areas would help stabilize the sandy soil, and perhaps improve McHenry County's long-term economic growth by establishing a timber industry. They envisioned a forest as large as 480,000 acres (750 square miles or 40 percent of the area of McHenry County) around the forest's current site near Denbigh.

The Denbigh Experimental Forest was established to determine: 1) which tree species grow well and are best suited for shelterbelt planting in the northern Great Plains; 2) which seed sources within species are best adapted for the northern Great Plains; 3) the best methods of soil preparation; and 4) which methods of tree establishment are most effective for shelterbelts.

The forest (really, no longer experimental) was established on a site that had been badly overplowed and overgrazed during the early part of the 20th century, resulting in bare wind-eroded sand dunes and economic hardship. The original overall plan envisioned by the Roosevelt administration called for planting a 100-mile-wide shelterbelt zone from North Dakota to the north Texas panhandle to reduce wind erosion, eliminate dust storms and provide local employment during the Great Depression via a jobs program. The shelterbelt zone never happened, but the Denbigh Forest remains today.

If you are in the area of the Denbigh dunes, be sure to take the time to visit the Experimental Forest. No staff is on site, but you can use the auto and hiking trails. The forested area provides excellent birding and short nature hikes. In addition to the 43 tree varieties in the forest, you may see wild turkeys, porcupines, elk and an occasional moose.

38 When I started mapping geology in North Dakota in the early 1960s, piles of erratics were much more common than they are today. Since then, many of the boulders have either been buried or removed to places like the shorelines of Lake Sakakawea and roads bordering Devils Lake to serve as riprap.

39 Passage is from *Blue highways: A Backroads Tour of Rural America,* William Least Heat Moon, 1982.

40 Most states have state rocks; North Dakota does not. We do, however, have a state fossil (Teredo-bored petrified wood), along with our state bird (Western meadowlark), state flower (wild prairie rose), state tree (American elm), state fish (northern pike), state grass (Western wheatgrass), state beverage (milk), and—since 2011—state insect (ladybug).

41 The Coghlan Castle was listed on the National Register of Historic Places in 2008. More information about the preservation of the structure is available from Save Coghlan Castle, Inc., PO Box 549, Rolla, ND 58676. The restoration of the Coghlan Castle is restricted to the exterior of the building; the inside will not be restored and it will not be open to the public.

42 Dr. Dibb reported discovering a "mountain of gold" in Montana a short distance east—within 50 miles—of the city of Butte, Montana, which did not yet exist in 1864. In spite of the efforts of many to re-discover the mountain of gold, it was never found again. I suspect that Dr. Dibb may have had a vivid imagination.

43 I am unsure whether this story of finding gold in the crops of turkeys is true. To say the least, it seems unlikely! I have butchered quite a few chickens, and a few turkeys and ducks, but I seldom checked the crops closely.

44 Personal communication, Kay A. Buri.

45 *Grand Forks Herald*, June 11, 1981.

46 In one instance, after a well drilled into an artesian aquifer near Grafton blew out, an attempt was made to plug the well by lowering a 30-foot section of railroad track into the wellbore. After a short time, the pressure in the water forced the track out of the well.

47 The *Sheldon Enterprise* newspaper article, quoted in the Sheldon, N.D. Centennial (1981) Book. The exact date of the news article is not given, but it apparently was in April 1906. The follow-up article, concerning the lawsuit against the Village of Sheldon, was published in the *Sheldon Progress* newspaper for May 8, 1908.

48 A scabby appearing whitish crust on dry, bare ground is usually due to salinity and sodicity. A few local terms used to describe such areas are "scab-land," "salt-land," "alkali," "alkaline," and "sour-ground." Three types of salts are found in North Dakota: sulfates, carbonates and chlorides. Soils containing sulfate salts are the most common.

Saline soils in the northern Red River Valley—the Grand Forks area (Kellys Slough, etc.)—are due to regional discharge of artesian ground water from the Dakota Formation. However, these salts are mainly chloride salts, uncommon for the northern Great Plains. As I noted, sulfate salts dominate saline soils in North Dakota.

Saline and sodic soils with similar properties can be characterized by geographic area. In most glaciated areas north and east of the Missouri River, such soils are associated with the edges of closed depressions or broad swales where discharge of ground water can occur. West and south of the Missouri River, in unglaciated areas, saline and sodic soils are not necessarily related to ground water discharge. These soils inherited the salts they contain from the sedimentary materials (shales, sandstones, etc.) on which they formed.

Chapter 6: The Devils Lake Basin

49 A closed drainage basin has no drainage outlet. Except for some ground water outflow, water escapes only by evaporation.

50 If Devils Lake does overflow, it will flow into the Tolna Coulee to the Sheyenne River. The flow of water into the coulee will be regulated by a control structure at the head of Tolna Coulee. The control structure, which was completed in 2012, is intended to ensure against rapid erosion and a catastrophic release of water into the Sheyenne River system.

51 This was a re-advance of the Late Wisconsinan glacier. Perhaps 2,000 years earlier, the ice margin had receded to some position north of the Devils Lake area

and, 12,000 years ago, in response to intensified glacial conditions, the glacier advanced over the area again.

52 As I write this, in July of 2013, the level of Devils Lake stands at 1,453.75 feet.

53 An exception to the rising/falling pattern might have occurred during the Altithermal Interval when the lake may have dried up, and remained dry, for prolonged periods of time.

54 A possible further explanation for the springs: at the time of the thrusting event that shaped the Devils Lake Basin, the glacier moved over the top of the older materials overlying the Spiritwood Aquifer in Towner County and northwestern Ramsey County. This movement may have caused small-scale thrusting. The layered beds overlying the aquifer may have been shredded, their edges peeled up by the glacier, opening conduits for water in the aquifer to move upward to the surface whenever the aquifer is sufficiently pressurized.

55 In the early 1990s, people in the Devils Lake area being impacted by the rising water of Devils Lake suggested that the level of Tolna Coulee, the lowest natural outlet from Stump Lake to the Sheyenne River, be lowered, thereby limiting the amount of potential future rise in the level of the (combined) Devils/Stump Lake. Geologists knew that Devils Lake had overflowed into the Sheyenne River several times in the past when it reached a level of 1,458 feet. However, the threshold level in the floor of the coulee (in the early 1990s) was 1,459 feet, and the lake would have to reach that level to overflow to the Sheyenne River. The extra foot was due to sediments that have accumulated in the coulee since the last time Devils Lake overflowed.

In response to the request for an opinion, the attorney general of North Dakota determined that the level of Tolna Coulee should be the same as it was when North Dakota became a state in 1889, a date that corresponds closely to the time when North Dakota was being settled. However, in the years since North Dakota became a state, erosional processes, due in part to agricultural practices, have partially filled in the floor of Tolna Coulee. Sediment has washed into the coulee, partly filling it in and making it shallower. The question that had to be answered was: How much sediment has been deposited in Tolna Coulee since statehood; how much shallower is the coulee than it was in 1889?

Early settlers to North Dakota brought the Russian thistle (*Salsoa iberica*) with them from Europe around the time of statehood. Prior to settlement, North Dakota had no Russian thistles. Therefore, the depth in the soil at which pollen from the Russian thistle first appears—the lowest (oldest) depth the pollen is present in the soil—has to correspond to the time it was introduced. Sediment on top of the thistle pollen had to have been deposited since the settlers arrived (since statehood). In addition to the Russian thistle, the pollen of cultivated grains such as wheat, rye and barley can be used to establish an upper date for the earliest farming (approximately 1889 in north-central North Dakota).

In March of 1997, the North Dakota Geological Survey conducted a study of Tolna Coulee, digging several trenches across the coulee and drilling test holes. More than 100 soil samples were collected from the surface, downward to depths of several feet, in the floor of the coulee. Using radiocarbon analyses, Survey geologists determined the age of the samples. Working with palynologists, our geologists determined that the deepest thistle pollen occurred at a depth of 1 foot. Given that information, the attorney general approved the removal of a foot of soil from the floor of the coulee, thereby lowering by a foot the potential highest level that Devils Lake can reach. Following the attorney general's opinion, the coulee was deepened from 1,459 feet to 1,458 feet so that its floor now corresponds more nearly to its original condition—essentially the same as it was at the time North Dakota became a state.

By limiting its highest possible elevation to 1,458 feet, the area that the lake can now flood is no more than 261,000 acres (408 square miles), compared with 275,000 acres (430 square miles) if it rose to 1,459 feet, the level of the coulee floor prior to excavations in 1997.

The information in this Endnote is derived from an article by Ann M.K. Fritz and Edward C. Murphy (Fritz, Ann M K., and Murphy, Edward C., 1997, Study of Devils Lake and Stump Lake Outlets, NDGS Newsletter, vol. 24, no. 2, p. 16–21).

56 A rise of 1 foot in the level of Devils Lake, from 1,425 to 1,426 feet elevation, requires the addition of 51,000 acre-feet of water and increases the area of the lake from 50,000 to 52,000 acres. A rise of 1 foot, from 1,450 to 1,451 feet elevation, requires the addition of 151,000 acre-feet of water and increases the area of the lake from 147,000 to 156,000 acres.

57 A caveat: The levees that have been built to protect the city of Devils Lake and other places along the lake exclude those areas from flooding, but they cause the rest of the lake to rise to a slightly higher level to compensate for the fact that the water cannot spread out over those areas.

Chapter 7: The Red River Valley

58 Keating, William H., 1824, *Narrative of an expedition to the sources of St. Peter's River, Lake Winnepeek, Lake of the Woods, &c. &c.*

59 Upham, Warren, 1896, *The Glacial Lake Agassiz.*

60 Born in Switzerland, Agassiz and his associates carefully traced the movements of the Aar glacier in the Alps. In his major work, *Études sur les Glaciers,* published in 1840, Agassiz concluded that "great sheets of ice, resembling those now existing in Greenland, once covered all the countries in which unstratified gravel (till) is found."

61 The elevation of the southern outlet of Lake Agassiz, the Browns Valley Outlet, used by both Lake Agassiz I and II, kept the water level from rising above the Campbell level. Water flowing through the outlet, which today contains Big Stone Lake and Lake Traverse, flowed over granite, which kept the outlet river from cutting down far enough to completely drain the lake.

62 Personal communication, Robert Marvinney, state geologist, Maine Geological Survey, February 7, 2012.

63 Meandering rivers follow winding paths to compensate for the low gradients over which they flow. The Meander River in Turkey provides the name for such rivers, but the Red River of the North is one of the best examples of a meandering river to be found anywhere. The straight-line distance from its headwaters near Wahpeton to where it empties into Lake Winnipeg in Manitoba is 251 miles, but the Red River flows 550 miles in that distance. The river miles are more than double the total distance. Throughout its total length, the river drops an average of 5 inches for every mile it flows, from its headwaters near Wahpeton to where it enters Lake Winnipeg in Manitoba. However, the gradient decreases northward,

from 1.3 feet per mile near Wahpeton to 2.5 inches per mile where it crosses the Canadian border.

64 Smith, G. Hubert, 1980, *Explorations of the La Verendryes in the Northern Plains, 1738–43.*

65 Owen, David Dale, et al., 1852, *Report of a Geological Survey of Wisconsin, Iowa, and Minnesota; and incidentally of a Portion of Nebraska Territory.*

Chapter 8: The Effects of Geology on the Human Environment

66 I was surprised to learn that Louis Agassiz, for whom Glacial Lake Agassiz is named, and who is considered to be the father of glacial geology and the first person to propose the idea of an Ice Age, had commented that "disturbance of electrical currents" caused by settlement would increase moisture, thereby making it better suited for agriculture. Even though it was a popular idea at the time, it had no basis in fact and people today regard the idea as quackery.

Chapter 9: North Dakota's Energy Resources

67 Quaife, M., and Will, G.F., 1926, *Vie militaire dans le Dakota, notes et souvenirs (1867–1869).*

68 Oihus, Colleen, A., 1983, *A History of Coal Mining in North Dakota, 1873–1982.*

69 In the autumn of 1964, I visited a small underground lignite mine in North Dakota. I do not remember the name of the mine, but it was one of the last, or possibly the last underground mine to operate in the state. It was operated by one man and his mule. The mine shaft was perhaps 5 feet 6 inches high and the mule pulled a hopper car in and out of the mine on a narrow rail line through the shaft. The man who operated the mine was around 5 feet 2 inches tall, and his mule was not as tall he was, so a 5-foot-6-inch shaft left him plenty of headroom.

Ted Freers, another geologist with the North Dakota Geological Survey, and I wanted to see the mine, so we followed the man and his mule for what seemed like a long walk into the mine. Ted is 6 feet 1 inch tall and I am 5 feet 11 inches. It was an interesting experience walking through the shaft keeping our heads down far enough to keep from hitting them on the ceiling (no hard hats then), all the while listening as the fellow talked about his mining operation.

When we arrived at some point deep in the mine, and the miner had told us all there was to say about his mine, he sat down and proceeded to give us a long lecture on politics. It was an election year. I do not recall who he was for or against, but I do remember he gave a colorful rendition, laced with marvelous profanity. I have not heard anything remotely approaching it since. We had no choice but to sit and listen; he had the only light, and neither Ted nor I even considered heading out in the dark. We were, perhaps, the classic example of a captive audience.

70 In 2010, the average cost to residential consumers for electricity in the United States was 11.53 cents/kilowatt hour. North Dakota residents paid the least, 7.98 cents/kwh. The most expensive electricity was in Hawaii (27.83 cents), Connecticut (19.3 cents), California (14.36 cents) and Alaska (15.12 cents).

71 Wind classes refer to average wind speeds, not extremes. Thus, class five areas have average speeds of 14 mph and class seven areas have average speeds of 21

mph (both at 30 feet off the ground).

72 In 1968, when I was mapping geology in northeastern North Dakota, I visited with a farmer near Neche, in Pembina County, who had fitted an electric pump to a water well. The apparatus was housed in a small wooden structure for protection from the weather. Sometimes, when the motor started, a spark ignited gas that had collected in the pump house, blowing the structure apart. The farmer remedied the problem by installing a hinged cover on the pump house so that, when the gas exploded, it merely blew the cover open. I determined that the gas was coming from decaying organic material in the same glacial deposits from which the water was being pumped.

73 Positive evidence that the Williston Basin exists was discovered in 1912 by geologist W. Taylor Thom Jr., who discovered fossil corrals in beds of clay and sandstone exposed along the banks of the Cannonball River near the South Dakota border. He concluded that the area had at one time been at the bottom of a sea, and he referred to the basin as the "Cannonball Sea Basin." The current name "Williston Basin" was first used formally in a scientific paper co-authored by Thom and C.E. Dobbin, and presented at a Geological Society of America meeting on December 28, 1923.

74 North Dakota's Bakken crude oil is "sweet," meaning it contains less than 0.5 percent sulfur. Oil with more than 1 percent sulfur is considered to be "sour." Sweet crude contains only small amounts of hydrogen sulfide and carbon dioxide gases.

75 American Petroleum Institute (API) gravity is a measure of how heavy or light a petroleum liquid is compared with water. Oil with an API gravity greater than 10 is lighter and floats on water; if less than 10, it is heavier and sinks. Gravity is used to compare the relative densities of petroleum liquids. For example, if one petroleum liquid floats on another and is therefore less dense, it has a greater API gravity. API gravity has no units, but is referred to as "degrees."

76 Proppants are sized (sieved) particles mixed with fracturing fluid to hold fractures open after a hydraulic fracturing treatment. Naturally occurring sand grains or man-made materials such as resin-coated sand or high-strength ceramic materials can be used.

Proppant materials are sorted for size and sphericity, so that when they are injected into a fracture, they will ensure an efficient conduit for the flow of fluid from the reservoir to the wellbore.

Currently, proppants being used in horizontal drilling in North Dakota are imported from other states, but studies are under way to find naturally occurring sand deposits that might be used, or clay deposits that could be used to manufacture ceramic beads. Kaolinite beds found in the Golden Valley Formation are being considered as a possible source for clay proppant materials.

77 The Bakken Formation shale, the Sanish sand, the Pronghorn Member and the Three Forks Formation are all considered to be part of the same oil play—the Bakken Play. The oil being produced from the Bakken/Three Forks system formed in the Lower Bakken and migrated into the adjacent zones (Middle Bakken and Three Forks).

78 The oldest (indirect) recognized date that I know of for the earliest glacial maximum in the North American midcontinent is a cosmogenic burial age of

2.41 million years for a paleosol buried beneath a till in Missouri. Presumably, a glacier advancing as far south as Missouri would have covered North Dakota. However, earlier glaciations, which did not reach so far south, may have affected North Dakota before then. A variety of marine climate and ice-rafted debris records date to 2.7 million years ago, and extensive glaciation dating to 3.5 million years ago at James Bay in Ontario has recently been confirmed.

79 In North Dakota, up to seven buried till units have been identified that are likely pre-Illinoian, but their ages are not known. As many as 10 pre-Illinoian stages have been identified in other places.

80 The Wisconsinan glaciation did not end with a smooth transition from a glacial climate to a warm climate. Tree-ring records from the Great Lakes area, which should have had a climate similar to North Dakota's as the Ice Age ended, show that temperatures were already warming by 16,000 years ago. They warmed rapidly until 14,500 years ago, but then reverted to colder glacial conditions, cooling again to a level comparable to 16,000 years ago, during the Younger Dryas. The Younger Dryas was a cold period that lasted from 12,800 until 11,500 years ago (the name "dryas" refers to a flowering alpine plant that tolerates cold and is one of the few plants found near the edges of glaciers). Following that, the climate warmed quickly, leveling off 9,000 years ago and leading into a prolonged period of warm climate, known as the Altithermal. Much of the Altithermal (7,500 until 4,000 years ago) was warmer and drier than our modern climate in North Dakota.

SOURCES AND RECOMMENDED READING

I used all of the following books and articles as sources of information in writing *North Dakota's Geologic Legacy*. Most were chosen because they contained pertinent information I wished to include in this book. Some of the books and articles listed below may make enjoyable casual reading.

Two of my own books, *The Face of North Dakota*, and *The 50th Anniversary of the Discovery of Oil in North Dakota*, provide additional information about important aspects of North Dakota's geology.

Other books or articles specifically relating to North Dakota include:

Guide to the Sheyenne National Grassland by Ann Fritz
Concretions and Nodules in North Dakota by Bob Biek
The Geology of Theodore Roosevelt National Park, Billings and McKenzie Counties, North Dakota by Bob Biek and Mark Gonzalez
Geology of the Lewis & Clark Trail in North Dakota by John Hoganson and Ed Murphy *Dinosaurs, Sharks and Wooly Mammoths: Glimpses of North Dakota's Prehistoric Past* by John Hoganson
Prairie Mosaic: An Ethnic Atlas of Rural North Dakota by William Sherman

While I was researching materials for this book, I noted several other books that may appeal to the general reader. I have marked some of these with asterisks; most of them should be readily available.

COMPILATION OF PERTINENT REFERENCES

Agassiz, Louis, 1840, *Études sur les glaciers.* Neuchâtel (Studies on glaciers: preceded by the Discourse of Neuchâtel—Carozzi, A.V., editor, 1967), Hafner Publishing Company, New York.

Anderson, F.J., 2011, *Dunes on the Wind-Swept Prairie: North Dakota's Eolian Sands,* North Dakota Department of Mineral Resources, Geo News, vol. 38, no. 2, pp. 3–7.

Arndt, B.M., 1977, *Stratigraphy of Offshore Sediment Lake Agassiz—North Dakota,* North Dakota Geological Survey Report of Investigation No. 60, North Dakota Geological Survey, Bismarck.

Atkinson, A., and Wilson, M.L., 1917, *Corn Planting and Cultivation in Montana,* University of Montana Agricultural Experiment Station: Bozeman.

*Anderson, B.G., and Borns, H.W. Jr., 1997, *The Ice Age World: An Introduction to Quaternary History and Research with Emphasis on North America and Northern Europe*, Universitetsforlaget, 2nd Edition, (in English).

Ashworth, A.C., 1999, *Climate Change in North Dakota Since the Last Glaciation—Review of the Paleontological Record,* Proceedings of the North Dakota Academy of Science, vol. 53, pp. 171–76.

Audubon, J.J. 1843, *Audubon and His Journals/The Missouri River Journals 1843 (continued) Episodes*, Scribner's Sons, New York.

Bagnold, R.A., 1941, *The Physics of Blown Sand and Desert Dunes,* Methuen, London.

Benton, T.H., 1937, *An Artist in America,* Robert M. McBride Co., New York.

*Biek, R.F., 1994, "Concretions and Nodules in North Dakota," NDGS Newsletter, vol. 21, no. 1, p. 6. North Dakota Geological Survey, Bismarck.

*Biek, R.F., and Gonzalez, M.A., 2001, *The Geology of Theodore Roosevelt National Park, Billings and McKenzie Counties, North Dakota*, North Dakota Geological Survey Miscellaneous Series 86, Bismarck.

Bluemle, J.P, 1967, *Ice-thrust Bedrock in Northwest Cavalier County, North Dakota,* North Dakota Academy of Science Proceedings, vol. 20, pp. 112–18, Grand Forks.

*Bluemle, J.P., 1988, *Geologic Highway Map of North Dakota,* North Dakota Geological Survey Miscellaneous Map 29, Scale 1:1,000,000, Grand Forks.

Bluemle, J.P., 1991, *The Little Missouri Badlands.* North Dakota Outdoors, North Dakota Game and Fish Department, vol. 54, no. 2, pp. 6–9, Bismarck.

Bluemle, J.P., 1993, *Hydrodynamic Blowouts in North Dakota,* Glaciotectonics and Mapping Glacial Deposits: Proceedings of the INQUA Commission on Formation and Properties of Glacial Deposits, Canadian Plains Research Centre, Regina, Saskatchewan, pp. 259–66.

Bluemle, J.P., 1996, *North Dakota's Badlands,* Geology Today, Blackwell Sciences Ltd., London, U.K., November-December, pp. 217–23.

*Bluemle, J.P., 2000, *The Face of North Dakota, Third Edition,* North Dakota Geological Survey Educational Series 26, North Dakota Geological Survey, Bismarck.

*Bluemle, J.P., 2001, *The 50th Anniversary of the Discovery of Oil in North Dakota,* North Dakota Geological Survey Miscellaneous Series 89, North Dakota Geological Survey, Bismarck.

Bluemle, J.P., Anderson, S.B., Andrew, J.A., Fischer, D.W., and LeFever, J.A., 1986, *North Dakota Stratigraphic Column,* Miscellaneous Series 66. North Dakota Geological Survey, Grand Forks, 3 charts.

Bluemle, J.P., and Clayton, Lee, 1982, *Permafrost Polygons in Southwestern North Dakota*, Geological Society of America Program Abstracts, Rocky Mountain Section, Annual Meeting, May, Denver.

Bluemle, J.P., Clayton, Lee, 1982, *Geologic Time in North Dakota*, North Dakota Geological Survey Educational Series 14, North Dakota Geological Survey, Grand Forks.

Bluemle, J.P., and Clayton, Lee, 1984, "Large-scale glacial thrusting and related processes in North Dakota," *Boreas*, vol. 13, p. 279–99, Oslo, Norway.

Bluemle, J.P., Hunke, N.T., Lord, M.L., 1993, "Exceptionally long, narrow drumlins formed in subglacial cavities, North Dakota," *Boreas*, vol. 22, Oslo, pp. 15–24.

Bluemle, J.P., and Jacob, A.F., 1972, *Geology Along the South Loop*

Road, Theodore Roosevelt National Memorial Park, North Dakota Geological Survey Educational Series 1, Grand Forks.

Bluemle, J.P., Reid, J.R., and Mitchell, K.J.R., 2004, *Glaciation of North Dakota, U.S.A.*, in Ehlers, J. and Gibbard, P.L. (Eds.), *Quaternary Glaciations—Extent and Chronology, Part II: North America,* Elsevier, Amsterdam, pp. 207–12.

Bluemle, J.P., Sable, J.M., and Karlén, Wibjörn, 1999, *The Rate and Magnitude of Past Global Climate Changes,* Environmental Geosciences, American Association of Petroleum Geologists, vol. 6, no. 2, pp. 63–75.

Bray, E.C., and Bray, M.C., 1993, *Joseph N. Nicollet on the Plains and Prairies: the Expeditions of 1838-39, With Journals, Letters and Notes on the Dakota Indians,* Minnesota Historical Press, St. Paul.

Brevik, E.C., 2013, *An Introduction to Soil Science Basics* in *Soils and Human Health* edited by Eric C. Brevik and Lynn C. Burgess, Taylor and Francis Group, CRC Press, pp. 1–28.

Brevik, E.C., Fenton, T.E., and Reid, J.R., 1998, "Soil Maps as a Tool in Mapping Poorly Preserved Landforms: A Case Study in Grand Forks County, North Dakota," *Soil Survey Horizons*, vol. 39, no. 3, pp. 61–67.

Brevik, E.C., and Reid, J.R., 2000, "Uplift-based limits to the thickness of ice in the Lake Agassiz basin of North Dakota during the Late Wisconsinan," *Geomorphology*, vol. 32, pp. 161–69.

Chamberlin, T.C., 1894, "Proposed Genetic Classification of Pleistocene Glacial Formations," *Journal of Geology*, vol. 2, pp. 517–38.

Clark, William, in *The Journals of the Lewis and Clark Expedition, March 21, 1805* in Reid, Russell, (ed), 1948, *Lewis and Clark in North Dakota*, State Historical Society of North Dakota, Bismarck, p. 181.

Clayton, Lee, and Freers, T.F., 1967, *Glacial Geology of the Missouri Coteau and Adjacent Areas*, in *North Dakota Geological Survey Miscellaneous Series 30,* North Dakota Geological Survey, Grand Forks, 170 p.

Clayton, Lee, Bickley, W.B. Jr., and Stone, W.J. 1970, "Knife River Flint," *Plains Anthropologist*, vol. 15, pp. 282–90.

Collier, A.J., and Thom, W.T. Jr., 1918, *The Flaxville Gravel and its Relation to Other Terrace Gravels of the Northern Great Plains,* United States Geological Survey Professional Paper 108-J, Washington, D.C.

De Saussure, H.B. 1786, *Voyages dan les Alpes*, vol. 1 and 2, 1779.

Dibb, W.D., *Diaries of Dr. William Dibb on Captain Fisk's Expeditions of 1862, 1863, and 1864, Overland Journeys to the Pacific* in White, H.M., 1996, *Hot For the Gold Fields*, Minnesota Historical Society,

St. Paul.

*Fast, D., and Huck B., 2007, *The Land Where the Sky Begins,* James Richardson & Sons Limited, Winnipeg.

Featherstonhaugh, G.W., 1847, *A Canoe Voyage up the Minnay Sotor; With an Account of the Lead and Copper Deposits in Wisconsin*, Schulze and Co., London.

*Fritz, A.M.K., 2001, *Guide to the Sheyenne National Grassland*. North Dakota Geological Survey Educational Series 27, North Dakota Geological Survey, Bismarck.

Galbraith, J.K., 1954, *The Great Crash, 1929*, Houghton-Mifflin, Boston.

Gao, C., McAndrews, J.H., Wang, X., Menzies, J., Turton, C.L., Wood, B.D., Pei, J., and Kodors, C., 2012, "Glaciation of North America in the James Bay Lowland, Canada, 3.5 Ma," *Geology*, vol. 40, no. 11, pp. 975–76.

Gilman, C., 1978, *Perceptions of the Prairie: Cultural Contrasts on the Red River Trails*, Minnesota Historical Society, St. Paul, vol. 46, no. 3, pp. 112–22.

Gonzalez, M.A., 2001, *Recent fluvial history and environmental change of some ephemeral streams in the Little Missouri Badlands of southwestern North Dakota,* North Dakota Geological Survey Report of Investigations 101, North Dakota Geological Survey, Bismarck.

Gourneau, P., 1993, *History of the Turtle Mountain Band of Chippewa Indians, Ninth Edition.*

Higgins, K.F., 1984, "Lightning Fires in North Dakota Grasslands and in Pine-Savanna Lands of South Dakota and Montana," *Journal of Range Management*, vol. 37, no. 2, pp. 100–103.

Harington, C.R., and Ashworth A.C., 1986, "A Mammoth (Mammuthus primigenius) Tooth from Late Wisconsin deposits near Embden, North Dakota, and Comments on the Distribution of Woolly Mammoths south of the Wisconsin Ice Sheets," *Canadian Journal of Earth Sciences*, vol. 23, pp. 909–18.

Harris, K.L., Howell, F.L., Wartman, B.L., and Anderson, S.B., 1982, *An Evaluation of Hydrothermal Resources of North Dakota, Phase III Final Technical Report, UND Engineering Experiment Station Bulletin No. 82-08-EES-01,* Conference: Geothermal energy exploration and resource assessment technical conference, Salt Lake City, Utah, April 5, 1982; p. 22.

Hitchcock, E., 1847, *Elementary Geology*. Mark H. Newman & Co., New York.

*Hoganson, J.W., 2007, *Dinosaurs, Sharks and Wooly Mammoths: Glimpses* *of North Dakota's Prehistoric Past*, North Dakota Geological Survey Educational Series 31, North Dakota Geological Survey, Bismarck.

*Hoganson, J.W. 2011, *Fossil Resources Reveal History of Ancient Life in North Dakota,* North Dakota Blue Book, North Dakota Secretary of State, Bismarck, pp. 2–42.

*Hoganson, J.W., 2011, *Mammalian Paleofaunas of North Dakota* in Seabloom, R.W., *Mammals of North Dakota*, North Dakota Institute for Regional Studies, Fargo, pp. 10–22.

*Hoganson, J.W., and Murphy, E.C., 2003, *Geology of the Lewis & Clark Trail in North Dakota*, Mountain Press Publishing Company, Missoula, Montana.

Hooke, R. LeB., and Jennings, C.E., 2006, *On the Formation of the Tunnel Channels of the Southern Laurentide Ice Sheet*, Quaternary Science Reviews 25, pp. 1,364–372.

Jackson, J.A., 1997, *Glossary of Geology, Fourth Edition*. American Geological Institute, Alexandria, Virginia.

Kantrud, H.A., Krapu, G.L., and Swanson, G.A., 1989, *Prairie Basin Wetlands: A Community Profile,* Biological Report 85, U.S. Department of Interior, Fish and Wildlife Service, Washington.

Keating, W.H., 1824, *Narrative of an Expedition to the Sources of St. Peter's, Lake Winnepeek, Lake of the Woods &c., &c., Performed in the Year 1823, by Order of the Hon. J. C. Calhoun, Secretary of War Under Command of Stephen H. Long, Major,* two volumes. H.C. Carey and I. Lea, Philadelphia.

Kehew, A.E., and Lord, M.L., 1989, "Glacial Lake Spillways of the Central Interior Plains, Canada-U.S.A.," *The Canadian Geographer*, vol. 33, pp. 274–77.

Kehew, A.E. and Kozlowski, A.L., 2007, *Tunnel Channels of the Saginaw Lobe, Michigan, USA,* in Johannsson, P., and Sarala, P. (eds.) *Applied Quaternary research in the central part of glaciated terrain,* Geological Survey of Finland, Special Paper 46, pp. 69–77.

Korom, Scott, 2013, *Underground Coal Gasification. What is it and what role does Hydrogeology Play?* Abstract Volume, Midwest Ground Water Conference, Bismarck, September 23–25, p. 11.

Krueger, A.M., 1981, *Sheldon's Outlaw Well* in *Sheldon Community History,* Sheldon, North Dakota Centennial, 1881–1981.

Kume, J., 1966, *The Dahlen Esker of Grand Forks and Walsh Counties, North Dakota,* North Dakota Geological Survey, Miscellaneous Series 32, Grand Forks.

Kupsch, W.O., 1962, "Ice-thrust ridges in western Canada," *Journal of Geology*, vol. 40, no. 5, pp. 582–94.

Larocque, François-Antoine, 1910, *Journal of Larocque from the Assiniboine to the Yellowstone, 1805,* Government Printing Bureau Ottawa, Canada.

Laskin, David, 1997, *Braving the Elements: the Stormy History of American Weather,* Anchor Press.

La Verendrye, Pierre Gaultier de Varennes, 1738, *Journals & Letters of Pierre Gaultier de Varennes la Verendrye and his Sons* in Burpee, Lawrence J. *The Publications of the Champlain Society,* Toronto, Canada.

*Least-Heat Moon, W. (William Trogden), 1982, *Blue Highways,* Little, Brown and Company, Boston.

*LeFever, J., 2014, "Source Beds" in North Dakota Department of Mineral Resources, *Geo News, vol. 41, No. 2,* pp. 17–22.

*LeFever, J., Bluemle, J.P., and Waldkirch R.P., 1999, *Flooding in the Grand Forks-East Grand Forks North Dakota and Minnesota Area,* North Dakota Geological Survey, Educational Series 25, Bismarck.

Lewis, Meriwether, in *The Journals of the Lewis and Clark Expedition, April 16, 1805* in Reid, Russell (ed), 1948, *Lewis and Clark in North Dakota*, State Historical Society of North Dakota, Bismarck, p. 241.

Link, Arthur, 1973, *When the Landscape is Quiet Again,* Speech in Mandan, North Dakota on October 11.

Long, S.H., 1824, *The Northern Expeditions of Stephen H. Long: The Journals of 1817 and 1823 and Related Documents,* published by the Minnesota Historical Society Press in 1978.

Lowenthal, D., 1961, *Geography, experience, and imagination: towards a geographical epistemology*, Annals of the Association of Geographers, American Geographical Society, vol. 51, New York, pp. 241–60.

*Manz, Lorraine, 2012, *Drilling for Energy on the North Dakota Capitol Grounds* in North Dakota Department of Mineral Resources, *Geo News,* vol. 39, No. 2, pp. 10–12.

Maximilian, Alexander Philipp, 1833, *The North American Journals of Prince Maximilian of Wied: May 1832–April 1833.*

McMillan, J.N., 1973, *Shelves of the Labrador Sea and Baffin Bay, Canada:* Canadian Society of Petroleum Geologists Memoir 1, pp. 473–515.

*McPhee, J.A., 1980, *Annals of the Former World,* Farrar, Straus and Giraoux, New York.

*McPhee, J.A., 1981, *Basin and Range*, Farrar, Straus and Giraoux, New York.

Murchison, R.I., 1839, *The Silurian System (Part I)*, John Murray, London.

*Murphy, E.C., 2014, *Knife River Flint and Other Siliceous Rocks in Western North Dakota*, North Dakota Department of Mineral Resources Geo News, vol. 41, No. 1, pp. 1–7

*Murphy, E.C., and Forsman, N.F., 1998, *Meteorites in North Dakota,* North Dakota Geological Survey, Educational Series 23, North Dakota Geological Survey, Bismarck.

*Murphy, E.C., Hoganson, J.W., and Forsman, N.F., 1993, *The Chadron, Brule, and Arikaree Formations in North Dakota—the Buttes of Southwestern North Dakota,* North Dakota Geological Survey, Report of Investigation No. 96, North Dakota Geological Survey, Bismarck.

*Oihus, C.A., 1983, *A History of Coal Mining in North Dakota, 1873–1982,* North Dakota Geological Survey Educational Series 15, North Dakota Geological Survey, Bismarck.

Owen, D.D., 1852, *Report of a Geological Survey of Wisconsin, Iowa, and Minnesota and Incidentally a Portion of Nebraska Territory: Made Under Instructions From the United States Treasury Department,* Lippencott, Grambo & Co., Philadelphia.

Patterson, C.J., 1997, *Southern Laurentide Ice Lobes Were Created by Ice Streams: Des Moines Lobe in Minnesota, USA* in *Sedimentary Geology,* vol. 111, pp. 249–61.

Powell, J.W., 1878, *Report on the Methods of Surveying the Public Domain, to the Secretary of the Interior at the Request of the National Academy of Sciences*, Government Printing Office, Washington.

*Raban, J., 1996, *Bad Land,* Pantheon Books, New York.

Reid, J.R., and Bluemle, J.P., 1997, *Geologic History of the Devils Lake Basin, North Dakota*. Proceedings of the North Dakota Academy of Science Annual Meeting, vol. 51.

Rubin, D.M., and Hunter, R.E., 1987, "Bedform alignment in directionally varying flows," *Science*, vol. 237, pp. 276–78.

Rutten, M.G., 1960, "Ice-pushed ridges, permafrost and drainage," *American Journal of Science*, vol. 258, pp. 293–97.

Sears, J.W., 2013, "Late Oligocene—early Miocene Grand Canyon: A Canadian connection," *GSA Today*, vol. 23, no 11, pp. 4–10.

Seelig, B.D., Richardson, J.L., and Barker, W.T., 1990, "Characteristics and Taxonomy of Sodic Soils as a Function of Landform Position," *Soil Science Society of America Journal*, vol., 54, pp. 1,690–697.

*Severson, K.E., and Sieg, C.H., 2006, *The Nature of Eastern North Dakota: Pre-1880 Historical Ecology,* North Dakota Institute for Regional

Studies, North Dakota State University, Fargo.

Shay, C.T. in preparation, *The Northeastern Plains* (personal communication), 2014.

*Sherman, W.C., 1983, *Prairie Mosaic: An Ethnic Atlas of Rural North Dakota,* North Dakota Institute for Regional Studies, Fargo, 152 p.

*Sherman, W.C., Henke, W.A., Kloberdanz, T.J., Pedeliski, T.B, Thorson, P.V., and Wilkins, R.P. 1988, *Plains Folk*, North Dakota Institute for Regional Studies, Fargo.

Sherrod, Neil, 1963, *Late Pleistocene Fish from Lake Sediments in Sheridan County, North Dakota,* North Dakota Geological Survey, Miscellaneous Series 21. Grand Forks.

Smith, G.H., 1980, *Explorations of the la Verendryes in the Northern Plains, 1738-43,* University of Nebraska Press, Lincoln. 1980.

*Steinbeck, John. 1962, *Travels With Charley, In Search of America*, Viking Press, New York.

Sveen, L.W., 1981, *Potpourri*, Grand Forks Herald, editorial, June 11.

Teller, J.T., and Clayton, Lee, 1983, *Glacial Lake Agassiz,* The Geological Association of Canada, Special Paper 26, University of Toronto Press, Toronto.

Thom, W.T. Jr., 1952, "How the Williston Basin Got its Name," *Oil and Gas Journal*, vol. 51, no. 10.

Thompson, David, 1797, *David Thompson's Narrative of His Explorations in Western America 1784–1812* in Tyrrell, J.B. *The Publications of the Champlain Society*, Toronto.

De Trobriand, Philippe Regis, 1982, *Military Life in Dakota: The Journal of Philippe Regis de Trobriand,* Translated by Lucile M. Kane, University of Nebraska Press.

Upham, Warren, 1895, *The Glacial Lake Agassiz*. United States Geological Survey Monographs of the United States Geological Survey Vol. XXV, 54th Congress, 1st Session, House of Representatives, Document No. 297, Washington.

Venetz, Ignaz, 1833, *Moire sur les Variations de la Temperature Dans les Alpes de la Suisse*, first publication of an 1821 address to the Swiss Academy.

Wegener, Alfred, 1912, *Die Entstehung der Kontiente und Ozeane (The Origin of Continents and Oceans)*, presented at the annual meeting of the German Geological Society, February 12, Frankfurt am Main, Germany.

Wheeler, J., *Weather Not Just Small Talk in North Dakota* in *2011–2013*

North Dakota Blue Book, North Dakota Secretary of State, Bismarck, pp. 566–69

Willard, D.E., 1904, *Second Biennial Report,* Agricultural College Survey of North Dakota, Fargo, 1903–04.

Willard, D.E., 1907, *The Story of the Prairies* or *The Landscape Geology of North Dakota, Fifth Edition,* Rand McNally, Chicago.

*Winchester, Simon, 2001, *The Map That Changed the World.* Harper Collins, New York.

Wright, H.E. Jr., 1973, *Tunnel Valleys, Glacial surges, and Subglacial Hydrology of the Superior Lobe , Minnesota,* in *The Wisconsinan Stage* ed. R. Black, R. Goldthwait, and H. Willman, Geological Society of America Memoir, pp. 136, 251–76.

Yansa, C.H., 2007, "Lake Records of Northern Plains Paleoindian and Early Archaic Environments: The "Park Oasis" Hypothesis," *Plains Anthropologist*, vol. 52, No. 201, pp. 109–44.

*Yergin, Daniel, 1991, *The Prize—The Epic Quest for Oil, Money, and Power,* Simon & Schuster, New York.

*Youngquist, W.L., 1997, *GeoDestinies—The Inevitable Control of Earth Resources Over Nations and Individuals,* National Book Company, Portland, Oregon.

Zachus, J., Pagani, M., Sloan, L., Thomas, E., and Billups, K., 2001, "Trends, Rhythms, and Aberrations in Global Climate, 65 Mya to Present," *Science*, vol. 292, pp. 686–93.

INDEX

Agassiz, Lewis, 224, 353n23
Altithermal Period, 176, 320
Andean/Saharan Glaciation, 312
aquifers, 141, 203
anticlines, 288
API Gravity, 362n75
Arikaree Formation, 14, 80
artesian aquifers, flows, 203
Atkinson, Alfred, 259

badlands,
- erosion, 69
- occurrence, 63
- origin, 68

Bakken Formation, 291
- extent, 292
- origin, 293
- crude oil (sweet/sour), 362n74

Bell River, 19
Benedict Esker, 154
Bentley Lake gold deposit, 200
Benton, Thomas Hart, 32
bentonite, 67, 68
Black Butte, 84, 85
blizzard, origin of name, 25
Blue Butte/Blue Buttes, 86, 87
"bottomless" pits, 204
boulder pavements, 183
Brevik, Eric, 28
Browns Valley outlet, 226, 360n61
Brule Formation, 14
buffalo boulders, 183
Bullion Butte, 82, 83, 84
Bullion Creek Formation, 14
buttes,
- caprock, 76
- erosion, 77
- origin, 76

Campbell level, 227, 237
Camel's Hump Butte, 84
Cannonball Formation, 13, 42
cannonball concretions, 58
Cannonball Sea, 13
caprock, 15, 76, 78, 90
Carlile Formation,
- age, 8
- exposures, 9

castellate erosion, 64
Central Lowlands, 34, 36
Chadron Formation, 14
chert, 191
Chimney Butte, 86, 87
Clark, William, 50, 188
climate, 23,
- characteristics in North Dakota, 24
- effect of topography, 24
- effect of vegetation, 27

climate change, 213
- effect of volcanic eruptions, 27
- long-term cooling, 14, 16
- modern fluctuations in climate, 26

past history, 176
recent effect on Devils Lake, 208, 215
warming, ending the Ice Age, 113, 117
clinker/clinker caprock, 50, 89, 90
Clouston, Robert, 257
Chalky Buttes, 15, 84
Chamberlin, T. C., 352n18
Coghlan Castle, 194, 357n41
collapsible deposits, 251
collapsed lake topography, 139
compaction ridges, 231, 233
concretions, 57
Initial Rock log concretion, 61
internal bedding, 62
ledge concretions, 59
log concretions, 59
spherical concretions, 58
continental,
deposits, 12, 23
divide, 23
Coteau Slope, 34, 36
Cow Butte,
silcrete-caprock surface, 91
view from Cow Butte, 92
Cretaceous Period, 5
Crow-Flies-High,
ancient lake, 105
view from hill, 104
crustal rebound,
effect on erosion, 350n7
effect on Lake Agassiz beaches, 242
Cryogenian Glaciation, 307

Dahlen Esker, 153, 154, 356n32
Dakota Formation aquifer,
"bottomless" pits, 204
saline soil in Red River Valley, 206, 235
well blowouts, 203
dead-ice moraine, 135, 136, 137
Dead Man till, 316
Denbigh gold, 198
Devils Heart, 157
Devils Lake, 207, 359n52
geology, 208
ice thrusting, 212, 213
origin, 210
dimension stone, 193
dinosaur fossils, 7, 12
disintegration ridges, 139
Dogden Butte, 148, 150
Dogtooth Buttes, 88
doughnuts, 138, 139
dragline, 273
drought
historic, 25
reasons for, 25
drumlins, 129

elevations,
highest and lowest in North Dakota, 31
Elk Valley Delta, 239
end moraines, 132
Eocene Epoch, 12, 14
Epilogue, 306
erosion, 14, 18, 77, 350n6, 351n8
amount of, 15, 70
rate in badlands, 70
erratics, 5, 180, 182, 357n38
buffalo boulders, 183
notable erratics, 184, 185
eskers, 152
extent of glaciation,
in North America, 111
in North Dakota, 35, 119

field stones, 180, 191
flared gas, 301
Flaxville gravels, 19
flood damage, 251
flooding, 252
flood plains, 251
Four Bears Gravel, 316
Fox Hills Formation, 42
age, 11
landscapes formed on, 12
Frankhauser Esker, 154

Garrioch, Peter, 257
geography [of North Dakota], 23
geologic formations in the Williston Basin, 268
geologic hazards, 250
  floods in North Dakota, 252
  landslides in North Dakota, 253
geologic map (of North Dakota), 119
geologic time,
  analogy, 2
  duration, 4
  events through time, 4
geothermal energy, 279
Gilman, Carolyn, 352n16
Girard Spillway, 166
Golden Valley Formation, 14
glacial,
  lakes, 163
  periods, 207
  sediment, 120
Glacial River Warren, 226
glaciated buttes, 95
Glaciated Plains, 34, 38, 134
glaciers, 113, 134, 354n27
  behavior, 117
  characteristics, 115
  definition, 114
  earliest, 110
  wasting modes, 117, 118, 119
glaciation,
  causes, 110, 112
  effects on North Dakota, 17, 21, 37, 39
  extent, 111, 119, 314
gold in North Dakota, 195, 196
  Denbigh area, 198
  Dibb, William, 198
  Hungry Gulch, 201
  Lake Souris area, 198
Gourneau, Patrick, 352n17
Great American Desert, 259
Great Dakota Boom, 259
Great North Dakota Gold Rush, 201
Great Plains, 32, 34
Greenhorn Formation, 7, 349n4
Griswold Gold Mine, 197
ground moraine, 125
groundwater temperature, 280

halophytes, 206
Hamar Dune Field, 174
Hell Creek Formation, 42
  age, 11
  badlands topography,13, 67
  dinosaur fossils, 12
Herman Beach,
  characteristics, 237
  crustal rebound effect, 242
  origin, 226
Herman (Beach) level, 226
Hogback Ridge, 130, 131
Holarctic Glaciation, 312
Holocene time, 319
hoodoos, 71, 73
horizontal drilling, 296
HT Butte Clinker, 55, 351n12
Hungry Gulch Gold Rush, 201
Huronian Glaciation, 311
hydraulic fracturing, 298
  environmental concerns, 299
hydrocarbons,
  early development, 287
  marsh gas. 287
  oil and gas pipeline delivery, 300
  recoverable resource, 303
  regulation of industry, 295
  reservoir rock, 295
hydroelectric power, 278

ICBM sites, 349n1
Ice Age, 17, 21, 109, 110
iceberg grooves, 229
ice thrusting,
  Anamoose thrust, 146, 355n29
  Devils Lake thrust, 150, 213
  ice-thrust mechanism, 141, 149
  ice-thrust theory, 355n31
  ice thrust [topography], 142, 145,
  occurrences, 142-48
  Prophets Mountains thrust, 148

ice-walled lake plains, 140
Illinoian Glacial Stage, 318
iridium marker bed in North Dakota, 12
isostatic rebound, 17, 242

James, Edwin (Stephen Long Expedition), 257

kames, 152, 155
Karoo Glaciation, 312
Keating, William (Stephen Long Expedition), 223, 244, 256
Kellys Slough, 203, 206, 235
kettle holes, 138
Killdeer Mountains, 15, 78, 79
Knife River Flint, 186, 188, 189

Lake Agassiz, 223
- bars and spits, 238
- beaches, 237
- extent, 225
- history, 224
- scarps, 236
- sediment, 228
- shorelines, 236

Lake Ardoch, 203, 206, 235
Lake Cando, 164, 167
Lake Dakota, 164, 166
Lake McKenzie, 168
Lake Minnewaukan, 164, 167
Lake Mose, 174, 238
Lake Oakes Hills, 167
Lake Sakakawea, 104
lake sediment, 122
Lake Souris, 164
landforms,
- discussion, 31, 36–39, 41
- glacial vs. nonglacial, 31–33, 39
- perceptions, 352n17

landslides, 251, 253
Laramide Orogeny, 67
Laskin, David, 23
Lava Creek B Ash, 316
lignite, 43, 269
- characteristics, 270
- mining reclamation, 275
- production history, 270

Least-Heat-Moon, William, 184
Lewis, Meriwether, 50
Little Ice Age, 27, 310
Little Missouri Badlands, 33, 63–65
- age, 66
- erosion, 69
- landscape, 71
- origin, 68

lineations,
- oil and gas, 298
- origin, 276

Link, Arthur A., 176
loess, 172, 176, 177
log concretions, 60, 61, 62
Lowenthal, David, 352n16
Ludlow Formation, 13

marine deposits, 5, 10
Marshall, Eugene, 37
marsh gas, 297
McKenzie Upland, 35
McLean River, 98, 101
meandering rivers, 243, 360n63
Medicine Hole, 81, 82
Milankovitch Cycle, 313
meltwater channels,
- examples, 161
- terraces, 160

Miocene Epoch, 13
Missouri Coteau, 37, 134
Missouri Escarpment, 132, 133, 134
Missouri Plateau, 33, 41
Missouri River, 97, 103
modern landscape, 19
moraine,
- dead-ice moraine, 135
- end moraine, 132
- ground moraine, 125–129
- ice-thrust moraine, 141
- moraine: origin of name and history, 353n23, 354n23
- washboard ridges, 126

natural gas, 295, 300, 302
Nicollet, Joseph, 128, 207
Nikiforoff, Constantin Constantinovitch, 29
Niobrara Formation,
- age, 8
- exposures, 9
- coloration, 8

nodules, 57

oil and gas,
- early history in North Dakota, 287
- recoverable oil, 303
- regulation, 295
- source rocks, 295

Oligocene Epoch, 12
outwash plains, 123, 159
Owen, David Dale, 223, 244

Paleocene Epoch, 12
paleosols, 124, 176, 354n24
peat, 44
Pembina Delta, 239
Pembina Escarpment, 244, 245, 247
perched lake plains, 139, 140
petrified forest, 47
petrified log, 49
petrified wood, 43, 45
physiography, 31
Pierre Formation, 42
- age, 10
- landscape, 11

piping, 72, 74
Pleistocene Epoch, 110
Pliocene Epoch, 13
Powell, John Wesley, 259
Prairie Coteau, 37, 132
precipitation, 25–26
- effect of topography, 26
- effect of volcanism, 27

preglacial drainage, 99
proppants, 299, 362n76
Prophet's Mountains, 147, 148, 149

Red River Valley, 17, 39
red samphire, 234
reservoir rocks (vs. source rocks), 295
rill erosion, 70
riprap, 357n38
river terraces, 160, 161
Rock Column (stratigraphic column), 6, 269
Rocky Butte, 94
Rocky Ridge, 91, 92

sand dunes,
- how and where formed, 171–73
- when formed, 175-76
- where located, 175

Sangamon Interglacial Stage, 318
Schwert, Donald, 2
secondary recovery, 303
Sentinel Butte, 84, 86
Sentinel Butte Formation, 14
Shay, C. Thomas, 28
shelterbelts, 357n37
- beaded, 229, 231
- history, 357n37

Sherman, William C., 261
Sheyenne Delta, 239
Sheyenne National Grasslands, 172, 175
silcrete caprock, 91
silcrete/siliceous rocks, 186, 187
Slide Butte, 84, 352n15
Slope Formation, 14, 72
- age, 14
- landscape, 72

soils, 28
- origin of, 30
- quality of soils, 124
- saline soils, 358n49

soil slides/soil creep, 254
solar energy, 284, 286
source rocks (vs. reservoir rocks), 294
South Heart Badlands, 66
spring pits in the Red River Valley, 234
Spiritwood Aquifer, 211, 214, 218
state symbols, 357n40
Steinbeck, John, 31

stratigraphic columns (rock columns), 6, 269
Sveen, Lloyd W. (Grand Forks Herald), 201
synclines, 288

Table Butte, 86, 87
temperatures,
  effect of topography, 26
  extremes, 25, 26
  fluctuations, 15, 24, 25
Tepee Butte, 84
Tertiary-age formations, 12
tertiary oil recovery, 303
Thompson, David, 281
till, 122, 353n21
  characteristics, 122
  extent, 111, 119
  thickness, 124
treppe, 255
tunnel valleys, 158
Turtle Mountain, 37, 95, 352n17

underground coal gasification, 277
Upham, Warren, 224
uranium in lignite, 272

varves, 228
veblens, 156, 157, 356n33
Verendrye, Pierre la, 256, 355n30

washboard ridges, 126, 127, 354n25
Western Interior Seaway, 5
White Butte, 84, 85
Williston Basin, 1, 3, 290, 362n73
  location and extent, 1, 292
  structural features, 3
wind,
  energy, 281
  power potential, 283
wind-shaped landforms, 171, 178, 179
Wiota Gravels, 19
Wisconsinan Glaciation, 112, 115, 314, 318
Wormy Marker Bed, 80

Younger Dryas, 130, 164